Modular Design Strategies for TADF Emitters: Towards Highly Efficient Materials for OLED Application

Zur Erlangung des akademischen Grades eines

DOKTORS DER NATURWISSENSCHAFTEN

(Dr. rer. nat.)

von der KIT-Fakultät für Chemie und Biowissenschaften

des Karlsruher Instituts für Technologie (KIT)

genehmigte

DISSERTATION

von

M. Sc. Fabian Hundemer

aus Landau in der Pfalz

Dekan: Prof. Dr. Manfred Wilhelm

Referent: Prof. Dr. Stefan Bräse

Korreferent: Prof. Dr. Joachim Podlech

Tag der mündlichen Prüfung: 11. Dezember 2019

Band 84
Beiträge zur organischen Synthese
Hrsg.: Stefan Bräse

Prof. Dr. Stefan Bräse
Institut für Organische Chemie
Karlsruher Institut für Technologie (KIT)
Fritz-Haber-Weg 6
D-76131 Karlsruhe

Bibliographic information published by the Deutsche Nationalbibliothek

The Deutsche Nationalbibliothek lists this publication in the Deutsche Nationalbibliografie; detailed bibliographic data are available in the Internet at http://dnb.d-nb.de

ISBN 978-3-8325-5054-7
ISSN 1862-5681

Logos Verlag Berlin GmbH
Comeniushof, Gubener Str. 47,
10243 Berlin
Tel.: +49 030 42 85 10 90
Fax: +49 030 42 85 10 92
INTERNET: http://www.logos-verlag.de

Für meine Eltern und Selin.

"Mehr Licht."

Johann Wolfgang von Goethe

Honesty Declaration

This work was carried out from December 1st 2016 through November 6th 2019 at the Institute of Organic Chemistry, Faculty of Chemistry and Biosciences at the Karlsruhe Institute of Technology (KIT) under the supervision of Prof. Dr. Stefan Bräse.

Die vorliegende Arbeit wurde im Zeitraum vom 1. Dezember 2016 bis 6. November 2019 am Institut für Organische Chemie (IOC) der Fakultät für Chemie und Biowissenschaften am Karlsruher Institut für Technologie (KIT) unter der Leitung von Prof. Dr. Stefan Bräse angefertigt.

Hiermit versichere ich, Fabian Hundemer, die vorllegende Arbeit selbstständig verfasst und keine anderen als die angegebenen Hilfsmittel verwendet sowie Zitate kenntlich gemacht zu haben. Die Dissertation wurde von der Fakultät für Chemie und Biowissenschaften des Karlsruher Instituts für Technologie angenommen und wurde bisher an keiner anderen Hochschule oder Universität eingereicht.

Hereby I, Fabian Hundemer, declare that I completed the work independently, without any improper help and that all material published by others is cited properly. This thesis has been accepted by the Faculty of Chemistry and Biosciences at the Karlsruhe Institute of Technology and has not been submitted to any other university before.

German Title of this Thesis

Modulare Design-Strategien für TADF-Emitter: Hin zu Hocheffizienten Materialien für OLED Anwendungen

Table of Contents

Honesty Declaration V

German Title of this Thesis VII

Table of Contents IX

Kurzzusammenfassung 1

Abstract 3

1   Introduction 5

   1.1   Organic Light-Emitting Diodes (OLEDs) 7

     1.1.1   Working Principle and Architecture of OLEDs 8

   1.2   Thermally Activated Delayed Fluorescence (TADF) 10

     1.2.1   Working Principle of TADF 13

     1.2.2   History of TADF 14

     1.2.3   Molecular Design Strategies for TADF Emitters 16

       1.2.3.1   Twist-Induced Charge Transfer 17

       1.2.3.2   Through-Space Induced Charge Transfer 18

       1.2.3.3   Multi-Resonance Effect 19

     1.2.4   Molecular Design Towards Efficient OLEDs 20

       1.2.4.1   Host-Free TADF-OLEDs 21

       1.2.4.2   Aggregation-Induced Emission 22

       1.2.4.3   Orientation Controlled Emitters 23

2   Objective 27

3   Results and Discussion 31

   3.1   Derivatization of the 4CzIPN TADF System 31

     3.1.1   Tetrazole Derivatives 34

       3.1.1.1   Synthesis 35

       3.1.1.2   DFT Calculations 39

       3.1.1.3   Basic Photophysical Characterization 42

       3.1.1.4   Advanced Photophysical Characterization 44

       3.1.1.5   OLED Fabrication 46

       3.1.1.6   Bioconjugation 46

     3.1.2   Oxadiazole Derivatives 55

       3.1.2.1   Synthesis 57

       3.1.2.2   DFT Calculations 64

       3.1.2.3   Basic Photophysical Characterization 66

3.1.2.4   Advanced Photophysical Characterization 69

3.1.2.5   OLED Fabrication.. 72

3.1.2.6   Exciton Transport... 73

3.1.3   Benzothiazole Derivatives.. 76

3.1.3.1   DFT Calculations 77

3.1.3.2   Synthesis 78

3.2   TADF Emitters based on the Phthalimide Acceptor System 81

3.2.1   Synthesis.. 84

3.2.1.1   Acceptormodification 84

3.2.1.2   Donormodification.. 89

3.2.1.2.1   Functionalization by an Organometallic Approach............................ 90

3.2.1.2.2   Functionalization by Friedel-Crafts Reaction 92

3.2.1.2.3   Functionalization by Suzuki Cross-Coupling 96

3.2.1.2.4   Pre-functionalization and other Donor Groups 98

3.2.2   DFT Calculations... 101

3.2.3   Basic Photophysical Characterization 105

3.3   Tristriazolotriazine (TTT) as Acceptor Core...................................... 112

3.3.1   Synthesis.. 114

3.3.1.1   Variation on the Donor Type, Count and Arrangement.................... 114

3.3.1.2   Variations on the Phenylene Spacer... 115

3.3.1.3   Synthesis of the Tetrazole Precursors 119

3.3.1.4   Synthesis of the TTT Emitters 120

3.3.1.5   Isomerization of the TTT Core 122

3.3.1.6   Expanding the Concept to MTT and BTT 123

3.3.2   Physical Properties 125

3.3.3   DFT Calculations.. 126

3.3.4   Structural Investigations 132

3.3.5   Photophysical Characterization 134

3.3.6   OLED Fabrication 142

4   **Summary and Outlook**... **145**

4.1   Derivatization of the 4CzIPN TADF System.. 145

4.1.1   Outlook for 4CzIPN Derivatives 146

4.2   TADF Emitters based on the Phthalimide Acceptor System 147

4.2.1   Outlook for Phthalimide-based TADF Emitters.............................. 148

4.3   Tristriazolotriazine (TTT) as Acceptor Core... 149

4.3.1   Outlook for Tristriazolotriazines 150

5   **Experimental Section** **153**

5.1 General Remarks 153

    5.1.1 Materials and Methods 153

    5.1.2 Advanced Optoelectronic Data 156

        5.1.2.1 Tetrazole Derivatives 156

        5.1.2.2 Oxadiazole Derivatives 159

        5.1.2.3 Phthalimide-based TADF Emitter Project 164

        5.1.2.4 Tristriazolotriazine (TTT) Project 166

    5.1.3 Reaction Procedures and Analytical Data 168

        5.1.3.1 Derivatization of the 4CzIPN TADF System Project 168

        5.1.3.2 Phthalimide-based TADF Emitter Project 223

        5.1.3.3 Tristriazolotriazine (TTT) Project 277

5.2 Crystal Structures 315

    5.2.1 Crystallographic Data Solved by Dr. Martin Nieger 315

    5.2.2 Crystallographic Data Solved by Dr. Olaf Fuhr 328

6 List of Abbreviations **341**

7 Bibliography **345**

8 Appendix **359**

8.1 Curriculum Vitae 359

8.2 List of Publications 361

8.3 Acknowledgements 362

Kurzzusammenfassung

In den letzten Jahren haben sich Organische Leuchtdioden (OLED) zu einer der vielversprechendsten Beleuchtungstechnologien entwickelt. Sie bestechen nicht nur durch ihre Effizienz und ihr Energieeinsparpotenzial, sondern vor allem durch die Tatsache, dass transparente und flexible Displays realisiert werden können. Die Optimierung von Flüssigprozessierungsverfahren ist hierzu aktuell eine der dominantesten prozesstechnischen Hürden für die Herstellung von großflächigen OLED Displays. Der vielversprechendste Ansatz für effiziente Lichterzeugung in OLEDs ist das Prinzip der *thermally activated delayed fluorescence* (TADF).

In dieser Arbeit wurden neue TADF Materialien für flüssigprozesierbare OLEDs untersucht. Bestehende TADF-Emitterstrukturen auf Basis des Isophthalnitril-Akzeptors wurden chemisch modifiziert und der Einfluss auf die optoelektronischen Eigenschaften studiert, um ein besseres Verständnis der Struktur-Eigenschaftsbeziehungen zu gewinnen. Abhängig von den Modifizierungen konnte die Emissionsfarbe zwischen 438 nm und 541 nm variiert werden. Die vielversprechendsten Emitter wurden in OLEDs getestet und zeigten eine maximale *external quantum efficiency* (EQE) von bis zu 16 %.

Der modulare Ansatz ermöglichte die Einführung einer Vielzahl an funktionellen Gruppen und Strukturelementen. Anküpfstellen wurden geschaffen, um die TADF-Emitter an weitere Funktionalitäten wie Biotransporter zu binden. So diente die Kombination eines TADF-Emitters mit einem Biotransporter oder monodispersen Stäbchenoligomeren zur Untersuchung von Fluoreszenzbildgebung und von photophysikalischen Phänomenen wie Ladungsträgertransport.

In einem weiteren Ansatz wurden Phthalimid-basierte TADF-Emitter gezielt an Akzeptor und Donoreinheiten modifiziert und der Einfluss auf die optoelektronischen Eigenschaften und die Löslichkeit im Hinblick auf Flüssigprozessierbarkeit untersucht. Die Emissionswellenlänge konnte inkrementell in bis zu 2 nm Schritten zwischen 437 nm und 522 nm eingestellt werden.

Zuletzt wurde der Tris[1,2,4]triazolo[1,3,5]triazin Kern (TTT) als neue Akzeptorgruppe für TADF-Emitter erfolgreich evaluiert. Dazu wurde eine Synthese entwickelt, die den TTT-Kern mit peripheren Donoreinheiten dekoriert. Die Anzahl und Stärke der Donorgruppen beeinflusst maßgeblich die Emissionsfarbe (398–466 nm), während der Diederwinkel entscheidend für den TADF Charakter dieser Systeme ist. Die ersten himmelblauen OLEDs zeigen eine maximale EQE von 5.8 % und 11 %.

Abstract

Organic light-emitting diodes (OLEDs) are one of the most promising lighting technologies of the recent years not only due to their efficiency and energy saving capabilities, but also for their unique features in the realization of transparent and flexible display solutions. The commercial realization of large-area OLED panels is still limited by the dominant and cost-intensive vacuum processing techniques and challenges in solution processed device fabrication. Thermally activated delayed fluorescence (TADF) is considered as the most promising mechanism for electroluminescence for OLED application as this allows a complete conversion of excitons to photons.

In this thesis, the development of TADF materials for solution processed OLED devices was investigated. Existing isophthalonitrile-based TADF emitters were chemically modified in order to study the structure-to-property relationships with regards to the optoelectronic behavior. Based on the chemical modifications to the parent emitter, the emission color was varied in the range of 438 nm to 541 nm. Most promising emitters were employed in OLED devices, resulting in external quantum efficiencies (EQE) of up to 16%. A modular approach allowed the introduction of various functional groups and structural motives. Binding sites on TADF emitters were created in order to permit conjugation to other molecules or biotransporters. Combined systems of TADF emitter and biotransporter or rod-like monodisperse oligomers were used for fluorescence imaging or for the investigation of photophysical phenomena such as exciton transport and charge separation.

Phthalimide-based TADF emitters were thoroughly investigated with derivatization on the acceptor and donor groups to adjust optoelectronic properties and solubility for solution processing. For the emission color, a range of 437 nm to 522 nm was covered with typically small increments of 2 nm.

Lastly, the tris[1,2,4]triazolo[1,3,5]triazine (TTT) core was successfully implemented as novel acceptor group for TADF emitters. Hence, a synthesis for the TTT core with peripheral donor groups was developed. The concept was successfully extended to related systems yielding mono-triazolo triazines and bis-triazolo triazines. The donor strength and count were identified as key drivers for the emission color (398–466 nm) and the dihedral angle as critical measure for TADF characteristics. The initial sky-blue OLED devices showed maximum EQEs of 5.8% and 11%, respectively.

1 Introduction

"Light is a unifying symbol that signifies wisdom and excites the imagination across the world. Paintings and murals in all cultures show how artists have used light, shade and colour to illustrate mood and create atmosphere. Light is used in some therapies to promote health, and in religious ceremonies as aid to worship and reflection. On the most fundamental level through photosynthesis, light is necessary to the existence of life itself. Light science has revolutionized medicine, agriculture and energy, and optical technologies are part of the basic infrastructure of modern communications. For these reasons and more, light sciences are a cross-cutting discipline in the 21st century. As we strive to end poverty and promote shared prosperity, light technologies can offer practical solutions to global challenges. They will be particularly important in advancing progress towards the Millennium Development Goals, achieving the future sustainable development goals and addressing climate change. [...] This will mean more light in homes, hospitals and enterprises – and that will translate into a safer, healthier and more productive future." This is how Ban Ki-moon who served as the eighth Secretary-General of the United Nations, illustrated the importance of this research field at the opening ceremony of the International Year of Light and Light-based Technologies 2015.[1,2]

The sun as the most natural source of energy served nature and humanity since their dawn and brought them warmth and light. Fearing darkness and needing a heat source, the human race tamed one of nature's most destructive elements, fire.[3] Uncompeted for millennia, fire and all of its derived lighting sources like torches, oil lamps or candles, were substituted with the development of the incandescent light bulb by Thomas Alva Edison based on the observations of Louis Jacques de Thénard who found that current-carrying metal wires emit light and heat.[4,5] As most of the energy is lost as unwanted heat, the efficiency of the incandescent light bulb is merely 5%. Given this major drawback, the use of these in Europe is prohibited since 2009. However, the following generation of halogen bulbs with an energy saving of up to 30% also found their end with the production ban since 2018. The compact fluorescent lamp was believed to be a promising successor, however, the necessity of mercury as well as the relatively long time until maximum luminance renders them as an interim solution at most.[6] All of the light sources discussed so far generate light through detours, either by harvesting solely a small share of the injected energy via glowing metal wires with typical emission maxima in the infrared region (1000 nm) in incandescent light bulbs or by converting the initially created UV light into visible light as in compact fluorescent lamps. The development of light-emitting diodes through the utilization of an exciton generated in organic or inorganic semi-conductors by applying an electrical current is the first technique which allows the direct conversion of electrical energy to light.[5,7]

Despite the rapid development of increasingly efficient lighting techniques, approximately 15% of the global power consumption is still accounted to the electricity used for lighting, which then accounts for roughly 5% of the global greenhouse gas emission.[8] Despite introducing recent technologies in more developed countries such as the United States of America, the total electricity consumed for lighting in the U.S. still amounts to 6%.[9] In the same manner as the energy savings through novel lighting technologies rose, the general light output drastically increased as well, as illustrated in Figure 1 for central Europe. As the production costs for light sources decreased at the same time, illumination of previously unlit areas is increasingly available for more safety, convenience or comfort.[10,11] Apart from the need of solutions for more targeted alignment of lighting to avoid light pollution, the growing demand for lighting sources has still to be tackled by satisfyingly energy efficient technologies as Ban Ki-moon stated.

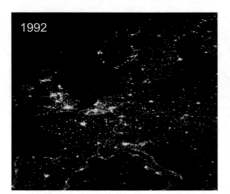 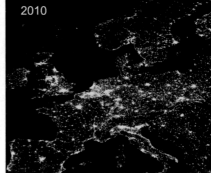

Figure 1. Nocturnal shot of central Europe from a satellite of the Defense Meteorological Satellite Program in 1992 (left) and in 2010 (right). Bright spots are caused by actual light emission on the ground. National borders were added afterwards for orientation purposes with red lines. Copyright: NGDC/DMSP/ESA.[10]

One of the most promising lighting technologies both for energy saving purposes as well as for innovative applications are organic light-emitting diodes (OLEDs). In contrast to the inorganic LED, where crystalline semiconductors are used as light source, the OLED is based on organic semiconducting emitters which can be processed in a thin film. The absence of crystalline materials allows solution-based processing techniques like inkjet printing,[12] spin-coating or photo-crosslinking[13] which are not only cost-efficient but also easily adoptable to a larger scale enabling the fabrication of large-area lighting solutions.[14] Already applied in commercial display solutions for smartphones or TV screens, OLEDs bring along another innovative solution: the realization of transparent[15] or flexible displays (Figure 2).[16] Besides these unique features, OLED displays are still impressive as they show high contrast, luminance and the realization of a true black. Hence, OLEDs fulfill modern demands for display solutions and also find application in the automotive industry as taillights[17,18] or by replacing the side-view mirror by a small camera and an indoor OLED display (Figure 2).[19]

Figure 2. OLED applications: replacement of the side-view mirror by an OLED display (left, copyright: Audi AG),[19] transparent display (center, copyright: LG Display),[15] flexible display (right, copyright: Samsung).[16]

1.1 Organic Light-Emitting Diodes (OLEDs)

The key phenomenon exploited in an OLED is the electroluminescence, which was first observed by Henry Joseph Round in 1907 when he witnessed light emission from silicium carbide upon the application of an electric current with a voltage of 10 V.[20] Later in 1955, this phenomenon was investigated by Bernanose *et al.* who measured phosphorescent dyes in a polymer matrix in dielectric cells,[21] while Pope and co-workers investigated the basic mechanisms on anthracene crystals in 1963.[22,23] Then, 24 years later, Tang and van Slyke reported the first OLED device based on the fluorescent tris(8-

hydroxyquinolinato)aluminum, which emitted green light at 550 nm with a brightness of 1000 cd m^{-2} at an external quantum efficiency of 1%.[24] Then, Heeger and co-workers' pioneering work on semiconducting and metal-like polymers was honored with the Nobel Prize in Chemistry in 2000, which paved the way for the commercial application of OLEDs.[25]

1.1.1   Working Principle and Architecture of OLEDs

The basic working principle of a simplified OLED, solely consisting of an organic emissive layer sandwiched between two electrodes is shown in Figure 3.[26,27] The work functions of the electrodes should be matched in such a fashion that the highest occupied molecular orbital (HOMO) is slightly lower in energy to the work function of the anode, while the lowest unoccupied molecular orbital (LUMO) needs to be slightly higher in energy to the work function of the cathode. This allows the injection of charge carriers into the organic material. Upon applying an external voltage to the electrodes, electrons are injected into the LUMO of the organic material at the interface from the cathode, while holes are generated by extracting electrons from the HOMO of the organic material at the Interface of the anode. Next, the charges migrate towards the electrodes of the opposite polarity due to the external electric field. This is believed to follow a hopping mechanism accounted to the distinct molecular orbitals.[28] As the conductivity is critical for the efficiency, π-stacking,[29] as well as conjugated π-systems were found to increase these processes.[30] When two opposite charge carriers approach each other closer than the Coulomb radius, they recombine to form an exciton located at an organic molecule. An exciton on a molecule, also described as excited state, can relax via different pathways. The emission of a photon by relaxation to the ground state is called electroluminescence.

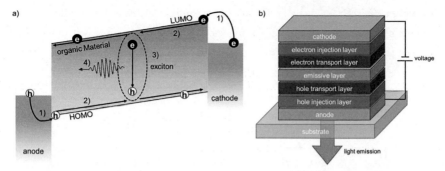

Figure 3. a) Simplified scheme of a single-layer OLED with the following elementary steps: 1) injection of charge carriers, 2) transport of charge carriers, 3) recombination of charge carriers, exciton generation, 4) light emission. b) Simplified OLED stack architecture. Figures recreated in accordance with literature.[31,32]

As the transport rate of charge carriers is different for holes and electrons, with the transport of holes known to be faster in organic semiconductors,[33] a multilayer device architecture with optimized layer thicknesses was found to be more efficient for OLEDs (Figure 3) to mitigate these issues. The **substrate** is needed to carry the OLED stack and typically consists of transparent polymers or glass. Like the substrate, the **anode** also must be transparent to allow light out-coupling. Transparent conductive oxides (TCO) like indium tin oxide (ITO, $(In_2O_3)_{0.9}(SnO_2)_{0.1}$) or non-indium-based TCOs are usually applied.[34] The **hole injection layer** (HIL) flattens out the usually rough surface of the anode material and facilitates the charge injection from the anode to the material by aligning the HOMO levels. PEDOT:PSS, a conductive polythiophene mixed with a sulfonic acid derivative is often used for this purpose.[5] As the name already suggests, the **hole transport layer** (HTL), facilitates the charge transport to the emissive layer. It is often built from electron rich heterocycles like triarylamines or carbazole derivatives. Furthermore, it acts as an **electron blocking layer** (EBL) preventing the electrons from entering this side of the stack. The **emissive layer** (EML) often consists of either solely the emitting molecule, or two or even three different materials forming a guest/host system. In this layer, the charges recombine in order to form excited emitter molecules (excitons) which then relax via emission of a photon. Like the HTL, the **electron transport layer** (ETL) simplifies the electron transfer via the LUMO while also preventing positive charge carriers from leaving the emissive layer. Lastly, the **electron injection layer** (EIL) as well as the **cathode** complete the device stack serving contrary tasks as the

HIL and anode. The cathode typically consists of readily available metals like magnesium, aluminum or different alloys.[5,32,35] Upon electrical excitation, the recombination of electron and hole results in the population of 25% excited singlet states and 75% excited triplet states according to the Fermi-Dirac statistics (Scheme 1).

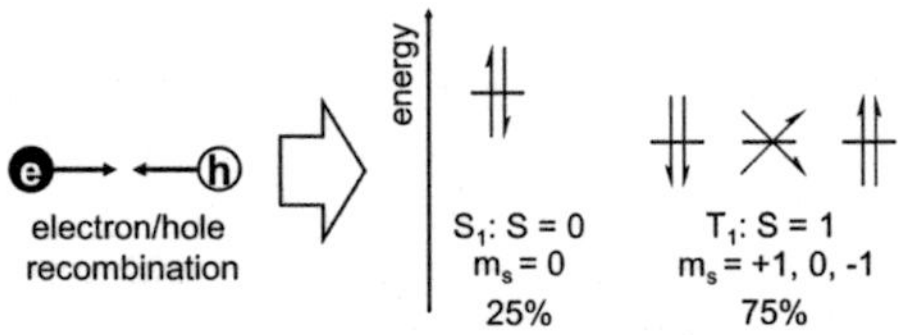

Scheme 1. Population of excited states upon charge carrier recombination.

1.2   Thermally Activated Delayed Fluorescence (TADF)

The relaxation from the excited singlet state S_1 to the ground state S_0 accompanied by the emission of light is called fluorescence (Scheme 2). Furthermore, the continuous population of the excited triplet state T_1 from S_1 via intersystem crossing (ISC) competes with the relaxation via fluorescence. As the excitons in the T_1 state cannot be used for light generation in conventional fluorescent emitters due to the forbidden $T_1 \rightarrow S_1$ transition, the maximum internal quantum efficiency is 25%. Typically, this is even lower as the ISC competes with fluorescence[36] and other non-radiative relaxation pathways.[37] Lastly, the populated triplet states can lead to additional unfavorable processes like triplet-triplet-annihilation (TTA) or radical mechanisms, which decrease the overall device lifetime. The first fluorescent molecule used in an OLED device, tris(8-hydroxyquinolinato)aluminum (Alq3, **1**) is a coordination complex of aluminum with three bidentate 8-hydroxyquinoline ligands (Figure 4) and emitted green light with a maximum of 550 nm in the first OLED device.[24]

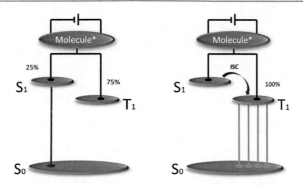

Scheme 2. Simplified Jablonksi diagrams for purely fluorescent emitter (left) and phosphorescent emitter (right). Applying voltage excites a molecule indicated with *. Basic pathways include $S_1 \rightarrow S_0$ (fluorescence), $T_1 \rightarrow S_0$ (phosphorescence) and $S_1 \rightarrow T_1$ (intersystem crossing (ISC)).

The second generation of OLEDs exploits that very triplet excitons by using heavy atom-based complexes to enhance not only the ISC from the excited singlet to the triplet state, but also to facilitate the $T_1 \rightarrow S_0$ transition accompanied by the emission of light (phosphorescence, Scheme 2) by spin-orbit coupling.[38,39] This class of metal complexes that are predominantly based on iridium,[39–42] but also on other heavy atoms like osmium[43,44], platinum[45,46], and ruthenium,[47,48] has found application in commercial electrophosphorescent devices. One of the most prominent phosphorescent dyes called Irppy3 (**2**)[36,39,49] is shown in Figure 4. It is often used as the standard for relative quantum yield measurements.[50] With this class of triplet-harvesting molecules, up to 100% of the generated excitons can be used for light generation. However, the precious metals needed for this emitter class is also considered as its bottleneck as they are cost-intensive, scarce and a limited resource. Lastly, the singlet excitons are converted to the triplet excitons which results in a loss of energy. As a consequence, the $T_1 \rightarrow S_0$ transition is slightly red-shifted in comparison to the $S_1 \rightarrow S_0$ transition.

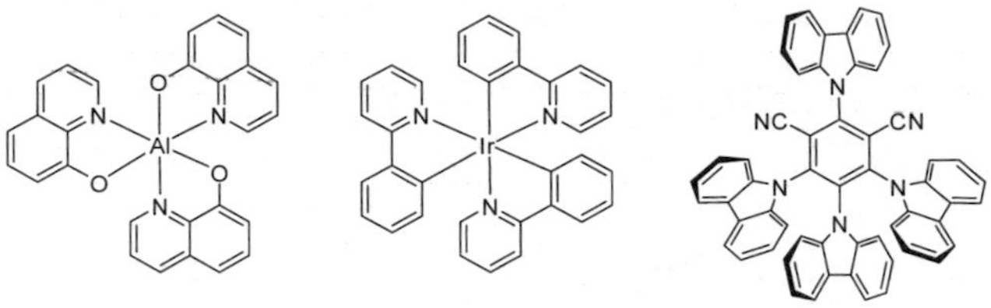

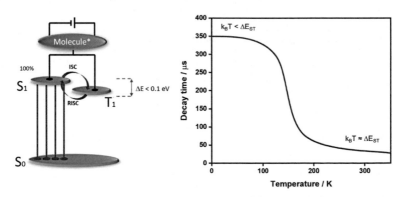

Figure 4. Fluorescent (left), phosphorescent (center), and TADF (right) dyes for OLEDs.

The concept of thermally activated delayed fluorescence (TADF) is considered to be one of the most promising mechanisms for electroluminescent devices. When the energy gap (ΔE_{ST}) between excited singlet S_1 and triplet T_1 state is sufficiently small, and the lifetime of T_1 long enough, the endothermic reverse intersystem crossing (RISC) from the T_1 to the S_1 state allows the harvesting of the triplet excitons via the fluorescent relaxation channel. The subsequent emission via the $S_1 \rightarrow S_0$ relaxation channel shows the same spectral distribution as the prompt fluorescence, but with a significantly longer decay time (Scheme 3). Developing TADF emitters that show these characteristics at ambient temperatures enables the usage of all electrically generated excitons in an OLED through singlet decay. This significantly increases the limit for fluorescent materials for internal efficiency from 25% to 100%.[51–54] Upon an increase in temperature, TADF emitters show a significant decrease of the fluorescence decay time indicating the efficient population of vibronic levels of T_1 followed by RISC (Scheme 3).

Scheme 3. Left: Simplified Jablonski diagram of a TADF emitter upon electrical excitation. Basic pathways include $S_1 \rightarrow S_0$ (fluorescence), $S_1 \rightarrow T_1$ (intersystem crossing (ISC)) and $T_1 \rightarrow S_1$ (reverse intersystem crossing (RISC)). Right: Temperature dependence of the emission decay time.

1.2.1   Working Principle of TADF

The working principle of TADF is under continuous investigation and has been summarized in numerous articles over the last years.[54-56]

The singlet-triplet energy gap (ΔE_{ST}) is the energy difference between the lowest excited triplet state (T_1) and lowest excited singlet state (S_1) and defines a system showing TADF characteristics. Thermal upconversion from the triplet state, followed by reverse intersystem-crossing (RISC), is possible when ΔE_{ST} is reasonably small (usually <0.1 eV).[51] Besides the prompt fluorescence from the S_1 to the S_0 state, TADF emitters usually show a second type of photoluminescence. The delayed fluorescence results from a triplet state converted to a singlet state via RISC prior to fluorescent emission.[51]

ΔE_{ST} is critical for efficient organic TADF materials as it directly influences the rate of RISC (k_{RISC}) according to the Boltzmann distribution (Equation (1), where k_B is Boltzmann's constant and T is the temperature).[37]

$$k_{RISC} \propto \exp\left(-\frac{\Delta E_{ST}}{k_B T}\right) \tag{1}$$

As seen in Equation (1), a large ΔE_{ST} results in a slow k_{RISC} which then results in a longer delayed fluorescence lifetime (τ_d) and also decreases the rate of delayed fluorescence (k_{DF} = 1 / τ_d).[57,58] The delayed fluorescence lifetime is further expressed in Equation (2)[59] which shows a decreasing τ_d upon an increasing rate of ISC (k_{ISC})[60] or RISC (k_{RISC}),[58,61] where the radiative and nonradiative decay rate constants of the singlet state are expressed by k_r^S and k_{nr}^S.

$$\frac{1}{\tau_d} = k_{nr}^T + \left(1 - \frac{k_{ISC}}{k_r^S + k_{nr}^S + k_{ISC}}\right) k_{RISC} \tag{2}$$

More importantly, the energy gap between the excited singlet (S_1) and triplet state (T_1) is dependent on the structure of the emitter and on the resulting exchange integral (J) as shown in Equation (3),[62] which itself is expressed by the density overlap between the frontier orbitals (HOMO and LUMO) under the assumption that the excited singlet (S_1) and triplet states (T_1) are dominated by a transition of LUMO to HOMO (Equation (4)). ϕ_{HOMO} and ϕ_{LUMO} are the spatial HOMO and LUMO distributions, and r_1 and r_2 are position vectors.[63] Decreasing the spatial overlap between the HOMO and LUMO results in a decrease of the exchange integral (J) and thus, ΔE_{ST}.

$$\Delta E_{ST} = E_S - E_T = 2J \tag{3}$$

$$J = \iint \phi_{HOMO}(r_1)\phi_{LUMO}(r_2)\frac{1}{|r_2 - r_1|}\phi_{HOMO}(r_2)\phi_{LUMO}(r_1)dr_1dr_2 \tag{4}$$

In theory, ISC and RISC processes in pure organic TADF emitters are spin-forbidden. However, due to a small ΔE_{ST} these transitions are allowed as shown in Equation (5).[51] Here, λ stands for the first-order mixing coefficient between the singlet and the triplet states and H_{SO} represents the spin-orbit interaction. In organic molecules, H_{SO} is typically small, a reasonably small ΔE_{ST} however, leads to efficient ISC and RISC processes.

$$\lambda \propto \frac{H_{SO}}{\Delta E_{ST}} \tag{5}$$

Another effect caused by a small ΔE_{ST} is the diminishment of El-Sayed's rule,[64] that states that there must be a change in symmetry of the excited states for an efficient ISC and RISC. As hyperfine-coupling (HFC) plays an important role for small ΔE_{ST}, El-Sayed's rule can be neglected in this case. Furthermore, different symmetries for the excited states are given, if T_1 is represented by a locally excited state (LE), while S_1 is represented by a charge-transfer state (CT). Typically, in TADF emitters, the excited states are neither pure CT nor LE states, but often mixed CT-LE states with varying shares. Hence, El-Sayed's rule can be attenuated, resulting in fast rates for ISC (k_{ISC}) and RISC (k_{RISC}).[65]

Furthermore, the heavy atom effect has often been utilized to enhance ISC and RISC. For this, heavier atoms have been implemented in the emitter. Predominantly sulfur,[66,67] but also other elements like selenium[68] have been reported. In addition, the manipulation of the device architecture using heavy atom based layer materials can lead to an enhancement of these rates due to an external heavy-atom effect.[69,70]

1.2.2   History of TADF

The TADF phenomenon, although initially reported as E-type delayed fluorescence, was first observed in 1961 by Parker and Hatchard for Eosin (**4**) (Figure 5) at elevated temperatures in ethanol.[71] Similar behaviors were later reported for benzophenone,[72] thioketones[73] and 9,10-anthraquinone.[74] In 1996, fullerene (5) was studied for its TADF properties and the rate equations for the description of the time-resolved processes of the TADF mechanism were derived (Figure 5).[75]

Until then, the TADF characteristics of the systems were not found to be efficient due to large ΔE_{ST} and the absence of design rules for efficient TADF systems. It were Yersin and Monkowius in 2008 that filed a patent claiming multinuclear complexes based on metals like iridium, palladium, platinum, rhodium and gold with small singlet-triplet energy gaps for application in OLEDs.[76] This novel mechanism of TADF for electroluminescence was first implemented in an OLED device by Adachi and co-workers in 2009 when they reported TADF properties for the Sn^{4+} porphyrin 6 shown in Figure 5 and employed this as the emitting material in a device.[77] One year later, the bis(phosphino)diarylamido dinuclear copper(I) complex 7 (Figure 5) was used as emissive material in an OLED with a high EQE of 16.1%.[65] This initial copper-based TADF emitter inspired many research groups like Yersin,[78,79] Bräse[80–82] and more.[83,84]

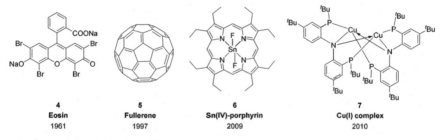

Figure 5. Historical outline of TADF emitters.

The breakthrough for TADF emitter design was in 2011, when Adachi and co-workers not only presented the first purely organic TADF molecule **PIC-TRZ** (8) (Figure 6) with a small ΔE_{ST} of 0.11 eV and an EQE of 5.3% in device, but also found the importance of reducing the overlap of the frontier orbitals to achieve a small ΔE_{ST} by quantum mechanical analysis. Hence, a molecular design was developed based on the combination of electron donor and acceptor groups in a twisted structure.[85] One year later, Adachi reported the prominent family of carbazolyl dicyanobenzenes (CDCB) by combining multiple carbazole groups as donor moieties with benzonitriles in different arrangements.[51] This class of emitters with its most prominent representative **4CzIPN**

(3) (Figure 6) with an outstanding performance with an EQE of up to 19.3% clearly demonstrated the harvesting process of both singlet and triplet excitons via the TADF pathway and demonstrated a significant breakthrough when compared to conventional fluorescence-based OLEDs. Since then, pure organic TADF emitters for application in electroluminescent devices are thoroughly investigated in literature not only for their potential to replace conventional fluorescent and phosphorescent emitting materials in already commercialized OLEDs but also for a better understanding of the mechanism and design principles for this special class of luminescent molecules, classified as "third generation emitters".[86]

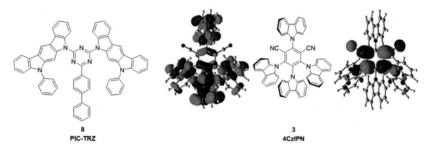

Figure 6. First pure organic TADF emitter (left) and **4CzIPN** (**3**) with its HOMO (left) and LUMO (right).

1.2.3   Molecular Design Strategies for TADF Emitters

Many design strategies for TADF emitters have been reported and were further refined since their advancement as OLED materials which can be followed in multiple review articles.[37,54–56,87] Most typically, the combination of donor and acceptor moieties is used to induce a charge transfer transition for efficient TADF characteristics. Most typically for donor groups, sterically demanding, annulated N-heterocycles like carbazole derivatives, triphenylamines or acridine-derived structures like phenoxazines, phenothiazine and 9,9-dimethyl-9,10-dihydroacridine (DMAC) are used. For acceptor groups, a vast plethora has already been reported. Predominantly however, benzonitriles, triazines, benzophenones, and sulfones are reported. Three distinct approaches have been proposed for molecular arrangement and design in order to create organic molecules with TADF properties.

1.2.3.1 Twist-Induced Charge Transfer

Based on the ground-breaking findings of Adachi and co-workers, one of the major design strategies uses the twist-induced charge transfer caused by a large steric hinderance between the acceptor and donor moieties. An increased dihedral angle between the acceptor and donor moieties separates the HOMO and LUMO orbitals and significantly reduces the exchange integral to a certain extent thus lowering the energy gap ΔE_{ST} as discussed earlier by localizing the HOMO on the donor groups and the LUMO on the acceptor groups.[56,88] The dihedral angle, mostly induced by the steric demand of the donor groups, can be varied depending on their relative arrangement. The prominent series of carbazolyl dicyanobenzenes (CDCB) by Adachi and co-workers squeezed up to four carbazole units and two nitrile groups to one central benzene unit, resulting in a highly twisted arrangement and separation of the frontier orbitals on the donor and acceptor groups (Figure 6).[51] Later, similar systems with up to five carbazole groups and one single acceptor unit such as **5CzBN (9)** were reported as efficient TADF emitters (Figure 7).[89,90]

In 2018, the group of Jun Yeob Lee investigated the dihedral angle control of two carbazoles linked via a phenyl unit to a diphenyltriazine acceptor. By investigating six isomers, it was found that the CT state, the singlet-triplet energy gap and the RISC rate could be affected by the dihedral angle. While the donor groups in 2-,6- and 2-,3- position showed the smallest energy gaps, substitution in 3-,4- and 3-,5- position was found to be not as efficient (**10**, Figure 7).[91] Another commonly used approach for increasing the dihedral angle is intentional installation of sterically demanding groups either on the donor group itself or on the respective spacer system. Often, the carbazole is methylated in the 1- and 8- position, or methyl groups are installed in *ortho* position to the donor group (**11**, Figure 7).[92] Furthermore, more profound modifications on the donor groups were also found to increase the dihedral angle and increase the TADF character of some molecules, like a triptycene modified carbazole[93] or the incorporation of the carbazole into a chiral [2.2]-paracyclophane (**12**) scaffold (Figure 7).[94]

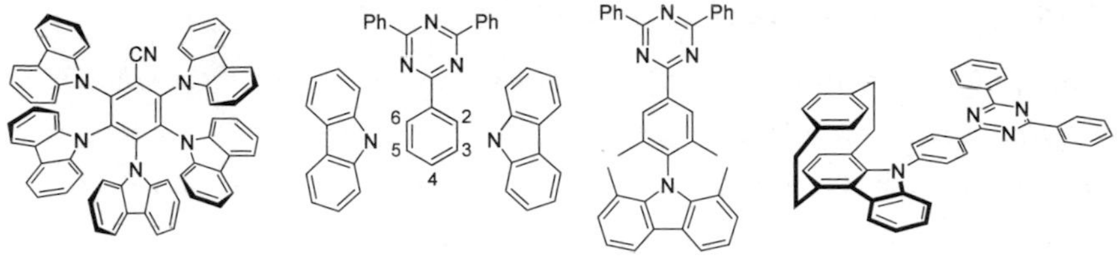

Figure 7. Different ways of adjusting the dihedral angle as crucial measure for a molecule showing TADF characteristics.

1.2.3.2   Through-Space Induced Charge Transfer

The second, less explored design principle for TADF molecules is based on a through-space induced charge transfer. Typically connected via a direct σ-bond or through twisted π-conjugation, this class separates its acceptor and donor groups and thus the frontier orbitals via a through space homojunction within a rigid framework. Hence, the charge transfer process happens through space via aromatic π-bonds or via a sp³-hybridized carbon center.[56]

Swager and co-workers designed triptycene-based emitters (**13**), where the acceptor and donor groups are separated on different fins of the triptycene scaffold. The frontier orbitals can still communicate via a homoconjugation (Figure 8).[95] Later, these systems were further investigated and a variation on the π-conjugation length and the donor units was conducted as a measure to optimize this class of TADF emitter.[96]

Adachi and co-workers reported the first TADF molecule implementing a spiro-center to separate the donor and acceptor groups.[97] The molecular structure of the acridine-based emitter (**14**) is shown in Figure 8. Here, the sp³-hybridized carbon center prevents an electronic communication, resulting in a minimal HOMO/LUMO overlap induced by the orthogonal arrangement and spatial communication. The concept of separation via a spirojunction was further explored with adjustments on the molecular setup.[98–100]

Spuling *et al.* utilized the unique [2.2]-paracyclophane scaffold once more to fix the benzophenone acceptor and diphenylamine as donor group in certain orientations (Figure 8). The two paracyclophane decks with a distance of 3.09 Å to one another[101] are closer in distance than the van der Waals distance between layers of graphite (3.35 Å).[102] Although the photoluminescence quantum yields of these emitters (**15**) was poor, blue

TADF emission via a communication of the donor and acceptor group through-space was demonstrated.[103]

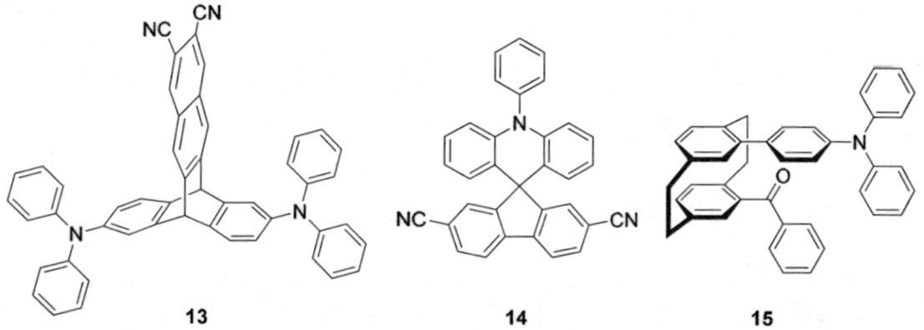

Figure 8. Emitters utilizing the through-space communication of acceptor and donor groups.

1.2.3.3 Multi-Resonance Effect

The pioneering work of Hatakeyama *et al.* in 2016 disregarded the conventional acceptor-donor design of TADF emitters and presented a class of organoboron-based emitters shown in Figure 9a.[104] In these molecules, the separation of the frontier orbitals is realized by "multi-resonance effects" where the HOMO and LUMO are alternated through the molecular framework. The boron atom localizes the center of the LUMO "mesh", while the HOMO is predominantly distributed on an alternative "mesh" centered at the nitrogen atoms. Both emitters, **DABNA-1 (16)** and **DABNA-2** show excellent optoelectronic characteristics such as large oscillator strengths of 0.205 and 0.415, excellent photoluminescence quantum yields (PLQY) doped as 1 wt% in mCPB of 88% and 90% and high EQEs of 13.5% and 20.2% with CIE coordinates of (0.13, 0.09) and (0.12, 0.13). The main advantage of this new class of emitters lies in their significantly thinner full width at half maximum (FWHM) of only 28 nm (typically 70–100 nm for conventional TADF emitters). This leads to a superior color purity than for conventional TADF emitters, facilitating the accessibility of the deep blue color region for industrial applications.[56]

Nakanotani *et al.* reported the light amplification abilities of **DABNA-2**,[105] while chemical modifications on the systems like extending the framework with additional boron and nitrogen atoms[106] or substitutions of donor groups on the framework,[107] mainly with the goal of developing a synthetic route and exploring the potential of this new TADF emitter class were conducted. The attachment of a carbazole group in *para* position to the boron atom conducted by Liang *et al.* led to a significant improvement of the efficiency of the resulting emitter (**17**) to be one of the best blue TADF-OLEDs up to date with an EQE of 32.1% (Figure 9b).[108] In depth theoretical investigations were conducted in 2019 by Pershin *et al.* to study the origin of the multi-resonance (MR) TADF emitters. A local alternating rearrangement of the electronic density upon excitation was found by using highly correlated quantum-chemical calculations.[109] Furthermore, a series of π-extended graphene-like structures doped with boron and nitrogen atoms were proposed as promising candidates for MR-TADF emitters (**18**) as shown in Figure 9c.

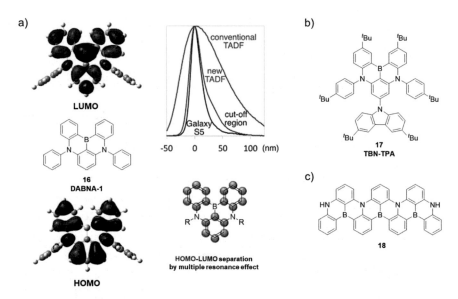

Figure 9. Developments of MR-TADF emitters: a) Chemical structure of **DABNA-1** (**16**) and its frontier orbitals, narrow emission peak compared to conventional TADF and the visualized HOMO-LUMO separation, (copyright: Wiley-VCH)[104] b) Substituted **DABNA-1** derivative (**17**) used in the most efficient blue TADF-OLED up to date, c) π-extended proposed structure based on theoretical investigations.

1.2.4 Molecular Design Towards Efficient OLEDs

Zheng *et al.* further reviewed on molecular design strategies in order to obtain efficient OLEDs.[56] Besides having a highly luminescent emitter with a high internal quantum efficiency, the out-coupling efficiency, strongly dependent on the device architecture is of crucial importance to achieve excellent external quantum efficiencies. Typically, the light outcoupling involves several channels for losses and is in the range of 20% to 30% for OLEDs.[110] The light outcoupling can be enhanced by deploying transparent electrodes and by targeted design of the refractive indices of the employed layers within the device. There are few possibilities for emitter design to enhance the light out-coupling, too.

1.2.4.1   Host-Free TADF-OLEDs

Typically, TADF emitters are doped in a host matrix to prevent exciton-exciton or exciton-polaron quenching. A less active emissive layer is also more gentle to the embedded molecules resulting in diminished efficiency roll-offs.[56] Recent studies found that the optimal doping concentration of TADF emitters is much larger than for fluorescent or phosphorescent emitting materials.[111] A fully host-free setup, solely using the pure emitter as emissive layer furthermore simplifies the device fabrication by preventing the need of depositing two materials simultaneously, resulting in lower costs in commercial applications and more reliable devices.[112-114] Still, it is necessary to suppress the non-radiative processes through interference of the molecular stacking or by largely twisted conformations.

The group of Adachi reported a set of sulfone and benzophenone based emitters (**19, 20**) (Figure 10) used in host-free OLEDs that were as efficient as the best doped OLEDs with maximum EQEs of 18.9% and 19.5%. They proposed that both, a large Stokes shift and weak $\pi-\pi$ stacking interactions, led to the suppression of the non-radiative processes.[115] The combination of the DMAC or phenoxazine donor with a triazine acceptor led to two similar TADF emitters, **DMAC-TRZ (21)** and **PXZ-TRZ (22)** (Figure 10). **DMAC-TRZ (21)** was not only found to result in highly efficient conventional doped OLEDS (EQE of 26.5%), but also in a device architecture employing a non-doped emitting layer with an EQE of 20%.[116] The possibility of achieving high efficiencies of one emitter in doped or non-doped OLEDs demonstrates the feasibility of device and fabrication simplification and ultimately cost reduction.

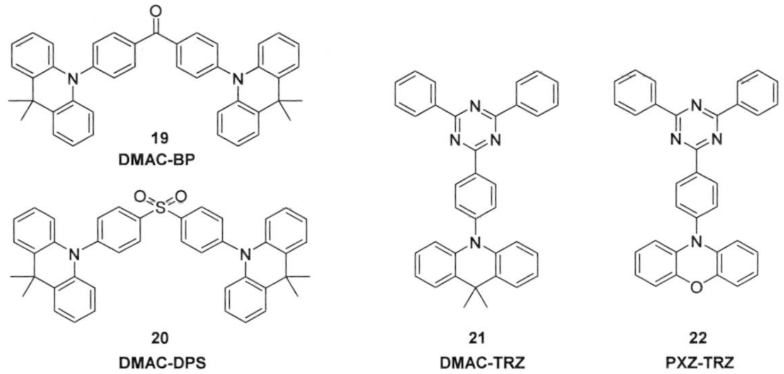

Figure 10. Efficient TADF emitters suitable for host-free device stacks.

1.2.4.2   Aggregation-Induced Emission

Another effect which was found and investigated to suppress concentration quenching was found in 2001 by Tang and co-workers and named aggregation-induced emission (AIE).[117–119] Molecules, showing this effect are non-luminescent in diluted solutions but show intense luminescence upon the formation of aggregates in solid state.

Wang *et al.* reported the first TADF emitter with AIE characteristics in 2014 by combining a thioxanthenone-dioxide acceptor with a *N*-phenyl carbazole and triphenylamine donor (Figure 11). The AIE properties were demonstrated by a significantly higher PLQY in solid state than in solutions. The investigation of the crystal structure and molecular packing gave evidence to understand the different emission behavior in solid state. As the molecules align themselves in a rigidified molecular conformation, the intramolecular rotation is hindered and non-radiative processes are suppressed. In solution, the emission is quenched by the enabled rotation of the donor units from the twisted conformation.[120,121] Although a non-doped device architecture was not reported, the doped OLEDs of **TXO-PhCz (23)** and **TXO-TPA (24)** showed maximum EQEs of 21.5% and 18.5%, respectively.

Groundbreaking results on AIE-based emitters were reported by Tang and coworkers in 2018 as they combined the 4-(phenoxazine-10-yl)benzoyl building block[122–125] with common host motifs like DCB, CBP, mCP and mCBP. Two of them, **DCB-BP-PXZ (25)** and **mCP-BP-PXZ (26)** are shown in Figure 11. The optoelectronic properties of this set of emitters were quite similar, mainly influenced by the 4-(phenoxazine-10-yl)benzoyl fragment. Almost non-fluorescent in solution, excellent PLQYs in non-doped films were reported, as well as excellent performing OLEDs with almost 100% exciton use, negligible efficiency roll-offs even at a luminance of 1000 cd m^{-2} and maximum EQEs of 22.6% for **DCB-BP-PXZ (25)** and 22.1% for **mCP-BP-PXZ (26)**. The emission quenching is suppressed by a highly twisted molecular geometry which also prevents the excitons at the central accepting benzoyl moiety from Dexter energy transfer. Lastly, the peripheral carbazole moieties are presumed to transfer excitons to the benzoyl moiety to further enhance the EL efficiencies.[126]

The combination of the benefits of host-free emissive layers for simplified device architectures and the outstanding performance of aggregation induced emission properties shows great potential for non-doped OLED devices with superior efficiency and stability.[56]

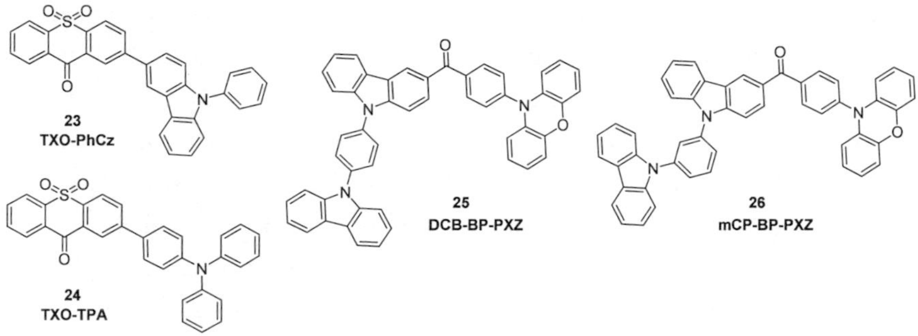

Figure 11. Emitters showing AIE properties: Thioxanthenone-dioxide based emitters (left) and the combination of common host groups with an emissive unit (right).

1.2.4.3   Orientation Controlled Emitters

The specific orientation of emissive molecules with respect to the device stacking direction is another tool for improving the device efficiency. The already mentioned out-coupling efficiency of 20% to 30%[110] is achieved upon an isotropic orientation of the emitting molecules. Studies have shown that a horizontal orientation of the emitters with respect to the substrate reduces the surface plasmon losses, and therefore significantly increases the out-coupling efficiencies.[127–130] The anisotropy factor Θ, defined as the share of vertically oriented dipoles (p_z) to the total dipole moments, as shown in Equation (6), can be used as a measure for the emitter orientation.[131]

$$\Theta = \frac{[p_z]}{[p_x] + [p_y] + [p_z]} \tag{6}$$

The group of Brütting investigated the previously reported triazine-based emitter **CC2TA** (**27**)[132] (Figure 12) and found a preferential horizontal orientation of its transition dipole moments of 92% when doped in DPEPO as host. Due to the TADF properties of the emitter, they observed an increase of the internal quantum efficiency by a factor of 2.24. However, the horizontal orientation also increased the external quantum efficiency due to an increased light out-coupling by a factor of 1.46 resulting in an experimental EQE of 11% that clearly exceeds the theoretical value (3.4%) for conventional fluorescent emitters.[133] Typically, the orientation is determined by analyzing angular photoluminescence spectra.[134]

Adachi and co-workers found that by adjusting the film fabrication temperature, the orientation of the triazine-based emitter **PXZ-TRZ** (**22**) could be controlled.[135] This is attributed to the molecular interaction of the emitting molecule and the mCBP host. Later in 2016, the group of Adachi demonstrated a correlation between the degree of emitter planarity and the preference for horizontal orientation. Hence, they designed a series of carbazole-based TADF emitters varying in their molecular planarity. Attributed to a strong intermolecular interaction between the emitter molecule and the host (DPEPO), the most planar emitters **BDQC-2** and **BDQC-4** (**28**) (Figure 12) showed the highest share in horizontal orientation, also effecting the EL performance of the resulting OLEDs.[136]

In conclusion, controlling the orientation of the emitting molecules either by molecular design or by fabrication techniques is a suitable tool to further enhance the efficiencies of OLED devices beyond the conventional limits.

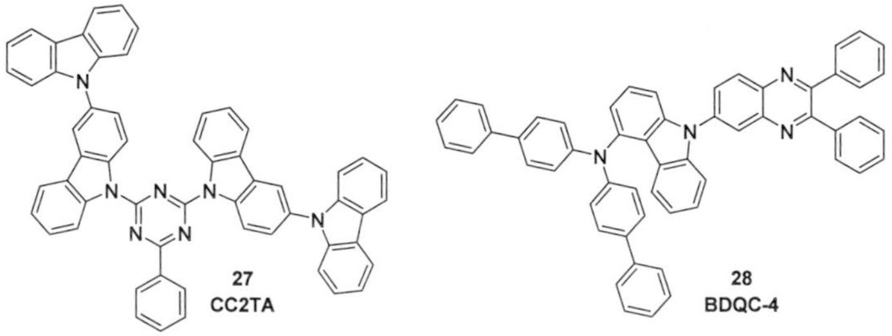

Figure 12. Reported molecules for horizontal orientation.

2 Objective

Lighting technologies are essential for working-around-the-clock modern society to provide basic services. Therefore, the continuous improvement towards higher efficiency is subject of lighting technology research to save energy, develop more environment-friendly processing techniques, and improve lifetimes of lighting devices to conserve resources.

In this regard, organic light-emitting diodes (OLEDs) have recently emerged as one of the most promising approaches, as they not only show favorable characteristics regarding large-scale fabrication and energy-saving capabilities, but in addition they possess unique advantages in the realization of novel display solutions such as flexible and transparent displays. In the field of OLEDs, the thermally activated delayed fluorescence (TADF) -based emitters are the most promising class for OLED application due to their excellent efficiencies and the absence of heavy metal ions in the molecular structures. Therefore, they are considered to be the 3rd generation of OLED materials. As not only the pure efficacy of an emitter is of great importance, but also the in depth understanding of the underlying fundamental mechanisms of TADF and structure-property relationships in order to fully understand the potential, increasing efforts have been devoted to explore the scope and limitations of TADF over the recent years.

The aim of this thesis was the synthetic development and in-depth characterization of three sets of TADF emitter classes for OLED application as well as for further less explored applications such as bioimaging and charge carrier transport investigations.

The rationale for the design is threefold: First, existing nitrile-based TADF emitters are chemically modified by transforming the nitrile acceptor groups into substituted tetrazoles or oxadiazoles as new accepting motifs. The modularity of this approach not only allows the adjustment of optoelectronic properties by altering the electron donating or withdrawing character, but the introduction of functional groups such as halides, alkynes, alkenes, and more offer pathways for follow-up chemistry and the attachment to other systems such as biomolecules or surfaces. Second, a phthalimide-based TADF emitter motif is thoroughly investigated through chemical modification with the aim to extensively explore structure-to-property relationships with the goal of realizing highly efficient, color tunable TADF emitters with enhanced properties for vacuum or solution processing. Lastly, the employment of the tristriazolotriazine (TTT) as acceptor motif is investigated in order to achieve a new family of TADF emitters. A synthetic protocol for the TTT core with peripheral donor groups is developed. The influence of donor strength, count and arrangement as well as variations on the dihedral angle between donor and acceptor groups on the optoelectronic properties is studied.

All promising TADF emitters that are synthesized are thoroughly characterized to assess their suitability for OLED application. The most promising candidates are then embedded as emitting materials in OLED devices.

TADF emitters featuring binding sites for follow-up chemistry are subjected to conjugation to molecular transporters for bioimaging and in the context of the Collaborative Research Centre 1176 "Molecular Structuring of Soft Matter" to rigid, monodisperse oligomers to investigate photophysical phenomena such as exciton transport and charge separation.

3   Results and Discussion

3.1   Derivatization of the 4CzIPN TADF System

In 2012, the Adachi group reported a set of highly efficient TADF emitters based on the combination of carbazole groups as donors and benzodinitriles as acceptors in different counts and arrangements.[51] To all three benzodinitrile isomers, namely phthalonitrile (1,2), isophthalonitrile (1,3) and terephthalonitrile (1,4), up to four carbazole units were attached, yielding the prominent set of carbazolyl dicyanobenzene-based TADF emitters **2CzPN (29)**, **4CzPN (30)**, **4CzIPN (3)** and **4CzTPN (34)** shown in Figure 13. By varying the count and arrangement of the carbazole donors, as well as by attaching methyl or phenyl groups in the 3- and 6-position of the carbazoles, the emission colors of the reported emitters in toluene ranged from 473 nm to 577 nm. Three of the reported emitters were employed in OLED devices, leading to sky-blue (**2CzPN (29)**), green (**4CzIPN (3)**) and orange (**4CzTPN-Ph (36)**) OLEDs. While the orange and sky-blue OLEDs showed maximum EQEs of 11.2% and 8.0%, the green OLED derived from **4CzIPN (3)** performed extraordinarily better with an EQE of 19.3%. Since then, all of these emitters were further investigated in literature, but predominantly the **4CzIPN (3)** system, which is mentioned in more than 115 patents and 200 journal articles to date, including a review article attributed solely to its derived application as a photocatalyst.[137]

Numerous modifications were conducted to change the optoelectronic properties of the emitter. Similarly to the initial modifications by Adachi, Lee *et al.* attached methyl and *tert*-butyl groups in the 3- and 6-position with the goal of enhanced solubility for solution-processed devices.[138] Besides the expected redshift due to the electron donating character of methyl and *tert*-butyl substituents, the latter was an efficient way in improving the solution processing of the emitting layer by stabilizing the film morphology, which was investigated by atomic force microscopy (AFM). The solution processed device of **4tCzIPN (32)** showed an improved EQE of 18.3% compared to 8.1% of the solution-processed **4CzIPN (3)** device.

Six years after the initial report of **4CzIPN (3)**, the Adachi group modified the carbazole donor units groups with electron withdrawing trifluoromethyl groups in the 3,6 and 1,8-position to introduce them to the **4CzIPN (3)** and **4CzTPN (34)** systems yielding **33** and **37**, respectively.[139] In contrast to the electron donating alkyl groups reported earlier, the electron withdrawing character of the CF₃-groups led to a significant blueshift in emission color, which is 40 nm to 55 nm shorter than for the unmodified parent emitters **4CzIPN (3)** and **4CzTPN (34)**.

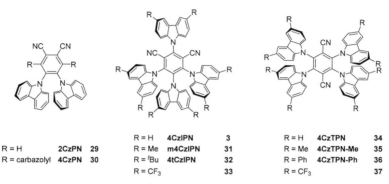

R = H		4CzIPN	3		R = H	4CzTPN	34
R = H	2CzPN 29	R = Me	m4CzIPN	31	R = Me	4CzTPN-Me	35
R = carbazolyl	4CzPN 30	R = ᵗBu	4tCzIPN	32	R = Ph	4CzTPN-Ph	36
		R = CF₃		33	R = CF₃		37

Figure 13. Carbazolyl dicyanobenzene series of TADF emitters first reported by Adachi and co-workers, and modifications to the carbazole donor groups.

Bunz and co-workers investigated the effects of attaching halides to the carbazole donors.[60] Each carbazole was modified with either one or two chlorides, bromides, or iodines, respectively. A significant shortening of the excited state lifetime, proportional to the weight of the halide atom was found. This decrease results from the heavy atom effect that facilitates intersystem crossing (ISC). The group of Waser used the effects of halide-substituted carbazoles for fine-tuning of the reduction potential, employing **4CzIPN (3)** derivatives as part of a photoredox catalytic system for the alkynylation of alkyl nitrile radicals upon oxidative ring opening of cyclic alkyl ketone oxime ethers.[140]

Zeitler and co-workers reported an impressive toolbox approach for the fine-tuning of the redox potentials of catalysts for photoredox chemistry starting from **4CzIPN (3)**.[141] They systematically investigated the effects of altering the carbazole group to diphenylamine or 3,6-dimethoxycarbazole. The count of donor and acceptor groups were varied, as well as their replacement by halides, mainly chlorides or fluorides. In total, eight catalysts were reported and tested, exhibiting balanced redox potentials and thus allowing a broad set of photoredox applications due to the tunable difference of the most reducing and most oxidizing potentials (Figure 14). These catalysts were successfully employed in the photocatalytic bromination of anisole, the sequential C-O-bond cleavage with subsequent reductive pinacol coupling, the detriflation of 2-naphthol triflate and more.

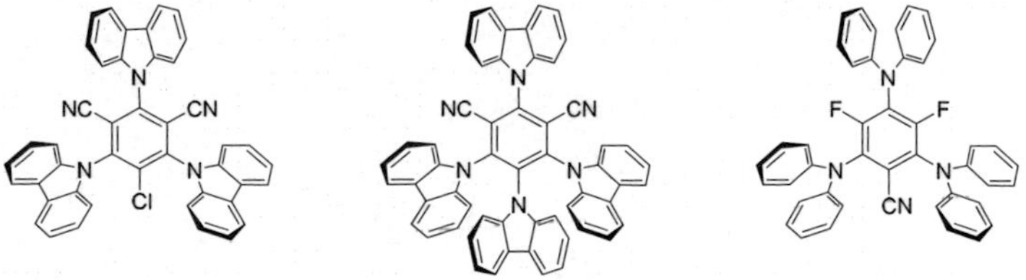

38	3	39
$E_{1/2}$ (PC$^{\bullet+}$/PC) = +1.79 V	$E_{1/2}$ (PC$^{\bullet+}$/PC) = +1.49 V	$E_{1/2}$ (PC$^{\bullet+}$/PC) = +1.24 V
$E_{1/2}$ (PC/PC$^{\bullet-}$) = -1.16 V	$E_{1/2}$ (PC/PC$^{\bullet-}$) = -1.24 V	$E_{1/2}$ (PC/PC$^{\bullet-}$) = -1.92 V

Figure 14. Structures and properties of the photocatalysts exhibiting the highest oxidizing and reducing potentials compared to **4CzIPN** (**3**).

4CzIPN (**3**), one of the most prominent TADF emitters in literature, was also the subject of this thesis. Despite numerous reported modifications of the donor units, little was reported on the derivatization of the nitrile acceptor groups. Therefore, the aim of this project was to find efficient derivatization strategies in order to establish an additional handle to fine-tune the TADF properties of this widely investigated system. The steric situation in **4CzIPN** (**3**) with regards to transformation of the benzonitrile groups shielded by the bulky carbazole units is very challenging, giving rise to the assumption that any modification that is achieved on this strongly demanding system can be transferred to any benzonitrile-based molecule.

DFT calculations were performed using the Turbomole program package. First, geometries of the molecular structures in the ground state in gas phase were optimized employing the BP86 functional and the resolution identity approach (RI). By employing Time-Dependent DFT (TD-DFT) methods, excitation energies were calculated using the (BP86) optimized structures. For orbital and excited state energies, the B3LYP functional and Def2-SVP basis sets and a m4-grid numerical integration were used. The molecular orbitals were visualized to get initial insights the localization of the frontier orbitals.

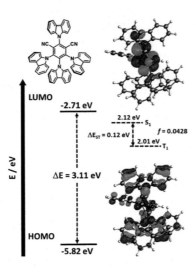

Figure 15. DFT calculations of **4CzIPN** (**3**). Results provided by cynora GmbH.

The relevant energy levels and frontier orbitals for **4CzIPN** (**3**) are shown in Figure 15. Similar to the reports in literature,[51] the HOMO is located on the carbazole-donors. Due to the presence of several carbazole groups, many carbazole-located molecular orbitals are formed partially delocalized over several carbazole moieties (HOMO, HOMO-1, HOMO-2, etc.). The LUMO is located on the benzonitrile-core. The HOMO and LUMO energies were calculated at –5.82 eV and –2.71 eV resulting in an optical band gap of 3.11 eV. The singlet-triplet energy gap was found sufficiently small with 0.12 eV.

3.1.1 Tetrazole Derivatives

Excerpts of this chapter have been already reported in the master thesis of the author. [142] This includes parts of the synthesis and basic photophysical characterization.

The core structure of **4CzIPN** (**3**) and its derivatives were synthesized by the reported methods starting from the perfluorinated isophthalonitrile precursor (**40**) and the respective carbazole groups via nucleophilic aromatic substitution using sodium hydride as base.[51] The synthesis and moreover the purification was slightly modified to transfer this method to a multi-gram scale to yield up to 36.4 g of **4CzIPN** (**3**) and 8.37 g of **4tCzIPN** (**32**) (Scheme 4).

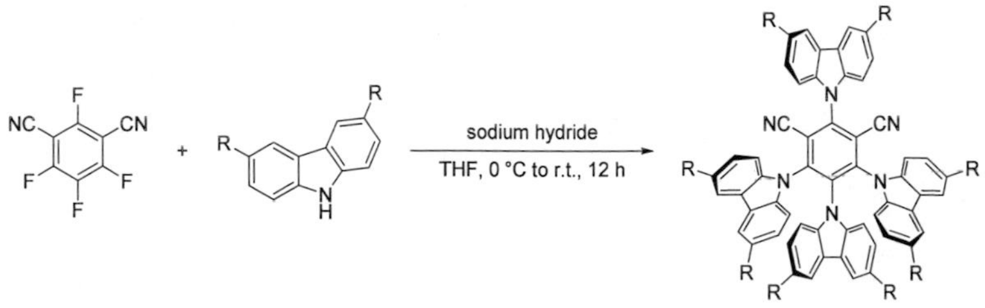

40		**41**	R = H		**3**	R = H	92%
		42	R = ^{t}Bu		**32**	R = ^{t}Bu	90%

Scheme 4. Synthesis of the TADF emitters **4CzIPN (3)** and **4tCzIPN (32)** via nucleophilic aromatic substitution.

3.1.1.1   Synthesis

In previous investigations it was determined, that the formation of tetrazoles is a suitable strategy in order to obtain weaker acceptors for TADF as well as a nucleophilic handle for further derivatization.[142]

For this, the nitrile groups were reacted with sodium azide to form tetrazoles using copper sulfate as a catalyst.[143] The outcome of the reaction can be varied between the mono-tetrazole derivative and bis-tetrazole derivative with respect to the equivalents of the added azide and catalyst. Both, **4CzIPN (3)** and **4tCzIPN (32)** were reacted to yield the respective mono- and bistetrazole derivatives of both cores as shown in Scheme 5. Single crystals, suitable for single crystal analysis were obtained for **4CzCNTl (43)** and **4CzdTl (45)** by slow evaporation of chloroform (Table 1). Due to the use of an excess of sodium azide and the need to acidify the reaction mixture during the workup, the maximum scale for this type of reaction was limited to two grams.

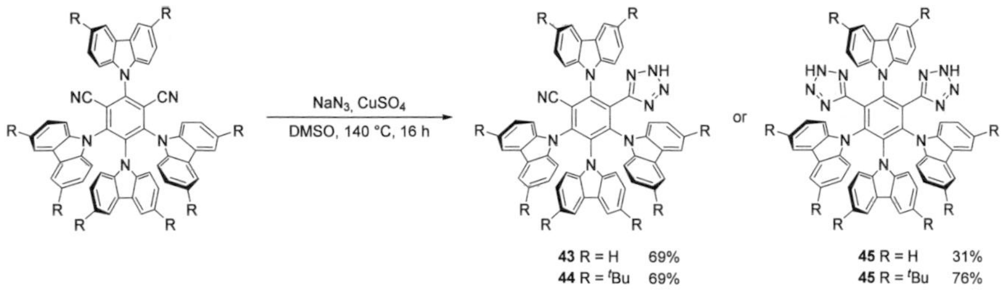

43 R = H   69%
44 R = ^tBu   69%

45 R = H   31%
45 R = ^tBu   76%

Scheme 5. Cyclization of the nitrile groups in **4CzIPN (3)** and **4tCzIPN (32)** with sodium azide to the respective tetrazole derivatives. The outcome was varied depending on the equivalents of reagents.

In a next step, alkylation of the tetrazole derivatives under mild reaction conditions using sodium carbonate as a base to deprotonate the tetrazole was investigated, followed by subsequent nucleophilic substitution using various alkyl halides.[144]

Using this method, a variety of structural motives- and functional groups such as esters, amides, halides, alkenes, alkynes and benzyls was introduced in fair to very good yields (Scheme 6). Due to the possible tautomerization of tetrazoles, alkylation can in principle take place in the 1- or 2-position. Both products of alkylation are known in literature and often separable by column chromatography.[145,146] Despite this, for **4CzCNTlallyl (47d)**, single X-ray analysis showed that the alkylation occurred exclusively in the sterically more accessible 2-position (Table 1).

In previous findings,[142] a highly regioselective 2-arylation of 5-substituted tetrazoles was found to be successful to attach arylboronic acids in the presence of [Cu(OH)(TMEDA)$_2$Cl$_2$] to yield the respective arylated tetrazole derivatives.[147] While this method did not allow the introduction of electron-deficient arenes, the non-substituted phenyl, and electron donating derivatives such as *p*-toluyl, and *p*-anisoyl groups could be successfully attached (Scheme 6). As reported in literature, the arylation in the 2-position was confirmed via single crystal analysis for **4CzCNTlOMePh (47k)**.

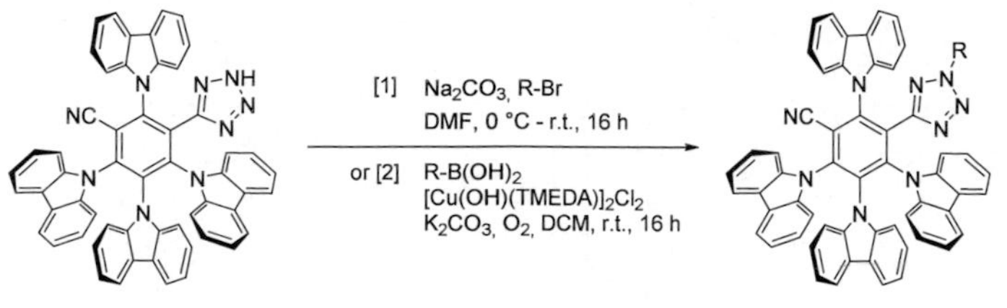

R =

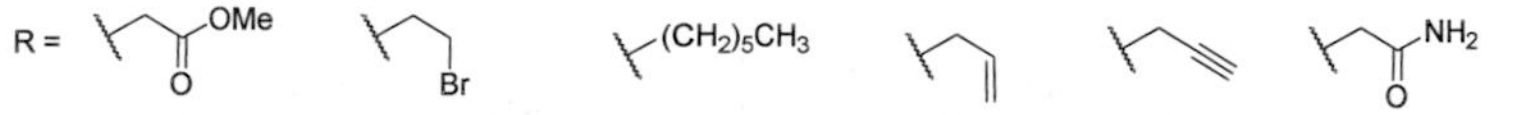

47a	47b	47c	47d	47e	47f
83%[1], a	78%[1], a	72%[1]	75%[1], a	70%[1], a	64%[1], a
4CzCNTICH₂COOMe	4CzCNTI(CH₂)₂Br	4CzCNTInHex	4CzCNTIallyl	4CzCNTIpgyl	4CzCNTICH₂CONH₂

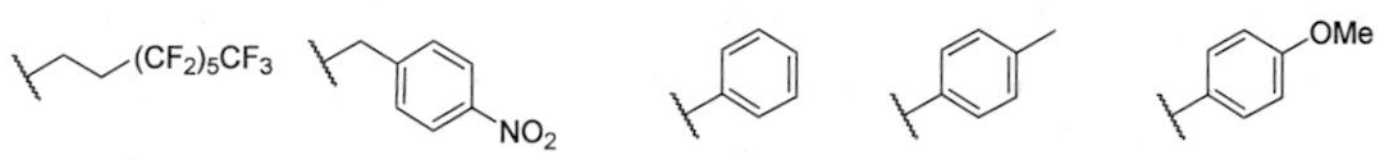

47g	47h	47i	47j	47k
63%[1], a	83%[1], a	61%[2], a	53%[2], a	67%[2], a
4CzCNTI(CH₂)₂(CF₂)₅CF₃	4CzCNTIBnNO₂	4CzCNTIPh	4CzCNTITol	4CzCNTIOMePh

Scheme 6. Synthesis and scope of the monotetrazole derivatives derived from **4CzIPN (3)** either via nucleophilic substitution [1] or by a copper catalyzed coupling reaction [2]. a: Result in the context of the authors master thesis.[142]

The molecular structure of the bis-arylated-tetrazole **4CzdTlPh (48c)** which was verified by X-ray analysis, showcases the absolute functionalization position in the tautomeric tetrazole system. Furthermore, the bis-tetrazole intermediate **4CzdTl (45)** was reacted in the nucleophilic substitution using the same conditions to yield **4CzdTlallyl (48a)** and **4CzdTlpgyl (48b)** in moderate yields (Scheme 7). Given the initial photophysical data with regards to TADF performance of both alkylated bis-tetrazole derivatives **4CzdTlallyl (48a)** and **4CzdTlpgyl (48b)**, (see 3.1.1.3 Basic Photophysical Characterization), **4tCzdTl (46)** with its slightly stronger donor groups was used as starting material for **4tCzdTlnHex (49)** to stabilize the HOMO orbital which results in an optimized emitter. The results and synthesized bis-tetrazole derivatives are shown in Scheme 7.

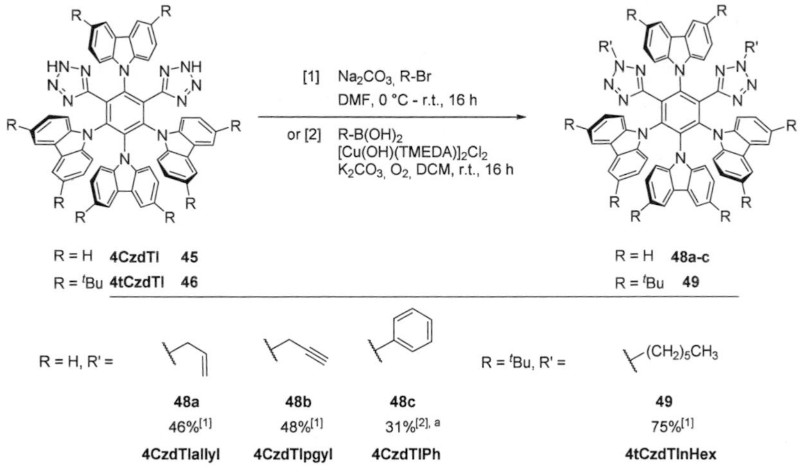

R = H   4CzdTl   45
R = ^{t}Bu 4tCzdTl   46

R = H   48a-c
R = ^{t}Bu   49

R = H, R' =

48a   48b   48c
46%[1]   48%[1]   31%[2], a

4CzdTlallyl   4CzdTlpgyl   4CzdTlPh

R = ^{t}Bu, R' =

49
75%[1]

4tCzdTlnHex

Scheme 7. Synthesis and scope of the bistetrazole derivatives derived from **4CzIPN** or **4tCzIPN** either via nucleophilic substitution [1] or by a copper catalyzed coupling reaction [2]. a: Result in the context of the authors master thesis.[142]

The molecular structures of tetrazole derivatives derived from **4CzIPN (3)** and the relevant data are shown in Table 1 and Figure 16. All six compounds show similar conformations with all four carbazole groups highly twisted in regard to the central benzene plane with dihedral angles between 60° and 87°. Interestingly, for all examples the top carbazole (Cz1), located in *ortho* position to both acceptor units (nitrile and tetrazole or both tetrazoles), is closer to orthogonal (70°–87°) than the other carbazoles units, although the acceptor units do not occupy as much space as carbazoles. The newly introduced tetrazoles can occupy a less twisted conformation due to their smaller size. Hence, the dihedral angles for the tetrazoles were measured to be between 47° and 60° with the exception of **4CzCNTl (43)** and **4CzdTlpgyl (47e)**. Upon a further arene substituent attached to the tetrazole such as in **4CzCNTlOMePh (47k)** and **4CzdTlPh (48c)**, the substituent was significantly less twisted with small torsion angles from 23° to 29°. Interestingly, **4CzdTlpgyl (47e)** showed significant variations compared to the other molecules, forcing all substituents of the central benzene unit to nearly orthogonal arrangements, with dihedral angles of up to 74° to 87°. On average, substituents in **4CzdTlpgyl (48b)** are 25° to 35° more twisted than in the other examples.

Table 1. Dihedral angles calculated from the molecular structures. Cz1 refers to the top carbazole located between tetrazole and nitrile or tetrazole and tetrazole. Cz2 to Cz4 refers to the next carbazoles when counting clockwise. Tl1 and Tl2 refer to the tetrazoles and if present, their substituted arene (Tl1R and Tl2R).

Entry	Name	#	Cz1	Cz2	Cz3	Cz4	Tl1	Tl1R	Tl2	Tl2R
1	**4CzCNTl**	**43**	81°	64°	67°	60°	69°	-	-	-
2	**4CzCNTlallyl**	**47d**	83°	71°	66°	73°	47°	-	-	-
3	**4CzCNTlOMePh**	**47k**	73°	65°	63°	62°	47°	25°	-	-
4	**4CzdTl**	**45**	70°	64°	67°	67°	60°	-	52°	-
5	**4CzdTlpgyl**	**48b**	87°	86°	77°	74°	85°	-	75°	-
6	**4CzdTlPh**	**48c**	73°	69°	67°	70°	49°	29°	50°	23°

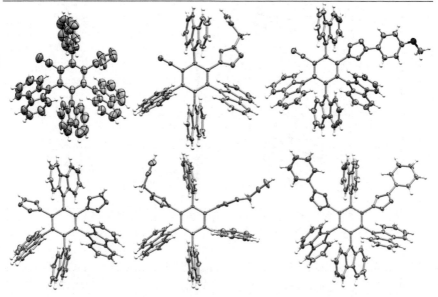

Figure 16. Molecular structure of **4CzCNTl** (**43**, top left), **4CzCNTlallyl** (**47d**, top center), **4CzCNTlOMePh** (**47k**, top right), **4CzdTl** (**45**, bottom left), **4CzdTlpgyl** (**48b**, bottom center) and **4CzdTlPh** (**48c**, bottom right); minor disordered parts omitted for clarity, displacement parameters are drawn at 50% probability level.

3.1.1.2   DFT Calculations

DFT calculations were performed using the Turbomole program package. First, geometries of the molecular structures in the ground state in gas phase were optimized employing the BP86 functional and the resolution identity approach (RI). By employing

time-dependent DFT (TD-DFT) methods, excitation energies were calculated using the
(BP86) optimized structures. For orbital and excited state energies, the B3LYP functional
and Def2-SVP basis sets and a m4-grid numerical integration were used. The molecular
orbitals were visualized to get initial insights on the influence of replacing the nitrile
acceptor with tetrazole groups.

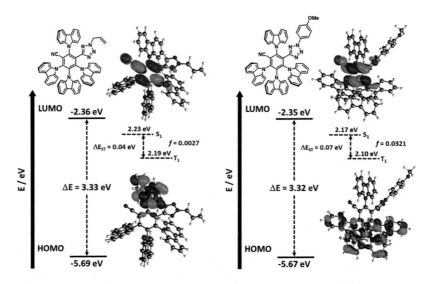

Figure 17. DFT calculations of the alkylated and arylated mono-tetrazole derivatives
4CzCNTIallyl (47d) and **4CzCNTIOMePh (47k)**. Results provided by cynora GmbH.

By comparing the results with **4CzIPN (3)**, some major changes can be observed. Similar
to **4CzIPN (3)**, the HOMO is located on the carbazole donors. Due to the presence of
several carbazole groups, many carbazole-located molecular orbitals are formed partially
delocalized over several carbazole moieties (HOMO, HOMO-1, HOMO-2, etc.). For **4CzIPN
(3)**, the LUMO is located on the benzonitrile-core. In case of the mono-alkylated tetrazole
derivative **4CzCNTIallyl (47d)**, the newly formed tetrazole is only partially part of the
LUMO orbital while most of the LUMO is still located on the benzonitrile moiety. This is
expected to destabilize the LUMO resulting in an increase of the optical band gap resulting
in a blue shift of the emission with respect to the reference emitter **4CzIPN (3)**. Indeed,
while the HOMO is also slightly increased from −5.82 eV to −5.69 eV, the LUMO is
significantly destabilized from −2.71 eV to −2.36 eV resulting in an overall increase in the

band gap of 0.22 eV. A similar situation is observed for the mono-arylated tetrazole **4CzCNTlOMePh (47k)**, resulting in very similar energy levels than for the LUMO of –2.36 eV for **4CzCNTlallyl (47d)** and –2.35 eV for **4CzCNTlOMePh (47k)**. Furthermore, there is minute influence on the HOMO level, resulting in similar band gaps of 3.33 eV and 3.32 eV, respectively. Hence, the substitution pattern on the mono-tetrazole derivatives barely has an influence on the resulting emission wavelength of the emitter. Both, the mono-alkylated and mono-arylated tetrazole derivatives show promisingly low ΔE_{ST} values of 0.04 eV and 0.07 eV which are beneficial for an efficient RISC process.

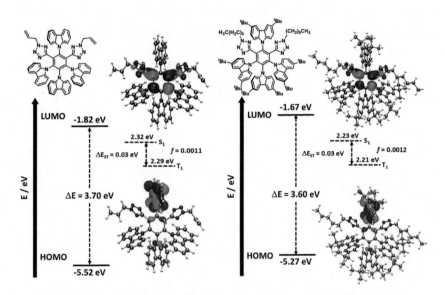

Figure 18. DFT calculations of the alkylated bis-tetrazole derivatives **4CzdTlallyl (48a)** and **4tCzdTlnHex (49)**. Results provided by cynora GmbH.

Replacing both nitriles with substituted tetrazole groups as in **4CzdTlallyl (48a)** and **4tCzdTlnHex (49)** destabilizes the LUMO even further, resulting in a significant increase of the optical band gap as shown in Figure 18. While the HOMO is slightly increased by 0.17 eV compared to **4CzCNTlallyl (47d)**, the LUMO is lifted by 0.54 eV to –1.82 eV. Here, the tetrazole starts to localize the LUMO orbital on its nitrogen atoms adjacent to the central benzene. After the initial optoelectronic results, the implementation of 3,6-di-*tert*-butylcarbazole as donor groups was used to stabilize the HOMO, thus leading to a more stable and efficient emitting molecule. As expected, the HOMO for **4tCzdTlnHex**

(**49**) is increased in comparison to the emitter bearing unsubstituted carbazoles by 0.25 eV due to the electron-enriched donor groups. However, the LUMO is further destabilized resulting in a similar band gap of 3.60 eV compared to 3.70 eV for the unoptimized structure **4CzdTlallyl** (**48a**). Due to the small calculated ΔE_{ST} of 0.03 eV for both emitters, an efficient RISC should be possible. This shows that tetrazole groups act as acceptor motives in TADF emitters indeed.

3.1.1.3   Basic Photophysical Characterization

Basic photophysical characterization was conducted by cynora GmbH. The photoluminescence spectra of the substituted tetrazole derivatives were recorded using the emitter as 10 wt% dopant in PMMA (Table 2). The unsubstituted tetrazole derivatives bearing acidic *NH* protons at the tetrazole groups show blue and blue/green emission maxima at 444 nm for **4CzdTl** (**45**) and 480 nm for **4CzCNTl** (**43**). The corresponding photoluminescent quantum yields are reduced compared to the initial **4CzIPN** (**3**) reference to 60% and 29%, respectively.

Alkylation of the mono-tetrazole derivatives drastically increased the photoluminescent quantum yields to a level similar to **4CzIPN** (**3**) ranging from 83% for **4CzCNTl(CH₂)₂Br** (**47b**) up to 94% for **4CzCNTlallyl** (**47d**). The substituent modestly influenced the maxima of emission ranging from 479 nm to 485 nm, allowing the modification of the variation of the emitter without affecting its emissive properties. All characterized mono-alkylated tetrazoles showed short excited state lifetimes for the delayed component of the emission from 7.9 µs to 12 µs.

Arylation of the tetrazole was not effective in terms of achieving high photoluminescent quantum yields. While the arylated tetrazole **4CzCNTlPh** (**47i**) showed a redshift in emission maxima up to 503 nm, the corresponding PLQYs drastically suffered, yielding 16%, 39% and 49%.

Similar to the monoarylation, bisarylation resulted in the poorly emissive molecule **4CzdTlPh** (**48c**) with 9% PLQY emitting at 474 nm. Due to the poor performance of mono- and bisarylated tetrazole derivatives, the arylation approach was discarded. Alkylation however, resulted in a significant blueshift in emission color to 438 nm for **4CzdTlpgyl** (**48b**) and 441 nm for **4CzdTlallyl** (**48a**), while maintaining moderate to good photoluminescent quantum yields of 37% and 53%. Both emitters showed moderate to long excited state lifetimes of 44 µs and 207 µs, respectively. In order to fine tune the emissive properties, a slightly stronger donor, 3,6-di-*tert*-butylcarbazole, was implemented to stabilize the HOMO, accepting the accompanying minimal redshift. As a result, emitter **4tCzdTlnHex (49)** was obtained emitting at a maximum of 448 nm with a photoluminescent quantum yield of 68% and a reasonably short excited state lifetime of 37 µs for the delayed component.

Table 2. Photophysical characterization of tetrazole-based TADF emitters.

Entry	Compound	#	$\lambda_{max, PL}$ [nm]	PLQY [%]	Excited state lifetime τ [µs]
1	**4CzIPN**	**3**	516	89	-
2	**4CzCNTl**	**43**	480	60	200
3	**4CzdTl**	**45**	444	29	32
4	**4CzCNTlPh**	**47i**	503	49	8
5	**4CzCNTlTol**	**47j**	503	39	5
6	**4CzCNTlOMePh**	**47k**	493	16	7
7	**4CzCNTlCH₂COOMe**	**47a**	485	91	12
8	**4CzCNTl(CH₂)₂Br**	**47b**	485	83	11
9	**4CzCNTlnHex**	**47c**	480	88	7.9
10	**4CzCNTlallyl**	**47d**	479	94	12
11	**4CzCNTlpgyl**	**47e**	485	87	10
12	**4CzCNTlCH₂CONH₂**	**47f**	487	87	11
13	**4CzdTlPh**	**48c**	474	9	-
14	**4CzdTlallyl**	**48a**	438	37	44
15	**4CzdTlpgyl**	**48b**	441	53	207
16	**4tCzdTlnHex**	**49**	448	68	37

The emission spectra of emitters **4CzCNTlallyl (47d)**, **4CzCNTlPh (47i)**, **4CzdTlallyl (48a)** and the optimized **4tCzdTlnHex (49)** are shown in Figure 19 concluding the effects of replacing either one or both nitriles with a tetrazole, and subsequent functionalization via arylation or alkylation and also the effect of fine-tuning the emissive properties by adjusting the donor groups. Regardless of the attached substituent, alkylation of tetrazole-modified emitters is a straightforward method to yield highly efficient emitting molecules.

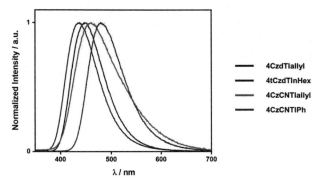

Figure 19. Emission spectra of selected mono- and bis-tetrazole derivatives doped in PMMA as host.

3.1.1.4   Advanced Photophysical Characterization

All experiments and analyses concerning the advanced photophysical characterization and OLED fabrication were conducted via collaboration with Lorenz Graf von Reventlow of the Dr. Colsmann group at the Light Technology Institute of the KIT.

In-depth photophysical studies were conducted for one emitter of this series, the mono-tetrazole derivative **4CzCNTlallyl** (**47d**) as this emitter showed the highest quantum yield, thus is most promising for an OLED application. To investigate if the modification still shows TADF characteristics, time- and temperature-dependent photoluminescence measurements were carried out. The photoluminescence decay of the emitter **4CzCNTlallyl** (**47d**) in tetrahydrofuran (0.1 g/L) showed two components, one assigned as the prompt (t_1 = 26 ns) and the other as delayed emission (t_2 = 6.5 µs). Upon flushing with oxygen for one minute, the delayed emission nearly vanished completely, indicating triplet states as the origin for the delayed emission (Figure 20a).[51]

Spin-coated thin films of **4CzCNTlallyl** (**47d**) in mCP as the host were subject to temperature-dependent PL measurements from 80 K to 300 K. Different lifetimes of the delayed emission were found with 1.2 µs in the host versus 6.5 µs in solution, attributed to the different molecular environment.[148] As shown in Figure 20b, the share of the delayed component significantly decreases with temperature and reaches 0% at around 100 K. This is a clear indication for the presence of the TADF mechanism, as a TTA mechanism would still show a delayed component at low temperatures. Also, the irradiation dependence of the emission can be fitted linearly (Figure S 1) in contrast to a TTA mechanism based case. The rate of intersystem crossing (k_{RISC}) can be calculated from the shares between prompt and delayed components I_{DF}/I_{PF}, while the temperature dependence yields the activation energy which can be approximated as the ΔE_{ST}. This gap was calculated at 62 meV, which is close to the energy of the reference emitter **4CzIPN** (**3**) (ΔE_{ST} = 83 meV)[51] (Figure 20c).

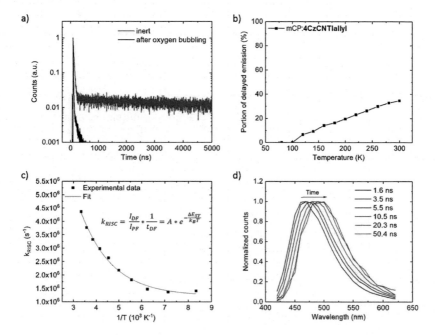

Figure 20. a) Emission decay of **4CzCNTlallyl** (**47d**) in THF in an oxygen-free environment (green line) and after bubbling with oxygen (black line). b) Temperature dependence of the delayed component in the emission of mCP:**4CzCNTlallyl** (**47d**). c) Reverse intersystem crossing rate k_{RISC} of mCP: **4CzCNTlallyl** (**47d**) in dependence on the inverse temperature. Inset: Formula to calculate ΔE_{ST}. d) Time-resolved emission spectra of mCP:**4CzCNTlallyl** (**47d**) after excitation with a 1 ns pulse at 340 nm. T = 0 ns is set to the time where the first photons of the pulse are detected at the PMT.

The emitter **4CzCNTlallyl** (**47d**) that is embedded in the host mCP showed two fast decay components in the range of prompt fluorescence (6 ns and 25 ns) unlike in solution. Hence, time-correlated single photon counting (TCSPC) measurements were recorded from 420 nm to 620 nm with changing emission wavelength in 10 nm steps. A clear redshift in the emission maximum was observed from 465 nm to 490 nm (Figure 20d).

This behavior was also observed for a similar class of emitters (see 3.1.2.4 Advanced Photophysical Characterization).

3.1.1.5   OLED Fabrication

The emitter **4CzCNTlallyl** (**47d**) not only possesses the highest PLQY among the synthesized set, but also other optoelectronic properties which are beneficial for an OLED application. The ionization potential of 5.9 eV was determined by photoelectron spectroscopy in air (PESA) and the optical bandgap was determined from the absorbance onset of 2.7 eV. With its small singlet-triplet gap ΔE_{ST} of 62 meV and the triplet energy approximated at 2.6 eV, **4CzCNTlallyl** (**47d**) was used as emitting material in OLED devices.

Solution-processed OLEDs with the architecture ITO/PEDOT:PSS/mCP: **4CzCNTlallyl**/BP4mPy/Liq/Al were built with a maximum EQE of 13.5% at 100 cd/m^2 with sky-blue emission at CIE1931 coordinates of (0.17;0.34). Unfortunately, the lifetime of the devices was very low with severe degradation during repeated J-V-L measurements and already accompanied by a redshift in emission and a drastic decrease in EQE (Figure S 3).

The molecular degradation of the tetrazole motif is suspected to either appear upon the increased temperatures within the devices or by electrical and optical excitation. Hence, the tetrazole groups possess excellent initial optoelectronic properties, although they were not found to be stable under electric excitation conditions used in OLED application.

3.1.1.6   Bioconjugation

Fluorescence imaging is a versatile method in investigating and understanding biological activities and systems *in vivo* and *in vitro*.[149] Besides being a minimally invasive technique, fluorescence imaging offers high resolution[150,151] which can also investigate deeper layers of tissue by spectroscopic measures.[152,153] One of the major limitations of this technique is the high autofluorescence of the tissue.[153–155] This issue can in principles be overcome by an increase of fluorescence efficiency and tuning of the emission wavelength of the fluorescent dye.

Several strategies have been developed to overcome the background autofluorescence. Near-infrared (NIR) emitters were used to isolate the target fluorescence from the autofluorescence.[155–157] However, the typically small Stokes shifts for organic NIR emitters often lead to reabsorption of emitted photons which results in weak emission.[158] When using emitters with long-lived fluorescence lifetimes, time-resolved fluorescence imaging (TRFI) can be used to separate the short-lived autofluorescence resulting in a high signal-to-noise ratio.[159–162]

Phosphorescent emitters based on heavy metal complexes possess reasonably long lifetimes in the order of milliseconds. Typically Ru(II)-, Ir(III)-, Pt(II)-, and Ln(III)-based complexes,[163–166] have been the predominant subject of interest for TRFI. One of the major drawbacks of these heavy metal based fluorophores is the accompanying toxicity of the ions.[167]

Similar to phosphorescent emitters, the long-lived excited state of TADF emitters is sufficient to be considered as a luminophore for TRFI. In 2014, Peng and co-workers reported **DCF-MPYM (50)**, a fluorescin derivative shown in Figure 21, which showed a long-lived luminescence of 22.1 µs.[154] The emitter was used to stain MCF-7 cancer cells, which were then investigated by time-resolved fluorescence imaging. **DCF-MPYM (50)** was found to be a successful biolabel for TRFI applications under enhanced cell penetration techniques. Furthermore, costaining with a lysosome-specific dye showed the localization of the emitter in lysosomes of living cells. Lastly, the low cytotoxicity renders **DCF-MPYM (50)** as a suitable TADF-based emitter for TRFI applications.

In 2017, Huang and co-workers combined the TADF emitter **CPy (51)** with **DSPE-PEG2000 (52)** to give **CPy**-based organic dots (Odots), which showed a long fluorescence lifetime of 9.3 µs in water at ambient conditions.[168] The **CPy-Odots** (Figure 21) were used for time-resolved and confocal fluorescence imaging of HeLa cells. Costaining with a typical cell membrane marker showed the preferential localization to be in the plasma membrane, although **DSPE-PEG2000 (52)** encapsulated Odots usually accumulate in the cytoplasm. The potential of **CPy-Odots** was further demonstrated in vivo by confocal microscopy and time-resolved fluorescence imaging experiments in living zebrafish. In these experiments they could successfully show that the fluorescence of the **CPy-Odots** was clearly distinguishable from the weak florescence of the organs by means of fluorescence lifetime imaging.

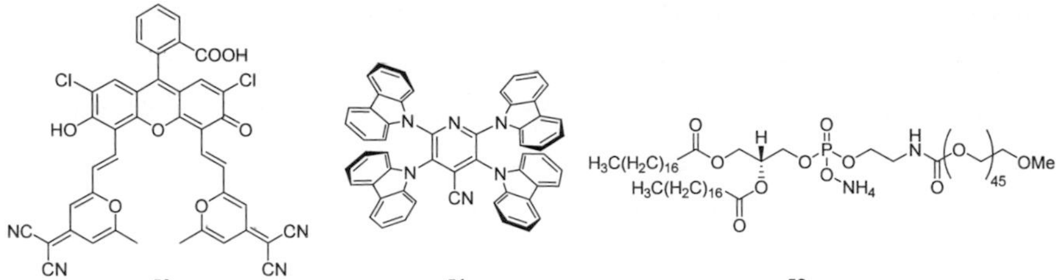

Figure 21. Fluorescin-derived TADF emitter (left) and the combination of TADF emitter **CPy** and **DSPE-PEG2000** (right) for time-resolved fluorescence imaging applications.

The TADF emitter **PXZT (53)** was first reported in 2018 and was rationally designed for time-resolved imaging experiments. It is based upon combining the phenoxazine donor group with a terpyridine as the acceptor moiety (Figure 22).[169] The terpyridine group not only serves as the acceptor group, but also as a binding site for zinc ions, which cause the quenching of the fluorescence. Hence, the release of zinc ions leads to a hydrophobic aggregation of the free **PXZT (53)**, which turns on the TADF emission and protects the emitter from the influence of oxygen. The zinc-containing molecule **ZnPXZT1 (54)** was employed in time-resolved fluorescence imaging of HeLa and 3T3 cells. After incubation, green luminescence was detected, which is attributed to the zinc-free **PXZT (53)**. This proves the decomplexation during the cell imaging process. The decomplexation is probably induced as zinc is an essential trace element for eukaryotes[170] and many zinc-based proteins.[163]

The group of Hu modified a prior developed phthalimide-based TADF emitter by attaching the mitochondria-targeting moiety triphenylphosphonium (TPP) and the lysosome-targeting group 2-morpholinethylamine which resulted in **AI-Cz-MT (55)** and **AI-Cz-LT (56)** as shown in Figure 22.[171] Fluorescence confocal imaging experiments were conducted with HepG 2 cells, demonstrating the ability to enter cells. By addition of Mito Tracker Green and Lyso Tracker Red DND-99, the organelle-targeting abilities of the two systems were validated. Lastly, time-resolved fluorescence imaging studies on HepG 2 cells were conducted and the fluorescence lifetimes were found with 16.7 μs and 28.0 μs, respectively.

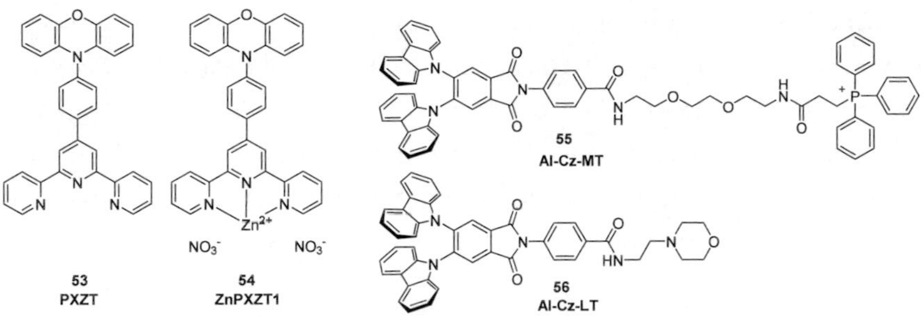

Figure 22. Phthalimide-based TADF dyes with mitochondria- and lysosome-targeting groups (right) and terpyridine-based TADF emitter (left).

Different approaches have been reported for the transportation of TADF dyes into cells for imaging applications. As already mentioned, the dye itself can possess a sufficient cell-penetrating ability like **DCF-MPYM (50)** and the zinc-containing **ZnPXZT1 (54)**. The TADF dye can also be embedded in an Odot (**CPy-Odots**) or conjugated to cell-penetrating peptides to be transported into cells.[172] Lastly, the covalent conjugation to a transporter moiety not only enables cell-penetrating properties, but the organelle-specific targeting groups like in **AI-Cz-MT (55)** and **AI-Cz-LT (56)** or similar ones reported by Yang and co-workers[173] act as a versatile tool for organelle-specific accumulation of the TADF luminophores.

For many years, the Bräse group has been investigating peptoids as analogous structures to peptides as pharmaceutically active agents and as molecular transporters. Molecular transporters are a class of molecules, that carry a covalently bound cargomolecule through the cytoplasmic membrane into the inner cell.[174] Peptoids based on lysine and arginine are known for their cell-penetrating properties.[175,176] Based on their side chains, the specific localization within the cell can be varied. To accumulate in mitochondria, peptoids need to possess a positive charge and a decent lipophilicity.[177]

All reactions and experiments concerning the peptoid and TADF-transporter synthesis were performed via collaboration with Dr. Stephan Münch of the Prof. Stefan Bräse group and have partially been reported in the respective doctoral thesis.[178] The following experiments and analyses for the bioimaging experiments were conducted by Dr. Ilona Wehl of the Prof. Ute Schepers group at the Institute of Toxicology and Genetics of the KIT and have partially been reported in the respective doctoral thesis.[179]

In order to combine the previously synthesized TADF emitters with peptoid transporters, the propargylic function of **4CzCNTlpgyl (47e)** was used for covalent conjugation via the copper catalyzed alkyne-azide cycloaddition (CuAAC). Three different peptoids were employed differing in the lipophilicity of the side chains, resulting in three TADF-transporter conjugates shown in Figure 23 where the TADF emitter is attached via the freshly formed triazole heterocycle to the peptoid transporters.

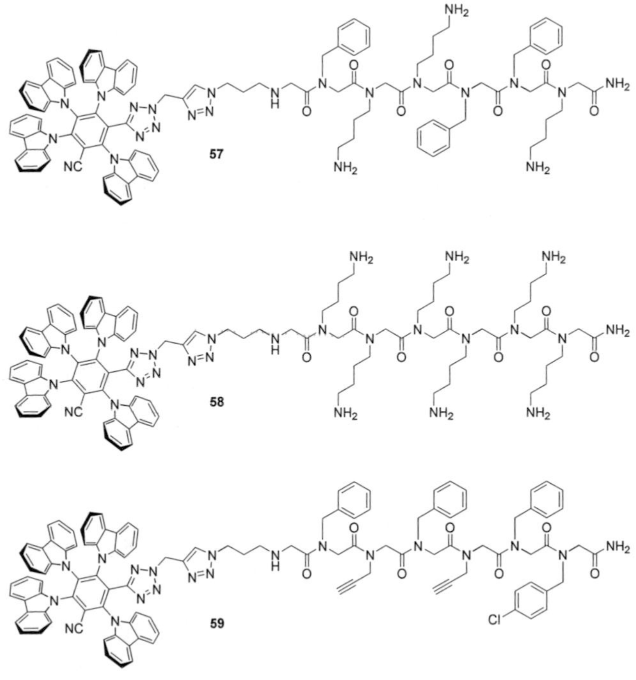

Figure 23. TADF-transporter conjugates with varying peptoid side chains, synthesized and characterized by Dr. Stephan Münch.

The synthesized TADF-transporter conjugates **57**, **58**, and **59** were investigated for biological application. Hence, toxicity, cellular uptake and intracellular localization was analyzed by Dr. Ilona Wehl in the group of Prof. Ute Schepers. The MTT assay for **57** and **58** is depicted in Figure 24 and demonstrated no toxicity to HeLa cells at low micromolar concentrations. For concentrations between 1 µM and 30 µM and 72 h incubation time, the LD_{50} values could not be determined.

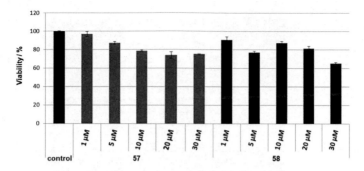

Figure 24. Cytotoxicity of TADF-transporter conjugate **57** and **58** to HeLa cells. Viability was determined for concentrations between 1 μM and 30 μM for 72 h using MTT assay.

To investigate the cellular uptake and intracellular localization, 1×10^4 HeLa and SK-MEL 28 cells and 2×10^4 RAW cells were incubated with the TADF-transporter conjugates **57** and **58** at different concentrations (1 μM, 5 μM and 10 μM) for 24 hours. Fluorescent confocal microscopy was used to analyze the cell-penetration properties. Both conjugates displayed strong uptake and fluorescence. The TADF-transporter conjugates were mainly accumulated in the endosomal-lysosomal system as seen in Figure 25 and Figure 26. The localization presumably resulted from the positive lysine side chain triggering the endocytotic uptake. No difference in uptake or localization for the different cell lines was detected.

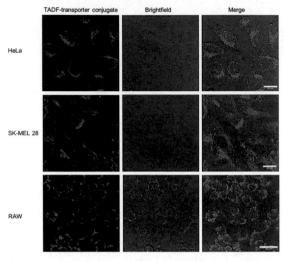

Figure 25. Cellular uptake of TADF-transporter conjugate **57** in HeLa, SK-MEL 28 and RAW cells. Cells were treated with 10 μM solution for 24 h at 37 °C. Accumulation was detected with fluorescence confocal microscopy Leica TCS-SPE. Excitation of the fluorophore at 405 nm; emission was recorded for the TADF-transporter conjugate at 420–550 nm at a scale of 20 μm.

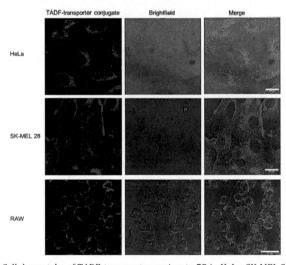

Figure 26. Cellular uptake of TADF-transporter conjugate **58** in HeLa, SK-MEL 28 and RAW cells. Cells were treated with 10 μM solution for 24 h at 37 °C. Accumulation was detected with fluorescence confocal microscopy Leica TCS-SPE. Excitation of the fluorophore at 405 nm; emission was recorded for the TADF-transporter conjugate at 420–550 nm at a scale of 20 μm.

The emission of the conjugates was measured between 410 nm and 550 nm upon excitation at 405 nm with a UV laser. The emission maximum was found to be at 480 nm which is in accordance with the photophysical characterization of the precursor emitter **4CzCNTlpgyl (47e)**. The conjugation to the peptoid transporter did not change the emission wavelength of the resulting TADF-transporter conjugate as the propargylic and triazole groups are insulated from the TADF acceptor system by one sp^3-hybridized center.

Furthermore, localization of the TADF-transporter conjugate **59** with the lipophilic peptoid moiety was tested in HeLa cells. The colocalization was supported by the addition of a Mito Tracker which is known to stain the mitochondria selectively. The overlapping yellow-orange areas in the merge image (Figure 27) resulted from randomly overlapping peptoid vesicles with mitochondria. In addition, a big share of endosomally localized TADF-transporter conjugates (green) is visible.

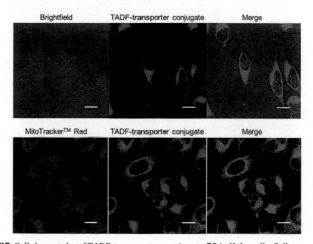

Figure 27. Cellular uptake of TADF-transporter conjugate **59** in HeLa cells. Cells were treated with µM solution for 24 h at 37 °C. Accumulation was detected with fluorescence confocal microscopy Leica TCS-SPE. Excitation of the fluorophore at 405 nm; emission was recorded for the TADF-transporter conjugate at 450–602 nm, and 637–697 nm for the MitoTracker at a scale of 15 µm.

To investigate the suitability of the TADF-transporter conjugates for applications in vivo, **58** was tested in zebrafish larvae. Hence, dechorionated 48 hpf (hours post fertilization) zebrafish embryos (Casper strain, mitfa$^{w2/w2}$; roy$^{a9/a9}$), anesthetized with 0.02% tricaine, were treated with 1 mM TADF-transporter conjugate **58** in E3-medium (5 mM NaCl, 0.17 mM KCl, 0.33 mM $CaCl_2$, 0.33 mM $MgSO_4$) solution by microinjection directly in the blood circulation into the caudal vein. One day after injection, fluorescence confocal microscopy was employed to investigate the accumulation on the zebrafish embryos. As seen in Figure 28, the TADF-transporter conjugate **58** strongly accumulated in the blood vessel system, mostly in endothelial cells and macrophages. Moreover, no noticeable toxic side effects on the zebrafish embryos were observed upon TADF-transporter conjugate treatment.

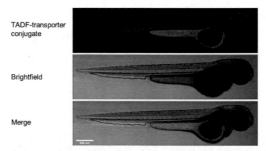

Figure 28. 3 dpf old zebrafish injected with 1 mM TADF-transporter conjugate **58**. Accumulation was detected with fluorescence confocal microscopy Leica TCS-SPE. Excitation of the fluorophore at 405 nm; emission was recorded for the TADF-transporter conjugate at 420–550 nm at a scale of 300 μm. The fluorescence detected from the yolk sac is not due to the TADF transporter but to autofluorescence.

The TADF-transporter conjugates showed selective cellular uptake and endosomal accumulation in vitro (HeLa, SK-MEL 28 and RAW cells) and in vivo in the blood vessel system, endothelial cells and macrophages in zebrafish embryos. For future optimization, the TADF emitter itself may be specifically designed to give a smaller molecule, diminishing its negative side effects like toxicity, lipophilicity and improve kinetic properties.

3.1.2   Oxadiazole Derivatives

Excerpts of this chapter have been already reported in a peer reviewed article of the author.[180] This includes parts of the synthesis, DFT calculations, basic and advanced photophysical characterization and OLED fabrication.

The electron accepting 1,3,4-oxadiazole is a thermally stable and weakly electron withdrawing group which is well known for its electron transporting and injection properties[90,181–185] and has been widely used as a part of electron-transporting materials for OLEDs.[186–188] Furthermore, oxadiazoles have been applied as part of conventional fluorescent emitters for OLEDs.[189–191]

In 2013, Adachi and co-workers developed one of the first oxadiazole based TADF emitters by combining the strong electron donating phenoxazine group via a phenyl spacer to the oxadiazole acceptor resulting in the D-A and D-A-D-type structures **PXZ-OXD (60)** and **2PXZ-OXD (61)** (Figure 29).[192] Both emitters were found to have similar PLQYs in toluene solution of 18.1% and 19.1%. However, **2PXZ-OXD (61)** showed more efficient TADF in the absence of oxygen, increasing its PLQY to 43.1% compared to 29.8% for **PXZ-OXD (60)**. Doped in DPEPO as host material, the PLQY of **2PXZ-OXD (61)** could be further increased up to 87% at an emission wavelength of 517 nm, resulting in a green OLED with an EQE of 14.9%.

Theoretical evaluation of a combination of DMAC with oxadiazole similar to previous findings, resulted in the two structures **ACR-OXD (62)** and **2ACR-OXD (63)**, (Figure 29) both showing reasonably low ΔE_{ST} values of 0.0076 eV and 0.0068 eV in silico and could therefore be considered promising oxadiazole-based TADF emitters.[193]

In a similar arrangement, the oxadiazole acceptor unit was combined with a 5,10-diphenyl phenazine donor to yield the emitter **PPZ-DPO (64)** which showed an emission maximum at 577 nm with a PLQY of 12%. Embedded in mCP as host, the PLQY could be further increased and resulted in a device with an EQE of 9.5%.[52]

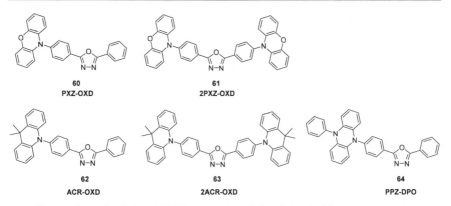

Figure 29. Oxadiazole based TADF emitters in a similar D-A or D-A-D arrangement with varying donor groups.

In 2017, Tao and co-workers combined oxadiazoles with carbazole groups as donor motives to develop **5CzOXD (65)** (Figure 30), a light blue TADF emitter with an emission peak at 496 nm and a PLQY of 58.2% in degassed toluene.[90] Besides its suitability in a light blue OLED with an EQE of 9.3%, its triplet energy level of 2.58 eV allowed this emitter to be used as a host material for green and yellow TADF OLEDs as well with EQEs of 16.2% and 17.1%. Lastly, a two-color WOLED implementing **5CzOXD (65)** and another emitter was fabricated resulting in an EQE of 7.2% with CIE coordinates of (0.35, 0.45).

During the work of this thesis, the Zysman-Colman group used the well-known **2CzPN (29)** emitter as starting material to convert the nitrile groups into oxadiazoles in an efficient two-step protocol.[194] It was shown that replacing the nitrile with weaker acceptor units significantly blue-shifted the emission wavelength of the parent emitter. In addition, different oxadiazole derivatives were reported to fine tune the photophysical properties (Figure 30). The doped emitters in thin-film showed emission maxima between 435 nm to 474 nm depending on the electron donating or withdrawing character of the substituent at the oxadiazole. The resulting sky blue (CIE of (0.17, 0.25)) and deep blue (CIE of (0.15, 0.11)) OLEDs of **2CzOXD4CF₃Ph (69)** and **2CzOXD4MeOPh (68)** with emission maxima at 474 nm and 446 nm showed maximum external quantum efficiencies of 11.2% and 6.6% respectively. Very recently, the Zysman-Colman group reported the same transformations starting from the isomer **2CzIPN**.[195]

A similar system was investigated by Kippelen and co-workers for the tetracabazolyl terephthalonitrile system (Figure 30). The resulting **TCZPBOX (70)** emitter derived from the well-known **4CzTPN** (**34**) with *tert*-butylphenyl groups attached to the oxadiazoles showed a photoluminescence emission maximum in toluene at 569 nm and a PLQY of 71% in neat film.[196] This led to the fabrication of an OLED with a host-free emissive layer with a maximum EQE of 21%. Further optimization of the device with the emitter doped in a host at 40 wt% led to an improved EQE of 28% in a yellow-green OLED.

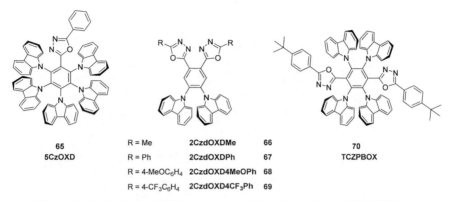

65	R = Me	2CzdOXDMe	**66**	**70**
5CzOXD	R = Ph	2CzdOXDPh	**67**	TCZPBOX
	R = 4-MeOC₆H₄	2CzdOXD4MeOPh	68	
	R = 4-CF₃C₆H₄	2CzdOXD4CF₃Ph	69	

Figure 30. Oxadiazole based TADF emitters derived from the well-known **5CzBN** (**9**) (left), **2CzPN** (**29**) (center) and **4CzTPN** (**34**) (right) systems.

In the very recent time, oxadiazole-based TADF emitters have been subject to many investigations focusing on the arrangement, count and type of donor used. Combining carbazole donors with one or two oxadiazoles have been found to result not only in efficient TADF emitters, but moreover possess the demonstrated potential to generate sky-blue to deep blue TADF emitters such as **5CzOXD** (**65**) or **2CzOXD4MeOPh** (**68**) and derivatives.

3.1.2.1   Synthesis

Many synthetic routes towards oxadiazoles are available, such as the condensation of a hydrazide with a carboxylic acid[197] or an aldehyde[198] or a intermolecular C-H functionalization of a hydrazide.[199] Still, the transformation of 5-phenyl-tetrazoles to 1,3,4-oxadiazoles upon reaction with acyl chlorides in pyridine at 60° to 100 °C, as reported by Huisgen and co-workers in 1960[200] and reused also by other groups[194,201] was found to be the most straightforward route, given the presence of the already installed tetrazole groups.

The proposed mechanism shown in Scheme 8 starts with the acylated tetrazole, whereas carboxamide resonance and aromatic tetrazole resonance are competing for the lone pair of the nitrogen in 2-position (**II**). After the ring opening, the tetrazole resonance is lost in the trait for resonance structures of **III** involving the carboxamide. Subsequent loss of molecular nitrogen yields intermediate **IV** which cyclizes to the 1,3,4-oxadiazole **V**.

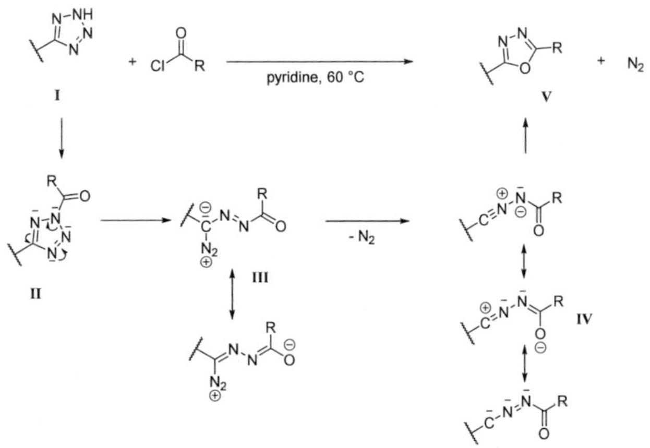

Scheme 8. Proposed mechanism for the formation of a 1,3,4-oxadiazole from a tetrazole upon reaction with acyl chlorides.

Although similar systems derived from **2CzPN**[194] (**29**), **2CzIPN**[195] and **4CzTPN**[196] (**34**) were reported, the respective **4CzIPN** (**3**) derived oxadiazoles have not been reported so far. Furthermore, solely the replacement of both nitrile moieties by oxadiazoles was reported. Given the availability of the **4CzIPN** (**3**) derived tetrazoles **4CzCNTl** (**43**) and **4CzdTl** (**45**) investigated in the previous chapter, transformation to mono- and bis-oxadiazole derivatives was investigated for this system. First the solvent was screened. Aside from the originally used pyridine reported by Huisgen, toluene, dioxane, tetrahydrofuran, chloroform and ethyl acetate were tested in a reaction of the mono-tetrazole **43** with acyl chloride at 100 °C over night. While only chloroform and pyridine showed full conversion, chloroform was selected as solvent for further investigations.

Furthermore, to monitor the progress of the reaction with the aim to investigate the mechanism, intermediates and reaction time, time resolved NMR spectroscopy was conducted (Figure 31). However, besides a conversion of the starting materials to the product, no intermediate was observed. Due to the little changes to the aromatic protons, barely any change was observed in the aromatic region between 6.5 ppm and 8.2 ppm. While the signal for the methyl protons of the acetyl chloride (2.4 ppm) slightly decreases over time, the signal of the continuously formed methyl group attached to the oxadiazole **71b** (1.59 ppm) could be successfully monitored reaching its maximum after 8 hours. This reaction time screening and excerpts of the following substrate scope were conducted in the context of a supervised bachelor thesis.[202]

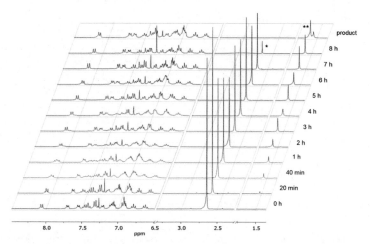

Figure 31. NMR screening: Monitoring of the reaction progress after 0, 0.33, 0.66, 1, 2, 3, 4, 5, 6, 7 and 8 hours and the final product. *acetyl chloride, ** product (**71b**).

The rationale for the choice of substituents at the oxadiazole is three-fold: First, the effect of implementing electron withdrawing and donating groups on the optoelectronic properties was studied (Scheme 9). Hence, alkyl chains (**4CzCNOXD(CH2)10CH3** (**71c**), **4CzCNOXDtBu** (**71a**)) and electron-rich arenes (**4CzCNOXD2,6OMePh** (**71i**), **4CzCNOXDMes** (**71h**)) were introduced. As electron withdrawing moieties, a perfluorinated alkyl chain (**4CzCNOXD(CF2)6CF3** (**71g**)) and electron-poor arenes (**4CzCNOXD4CNPh** (**71q**), **4CzCNOXDC6F5** (**71n**)) were used. Second, in order to show the feasibility of this method, the steric hindrance from the acyl chlorides in addition to the already present steric hindrance from the shielding carbazole moieties was investigated. Sterically demanding acyl chlorides like pivaloyl chloride, 2,4,6-trimethylbenzoyl chloride and 2,6-dimethoxybenzoyl chloride were used to yield the respective oxadiazole derivatives in good to high yields. Lastly, the tolerance of functional groups and structural motives was examined to open pathways for follow up chemistry. Many functional groups like ester, ether, nitrile, alkenes, halides and heterocycles were tolerated and successfully conjoined to the respective oxadiazole derivatives. As seen later in the photophysical investigations, (see 3.1.2.3 Basic Photophysical Characterization) **4CzCNOXD(CF₂)₆CF₃ (71g)** was found to show the strongest redshift in emission color amongst the tested substituents. Hence, it was combined with the slightly stronger 3,6-di-*tert*-butylcarbazole donor to further extend the limits of this method, resulting in **4tCzCNOXD(CF₂)₆CF₃ (72)**.

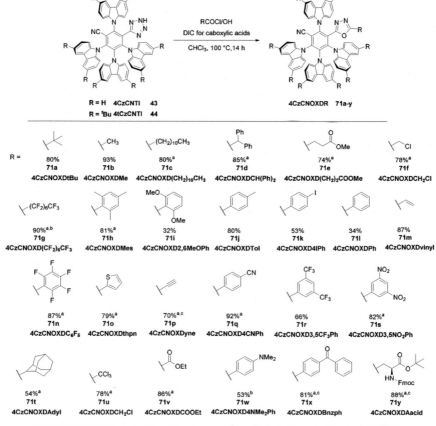

Scheme 9. Synthesis and scope of the mono-oxadiazole derivatives derived from **4CzIPN** or and **4tCzIPN (32)**. a: result from a supervised bachelor thesis[202] b: yield for **4tCzCNOXD(CF₂)₆CF₃ (72)** = 79%. c: reaction starting from the respective carboxylic acid.

Furthermore, carboxylic acids can also be used to transform a tetrazole to the corresponding oxadiazole derivatives by addition of a carbodiimide reagent such as *N,N'*,diisopropylcarbodiimde (DIC).[203] Using this method, a vast plethora of functional groups and structural motives can be added to the emitters that is otherwise expensive or unavailable by the acyl chloride route. Using the DIC approach, an alkyne as well as a protected amino acid were successfully attached to the oxadiazole showing the compatibility with larger biomolecules which opens pathways for follow up chemistry such as conjugation to biomolecules.

Four examples with the transformation of both tetrazole groups into oxadiazoles are shown in Scheme 10. Besides the simple methyl and phenyl groups, the strongest electron donating and withdrawing groups from the mono-functionalization were implemented to demonstrate the broad applicability of this synthesis. Alkylated and arylated emitters **4CzdOXDMe (73b)**, **4CzdOXDPh (73c)** and **4CzdOXDtBu (73a)** showed a stronger blueshift in emission color than the respective mono-oxadiazole derivatives.

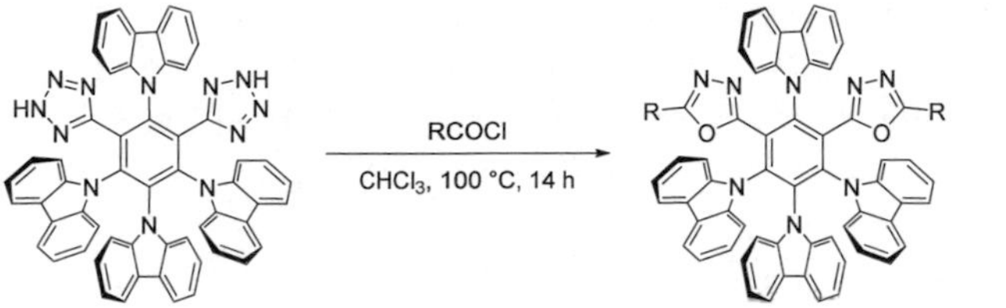

45 4CzdTI		
73a (4CzdOXDtBu)	R = ^{t}Bu	37%
73b (4CzdOXDMe)	R = Me	76%
73c (4CzdOXDPh)	R = Ph	67%
73d (4CzdOXD(CF$_2$)$_6$CF$_3$)	R = (CF$_2$)$_6$CF$_3$	34%

Scheme 10. Synthesis and scope of the bis-oxadiazole derivatives derived from **4CzIPN (3)**.

Five molecular structures of oxadiazole based emitters (**4CzCNOXDMe (71b)**, **4CzCNOXD2,6OMePh (71i)**, **4CzCNOXDMes (71h)**, **4CzCNOXDC$_6$F$_5$ (71n)** and **4CzCNOXDCH$_2$Cl (71f)**) were determined by X-ray analysis and are shown in Figure 32 and Table 3. The single crystals, suitable for X-ray analysis were obtained by slow evaporation of different solvents.

Similar to the reported tetrazole derivatives of **4CzIPN (3)** (3.1.1 Tetrazole Derivatives), the conformation of the oxadiazole derivatives was similar to all four carbazole groups in a highly twisted arrangement with regards to the central benzene plane. Neither the dihedral angles, nor the angles between the different carbazole donors within one molecule varied much when compared to the different emitters. The dihedral angles range from 64.4° to 70.1° with the exception of the top carbazole in **4CzCNOXDCH₂Cl** (**71f**) which is twisted at an angle of 75.1°. The oxadiazole heterocycles in these five emitters were less twisted when compared to the carbazoles due to their smaller size, resulting in less steric repulsion with the neighboring groups and the central benzene core, showing dihedral angles between 41.9° and 54.4°. In **4CzCNOXD2,6OMePh** (**71i**) and **4CzCNOXDMes** (**71h**) the arene substituents on the oxadiazoles were twisted with dihedral angles of 62.3° and 59.8° respectively to the oxadiazole plane, due to the steric repulsion of the methoxy and methyl groups in ortho position to the oxadiazole. In **4CzCNOXDC₆F₅** (**71n**), the pentafluorophenyl was oriented almost coplanar at an angle of 14.7° due to little steric repulsion also indicating a significant degree of conjugation between these ring systems.

Table 3. Dihedral angles of the twisted carbazoles and oxadiazoles with regards to the central benzene ring and dihedral angle of the aryl substituents on the oxadiazoles calculated from the molecular structures. Cz1 refers to the top carbazole located between oxadiazole and nitrile. Cz2 to Cz4 refers to the next carbazoles when counting clockwise. OXD refers to the oxadiazole group and if present, its substituted arene (OXDR).

Entry	Name	#	Cz1	Cz2	Cz3	Cz4	OXD	OXDR
1	4CzCNOXDMe	71b	67.0°	70.1°	66.5°	67.8°	43.6°	-
2	4CzCNOXDCH₂Cl	71f	75.1°	61.9°	66.0°	64.4°	52.5°	-
3	4CzCNOXD2,6OMePh	71i	69.4°	66.1°	66.9°	62.8°	41.9°	62.3°
4	4CzCNOXDMes	71h	69.6°	65.9°	66.9°	65.5°	54.4°	59.8°
5	4CzCNOXDC₆F₅	71n	64.5°	69.8°	68.7°	69.0°	45.5°	14.7°

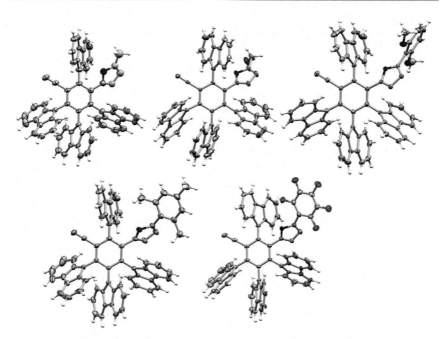

Figure 32. Molecular structures of **4CzCNOXDMe** (**71b,** top left), **4CzCNOXDCH₂Cl** (**71f,** top center), **4CzCNOXD2,6OMePh** (**71i,** top right), **4CzCNOXDMes** (**71h,** bottom left) and **4CzCNOXDC₆F₅** (**71n,** bottom right); minor disordered parts omitted for clarity, displacement parameters are drawn at 50% probability level.

Next, an iterative approach (Scheme 11) was used to combine tetrazole and oxadiazole moieties within one emitting molecule. With this strategy an approach to nitrile-free, but disparately substituted TADF emitters was evaluated. Starting from the mono-tetrazole **43**, the oxadiazole using acetyl chloride was formed first, followed by subsequent cyclization of the remaining nitrile to a tetrazole (**74**) which was then alkylated following the known protocol to yield the mixed emitter **4CzTlnHexOXDMe** (**75**).

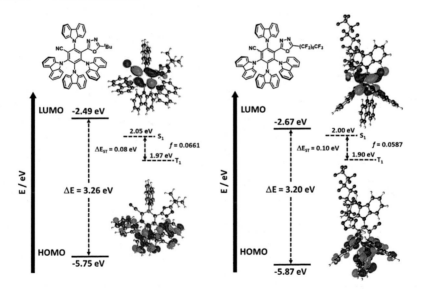

Scheme 11. Iterative approach towards an unsymmetrically substituted emitter with one tetrazole and one oxadiazole as an exemplary emitting system.

Verifying the proof of principle, this approach can be utilized to implement different functional groups to the emitter allowing orthogonal follow-up chemistry with two distinct binding sites. Furthermore, this iterative sequence can be used to install two oxadiazoles or tetrazoles with different groups attached.

3.1.2.2 DFT Calculations

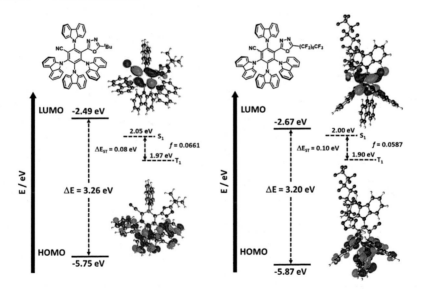

Figure 33. DFT calculations of the mono-oxadiazoles **4CzCNOXDtBu** (**71a**) and **4CzCNOXD(CF₂)₆CF₃** (**71g**). Results provided by cynora GmbH.

Oxadiazoles are known to be weaker acceptor groups than nitriles.[181,194,204] Hence, the replacement of nitriles by oxadiazoles is expected to destabilize the LUMO level, raising its energy and thus increasing the optical band gap. For the mono-oxadiazole derivatives, the LUMO is partially located on the oxadiazole moiety, while the benzonitrile still locates a large share due to the stronger acceptor properties of the nitrile group. In the case of the bis-oxadiazoles **4CzdOXDtBu** (**71a**) and **4CzdOXD(CF₂)₆CF₃** (**71g**) the LUMO is mostly located on the oxadiazole heterocycles and the respective benzene core as shown in Figure 34.

As expected, due to the lower electron withdrawing character of the *tert*-butyl-substituted oxadiazoles, **4CzCNOXDtBu** (**71a**) and **4CzdOXDtBu** (**73a**) show a significant destabilization of the LUMO orbital compared to **4CzIPN** (**3**) from –2.71 eV to –2.49 eV to –2.20 eV. Resulting in increased optical band gaps within this series from 3.11 eV for **4CzIPN** (**3**) to 3.26 eV for **4CzCNOXDtBu** (**71a**) and 3.46 eV for **4CzdOXDtBu** (**73a**), hence a blueshift in emission color (Figure 33 and Figure 34).

Interestingly, this effect slightly opposed when the oxadiazoles are substituted by the electron withdrawing perfluorinated alkyl chain. Here, although replacing both nitriles by the weaker oxadiazole groups, similar LUMO levels were calculated for **4CzCNOXD(CF₂)₆CF₃** (**71g**) with –2.67 eV and for **4CzdOXD(CF₂)₆CF₃** (**73d**) with –2.74 eV compared to –2.71 eV for **4CzIPN** (**3**) also at very similar HOMO levels. This results in a slight increase of the optical band gap for **4CzCNOXD(CF₂)₆CF₃** (**71g**) at 3.20 eV, while the band gap for the bis-oxadiazole **4CzdOXD(CF₂)₆CF₃** (**73d**) was calculated at 3.11 eV, the same as for **4CzIPN** (**3**) (Figure 34). Oxadiazoles substituted with strongly electron withdrawing groups appear to show similar acceptor strength as nitrile groups. All calculated ΔE_{ST} values are low in the range of 0.07 to 0.13 eV and are therefore well in the area needed for efficient TADF. No distinct trend of the absolute values is notable.

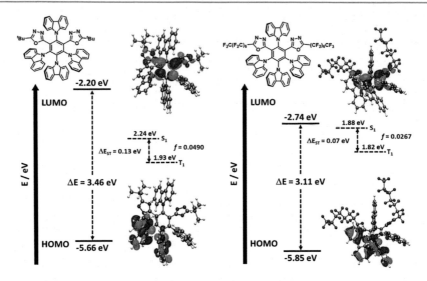

Figure 34. DFT calculations of the bis-oxadiazoles **4CzdOXDtBu** (**73a**) and **4CzdOXD(CF₂)₆CF₃** (**73d**). Results provided by cynora GmbH.

3.1.2.3   Basic Photophysical Characterization

Basic photophysical characterization was conducted by cynora GmbH. The photoluminescence spectra of the substituted oxadiazole derivatives were recorded using the emitter as 10 wt% dopant in PMMA. For the mono-oxadiazole derivatives, a trend on the emission maxima was observed with regards to the attached substituent between 498 nm for **4CzCNOXDtBu** (**71a**) and 527 nm for **4CzCNOXD(CF₂)₆CF₃** (**71g**). As expected, and shown by DFT predictions, the emission color is clearly dependent on the electron donating or withdrawing character of the substituent. Independent of the hybridization state of the attached group, the implemented electron donating groups (alky, mesityl, tolyl, phenyl) in **4CzCNOXDtBu** (**71a**), **4CzCNOXD(CH₂)₁₀CH₃** (**71c**), **4CzCNOXDMes** (**71h**) and **4CzCNOXD2,6MeOPh** (**71i**) lead to a blueshift in emission wavelength compared to electron withdrawing functions (halides, perfluorinated alkyl and aryl groups, nitriles) in **4CzCNOXD(CF₂)₆CF₃** (**71g**), **4CzCNOXDC₆F₅** (**71n**) and **4CzCNOXD4CNPh** (**71q**). All mono-oxadiazole derivatives showed high photoluminescence quantum yields in the range of 66% to 85% with the exception of **4CzCNOXDvinyl** (**71m**) (21%) and **4CzCNOXD3,5NO₂Ph** (**71s**) (18%). The latter emitter shows this poor PLQY due to the known quenching effect of the nitro group.[205–207] The emitters showed short excited state lifetime from 3.4 µs to 25.6 µs with the exception of entries 12, 17, 22 and 24.

Entries 22–25 show the photophysical properties of bis-oxadiazoles. According to the DFT calculations, replacing both nitriles with oxadiazole groups further destabilizes the LUMO orbital, resulting in a blueshift of the emission wavelength. This was confirmed for **4CzdOXDtBu (73a)**, **4CzdOXDMe (73b)** and **4CzdOXDPh (73c)** emitting at 473 nm, 482 nm and 490 nm with quantum yields of 48% to 74%, respectively, when compared to the respective mono-oxadiazoles **4CzCNOXDtBu (71a)**, **4CzCNOXDMe (71b)** and **4CzCNOXDPh (71l)** emitting at 498 nm, 499 nm and 506 nm. Replacing both nitriles by oxadiazoles that carry electron withdrawing substituents like in **4CzdOXD(CF₂)₆CF₃** (**73d**) did not result in a further redshifted emission when compared to the mono derivative **4CzCNOXD(CF₂)₆CF₃ (71g)** as predicted by DFT calculations. Here the bis-oxadiazole derivative is still slightly blueshifted by 4 nm to an emission wavelength of 523 nm. The substituted oxadiazole acceptor in **4CzCNOXD(CF₂)₆CF₃ (71g)** barely influenced the photophysical properties upon second transformation of the nitrile to the oxadiazole, giving rise to the assumption that nitrile and the electron withdrawing substituted oxadiazole are similarly strong acceptors.

Similar to the tetrazole approach, 3,6-di-*tert*-butylcarbazole was implemented and combined with a mono-oxadiazole derivative showing the strongest redshift in emission wavelength, yielding **4tCzCNOXD(CF₂)₆CF₃ (72)** with an emission color of 541 nm. By adjusting the donor strength, a redshift of 12 nm was achieved while maintaining a fairly high quantum yield of 66% and short excited state lifetime of 3.4 µs.

It is noteworthy, that substituted oxadiazoles bearing functional groups for follow up chemistry like **4CzCNOXD(CH₂)₂COOMe (71e)**, **4CzCNOXDCH₂Cl (71f)**, **4CzCNOXD4IPh (71k)** and **4CzCNOXDyne (71p)** result in highly efficient emitters with photoluminescent quantum yields between 66% and 82%

The emission spectra of selected emitters of the oxadiazole series are depicted in Figure 35 concluding the feasibility of adjusting the emission wavelength by acceptor substitution. Electron donating groups attached to the oxadiazole lead to the expected blueshift in emission wavelength, amplified by replacing the remaining nitrile with the bis-oxadiazole acceptor motif. Electron withdrawing groups lead to a redshift in emission wavelength, which can further be extended by donor adjustment. Referring to the parent emitter **4CzIPN (3)**, which itself shows an emission maximum at 516 nm with a PLQY of 89%, the reported modifications allow a blueshift as well as a redshift in emission color, covering 473 nm to 541 nm thus varying the emission color by 66 nm solely by adjusting the acceptor groups and slightly modifying the carbazole donors.

Table 4. Photophysical properties of oxadiazole-substituted TADF emitters doped in 10wt% in PMMA.

Entry	Compound	#	$\lambda_{max, PL}$ [nm]	PLQY [%]	FWHM [eV]	Excited state lifetime τ [μs]
1	4CzIPN	3	516	89	-	-
2	4CzCNOXDtBu	71a	498	85	0.42	10.3
3	4CzCNOXDMe	71b	499	80	0.43	5.3
4	4CzCNOXD(CH$_2$)$_{10}$CH$_3$	71c	499	80	0.43	8.9
5	4CzCNOXDCH(Ph)$_2$	71d	500	83	0.42	8.5
6	4CzCNOXD(CH$_2$)$_2$COOMe	71e	502	82	0.43	8.7
7	4CzCNOXDCH$_2$Cl	71f	508	81	0.44	7.1
8	4CzCNOXD(CF$_2$)$_6$CF$_3$	71g	527	71	0.43	4.2
9	4tCzCNOXD(CF$_2$)$_6$CF$_3$	72	541	66	0.42	3.4
10	4CzCNOXDMes	71h	499	83	0.42	8.3
11	4CzCNOXD2,6MeOPh	71i	499	84	0.42	10.7
12	4CzCNOXDTol	71j	505	85	0.42	38.0
13	4CzCNOXD4IPh	71k	505	66	0.41	25.6
14	4CzCNOXDPh	71l	506	85	0.42	19.3
15	4CzCNOXDvinyl	71m	509	21	0.44	5.6
16	4CzCNOXDC$_6$F$_5$	71n	509	83	0.42	6.3
17	4CzCNOXDthpn	71o	510	77	0.42	86.4
18	4CzCNOXDyne	71p	515	66	0.43	6.4
19	4CzCNOXD4CNPh	71q	516	80	0.41	17.4
20	4CzCNOXD3,5CF$_3$Ph	71r	516	82	0.41	16.9
21	4CzCNOXD3,5NO$_2$Ph	71s	523	18	0.42	-
22	4CzdOXDtBu	73a	473	50	0.45	52.9
23	4CzdOXDMe	73b	482	48	0.45	11.0
24	4CzdOXDPh	73c	490	74	0.43	64.2
25	4CzdOXD(CF$_2$)$_6$CF$_3$	73d	523	62	0.46	3.5

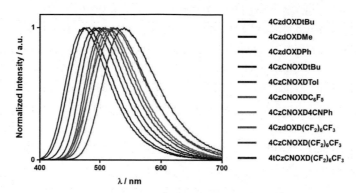

Figure 35. Emission spectra of selected oxadiazole emitters doped in PMMA as host.

3.1.2.4 Advanced Photophysical Characterization

All experiments and analyses concerning the advanced photophysical characterization and OLED fabrication were conducted via collaboration with Lorenz Graf von Reventlow of the Dr. Colsmann group at the Light Technology Institute of the KIT.

In-depth photophysical studies were conducted for two emitters of this series, the mono-oxadiazole **4CzCNOXDtBu (71a)** and the bis-oxadiazole **4CzdOXDtBu (73a)**. To investigate if both modifications still retain their TADF characteristics, time- and temperature dependent photoluminescence studies were conducted on solution processed thin films with the emitter as 10 wt% dopant in mCP. In contrast to the high PLQYs observed in PMMA of 85% and 50%, respectively, the PLQYs in mCP is reduced to 57% for **4CzCNOXDtBu (71a)** and 29% for **4CzdOXDtBu (73a)**. It is known that the PLQY of charge-transfer states is strongly dependent on the molecular environment making the polarity-based choice of a suitable emitter crucial.[148,208,209] When using mCBP as host for example, the PLQY is reduced even more drastically (34% and 15%). Embedded in mCP, both emitters show fast and slow emission decay components (Figure 36a). From this ratio the reverse intersystem crossing (RISC) rate is calculated. The degree of temperature dependence allows an Arrhenius fit over the inverse temperature to calculate the activation energy of the RISC process (Figure 36b). This activation energy is estimated to be the singlet-triplet energy gap ΔE_{ST}. For both oxadiazole emitters **4CzCNOXDtBu (71a)** and **4CzdOXDtBu (73a)**, ΔE_{ST} was calculated at 52 ± 10 meV and 59 ± 10 meV, respectively which is in good agreement with the predicted values by DFT calculations (80 meV and 130 meV). Furthermore, both oxadiazole-based emitters show similar ΔE_{ST} values as the reference emitter **4CzIPN (3)** with a ΔE_{ST} of 83 meV.[51] As none of the investigated oxadiazole-based emitters showed any observable delayed emission below 120 K, TADF is the main mechanism for the delayed emission at higher temperatures. Also, the structure of the intensity dependence of PL at room temperature rules out triplet-triplet annihilation (TTA) as the origin for delayed emission (Figure S 5). In order to clarify the cause of the much lower PLQY of the bis-oxadiazole **4CzdOXDtBu (73a)** compared to **4CzCNOXDtBu (71a)** and **4CzIPN (3)**, temperature-dependent PLQY measurements of **4CzdOXDtBu (73a)** doped in mCP were conducted. It was found that the emitter shows the maximum PLQY of 58% at 140 K with only a small share of delayed fluorescence. This clearly indicates that the low PLQY is induced by non-radiative decays from the singlet states which is caused by vibration of the emitter and not due to an inefficient RISC. The reference emitter **4CzIPN (3)** in contrast showed a decrease of the PLQY with decreasing temperature at the same rate as the share of delayed emission declines (Figure S 6).

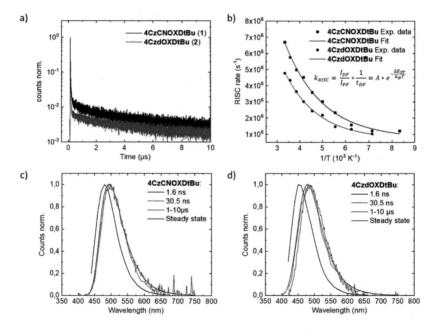

Figure 36. a) Photoluminescence decay of compound **4CzCNOXDtBu (71a)** and **4CzdOXDtBu (73a)** in mCP at room temperature. The samples were excited at 340 nm for all measurements and PL-decays were observed at 500 nm (**4CzCNOXDtBu (71a)**) and 480 nm (**4CzdOXDtBu**

(73a)). b) Arrhenius plot of RISC rate over the inverse temperature to calculate the activation energy for the TADF process. Time resolved and steady state emission spectra for compound **4CzCNOXDtBu (71a)** (c) and **4CzdOXDtBu (73a)** (d).

Surprisingly in solid state, both oxadiazole derivatives showed two fast decay components in the short ns-range not visible when dissolved in tetrahydrofuran (Figure S 7). Time-resolved emission maps (TRES) were recorded with the time-correlated single photon counting (TCSPC) technique in order to investigate this phenomenon. A TCSPC measurement was recorded for different wavelengths to resolve the spectra at different times. There was a clear red shift of the emission maximum for **4CzCNOXDtBu (71a)** from 480 nm to 498 nm within the first nanoseconds. The emission stabilizes after 30 ns and the subsequent delayed component (1–10 µs) shows the same spectral shape as the second prompt component (Full TRES data is shown in Figure S 7). As the intensity of the first component is rather low, the steady state emission nearly matches the second prompt and delayed emission. Similar observations were reported for D-A-D TADF molecules[210] and in TADF exciplex systems.[211] In the case of the oxadiazole-based systems the following approach is suggested: As the emitter concentration of **4CzCNOXDtBu (71a)** is quite low (10 wt%), singlet excitons on mCP are generated mostly, which are then transferred to the emitter via Förster transfer. The energy transfer to a local excited singlet state (^{1}LE) seems to be more favorable than the transfer directly to the charge transfer singlet state (^{1}CT) followed by subsequent transfer to the lower lying ^{1}CT state. This process is rather slow, as the highly twisted structure of the TADF emitter leads to a low coupling.[210] This results in the competition with the fluorescence from the ^{1}LE state which is visible in Figure 36c for times below 30 ns. Due to the higher degree of freedom in solution, an increased coupling of ^{1}LE and ^{1}CT prevents this effect in solution. The same measurements were conducted with another host (mCBP) in order to show that the fast fluorescence does not come from a CT state between host and guest. The spectra of **4CzCNOXDtBu (71a)** in mCBP perfectly match the spectra in mCP (Figure S 7).

Similar observations were found for the bis-oxadiazole **4CzdOXDtBu (73a)** (Figure 36d). Sandanayaka et al. also observed a similar behavior for **4CzIPN (3)** at <500 nm emission for 4CzIPN:m-CBP composites but no further investigation was conducted.[212] As the PL decays show more than one fast component, they may have observed the same effect.

3.1.2.5   OLED Fabrication

Both oxadiazole-based TADF emitters showed very high solubility (>3 g/L in THF), rendering them suitable for solution-processed OLEDs. The commonly used phosphorescent iridium complexes often show limited solubilities (<1 g/L) which result in strong molecular aggregation and quenching of the triplet states.[213] mCP, which is a commonly used host material for blue TADF OLEDs,[55] was chosen as it led to the highest PLQY for both emitters. The OLED stack chosen consisted of PEDOT:PSS as hole transport layer from the ITO anode, while BP4mPy acts as evaporated electron transport layer from the Liq/Aluminum cathode. The optimized OLEDs achieved an EQE of 16% at 100 cd m^{-2} for **4CzCNOXDtBu (71a)** and 5.9% at 100 cd m^{-2} for **4CzdOXDtBu (73a)**. The OLED using **4CzIPN (3)** in the same stack resulted in an EQE of 17% at the same luminance. Figure 37 shows the main optoelectronic characteristics. Barely any efficiency roll-off was observed up to 300 cd m^{-2} in every device which is in the commercially relevant brightness range of OLED displays. The lifetimes of all OLEDs are rather short (LT50 in the range of minutes at 300 cd m^{-2}) as shown in Figure S 9, which can be attributed to the instable solution processed architecture, as **4CzIPN (3)** typically has very long lifetimes in well optimized evaporated stacks.[214] The fast degradation may be induced by the acidity of PEDOT:PSS or by solvent residues in the layers.

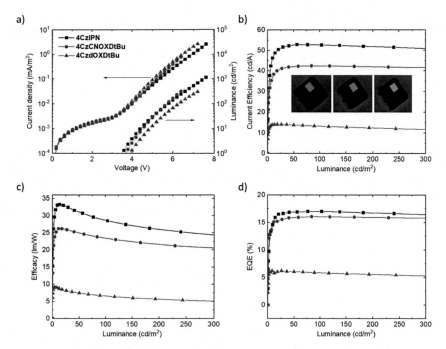

Figure 37. a) j-V (left axis) and L-V characteristics (right axis) of solution processed OLEDs containing **4CzIPN (3)**, **4CzCNOXDtBu (71a)** and **4CzdOXDtBu (73a)** as emitter. b) Current efficiency. Inset: Pictures of OLEDs containing (from left to right) **4CzIPN (3)**, **4CzCNOXDtBu (71a)** and **4CzdOXDtBu (73a)** as emitter. The shift in color is clearly visible. c) Efficacy and d) EQE. All curves display the second measurement of the devices, and current densities were kept low to prevent degradation of the OLEDs during the measurement.

3.1.2.6   Exciton Transport

All reactions and experiments concerning the oligomer and TADF-oligomer conjugate synthesis were conducted via collaboration with Dr. Rebekka Schneider of the Prof. Meier group at the Institute of Organic Chemistry of the KIT and have partially been reported in the respective doctoral thesis.[215] The following photophysical characterization has been conducted in the Prof. Lemmer group at the Light Technology Institute of the KIT. Excerpts of this chapter have been already reported in a peer reviewed article.[216]

Within the context of the Collaborative Research Centre 1176 "Molecular Structuring of Soft Matter", this thesis is part of the project A4 "Tailor-Made Sequence-Defined Oligomer-Dye Conjugates for Controlling Exciton Dynamics". The aim of this project is the creation and investigation of rigid, monodisperse conjugated oligomers, that can be attached to a chromophore (TADF emitter) to investigate photophysical phenomena such as exciton transport and charge separation in a precisely defined system. This is of high interest for a better understanding of the elemental processes for applications like organic solar cells and OLEDs.

In the model system (Figure 38), two sequence-defined oligomer chains that are each functionalized with either electron donating or accepting groups are attached to a TADF molecule in the center. Upon excitation, the exciton is expected to migrate towards the TADF molecule. As discussed in the introduction, TADF molecules possess a small energy splitting between the excited singlet and triplet state which allows an efficient transfer between the two without a significant loss in energy. It is then expected that the exciton is separated by the TADF molecule into an electron hole and an electron that moves backwards along the respective oligomer chain (hole to the donor chain, electron to the acceptor chain).

1. Absorption of photons
2. Excitation-transfer
3. Charge separation

Figure 38. Oligomer chains connected to a TADF dye to investigate exciton transport, charge separation (left) and hyperfluorescence (right). Figures from the CRC1176 project plan.

In a second system, hyperfluorescence, a promising concept for highly performant OLEDs with sharp emission peaks, shall be investigated. Hyperfluorescence occurs upon excitation of a TADF molecule that transfers the singlet excitations to a nearby fluorescence emitter. As depicted in Figure 38, the combination of a TADF emitter and a fluorescence emitter on a rod-like oligomer serves as a tool to precisely control the spatial distance of the TADF unit to the fluorescent part within the system in order to find the optimal distance and arrangement.

The rod-like oligomers were developed and synthesized by Dr. Rebekka Schneider from the Meier group.[215,217] In order to obtain the respective systems, the TADF emitter **71k** was coupled to oligomers of different lengths of 1, 3 and 5 repetition units via Sonogashira reaction. These model systems shown in Scheme 12 represent the TADF center combined with the donor chains of different lengths in order to get an initial understanding of the essential reaction steps and for the basic understanding of the photophysics.

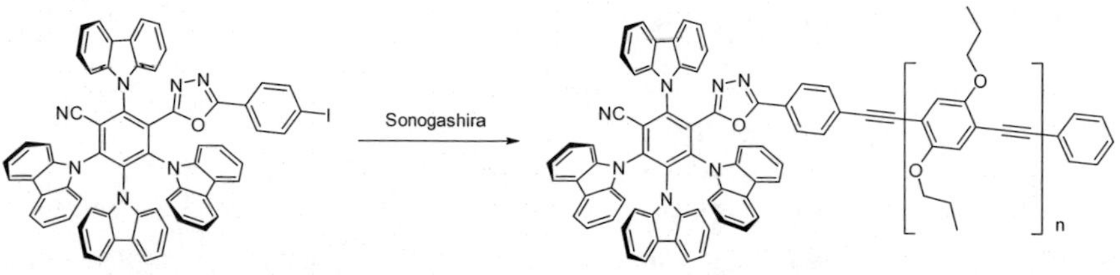

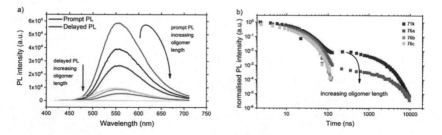

71k
4CzCNOXD4IPh

76a n = 1
76b n = 3
76c n = 5

Sonogashira

Scheme 12. Attachment of the rod-like oligomers of different length to the TADF molecule via Sonogashira reaction.

The initial TADF emitter **4CzCNOXD4IPh (71k)** as well as the respective TADF-oligomer conjugates were investigated photophysically. The total prompt and delayed photoluminescence were recorded in solution in chloroform and are shown in Figure 39a. The excitation and measurement conditions were identical, so that the absolute value of the curves can be compared. The decrease in prompt photoluminescence with increasing oligomer length correlates with an increase in the absorption of the 355 nm excitation pulse due to the absorption of the oligomer-chain acting as an antenna followed by an energy transfer to the TADF center. Figure 39b shows that the prompt lifetime did not change with an increase in oligomer length. The delayed component of emission however, drastically decreased upon extending the oligomer length and was not observable anymore for 3 and 5 repetition units in **76b** and **76c**.

Figure 39. a) Photoluminescence spectra and b) lifetime of the TADF-oligomer conjugates.

The causes for the drastic decrease of delayed fluorescence with increasing oligomer length is suspected to be manifold and should be further investigated. The exciton in the triplet state might be quickly quenched by either charge or triplet transfer. Also the attachment of the oligomer might increase the singlet-triplet splitting, turning off the reverse intersystem crossing process.

3.1.3   Benzothiazole Derivatives

Similarly to the investigations on oxadiazole acceptors, Adachi and co-workers developed donor-acceptor-donor-type and donor-acceptor type structures, implementing benzothiazoles as acceptors and combined them with phenoxazines as donor groups[218] (Figure 40). As is commonly observed in TADF design, the D-A-type structure **BT (78)** is a less efficient TADF emitter than the D-A-D-type molecule **BT2 (77)** due to a larger ΔE_{ST} resulting in a less efficient $T_1 \rightarrow S_1$ upconversion and a smaller oscillator strength. This effect was also visible in the performance of the resulting OLEDs with an EQE of 14.0% for **BT2 (77)** and 12.1% for **BT (78)**.

In 2015 Adachi *et al.* furthermore investigated the 5,10-dihydrophenazine donor group by combining it with different acceptor groups in an acceptor-donor-acceptor-type structure.[57] Besides common acceptor groups like benzimidazoles, benzonitriles and triazines, the benzothiazole group, which is also known for its application in electroluminescent[219,220] and electron-transporting materials[221] was evaluated as an acceptor group. The resulting emitter **DHPZ-2BTZ (79)**, shown in Figure 40 showed a yellow peak maximum at 577 nm when doped in mCP as host material and a photoluminescence quantum yield of 33%. The resulting device emitted at an emission maximum of 601 nm with an EQE of 5%.

Also in 2015, Adachi and co-workers combined the benzothiazole acceptor with bisdiphenylamine substituted carbazole to access the sky-blue emitter **DAC-BTZ (80)** shown in Figure 40 with an emission peak maximum at 496 nm in toluene solution with a PLQY of 45% which was increased to 65% upon nitrogen bubbling, already indicating its TADF characteristics. The ΔE_{ST} was found to be small with 0.18 to 0.22 eV. The fabricated device containing **DAC-BTZ (80)** as emitter showed an emission maximum at 493 nm with a maximum EQE of 10.3%.[222]

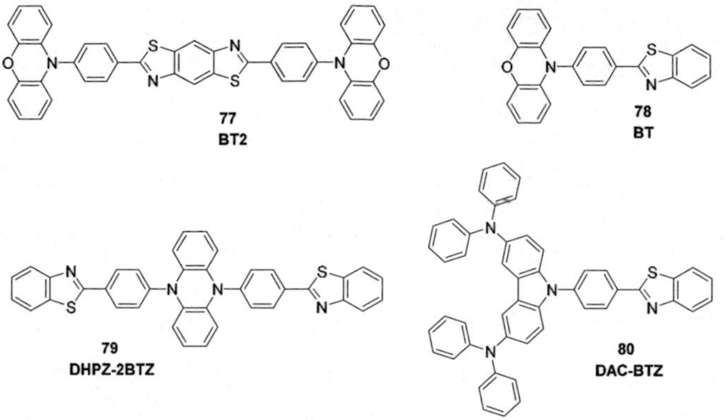

Figure 40. Benzothiazole based TADF emitters in literature.

3.1.3.1   DFT Calculations

Benzothiazoles can be synthesized from nitriles upon reaction with 2-aminobenzenethiol under copper catalysis[223] and were therefore also investigated for the derivatization of the **4CzIPN (3)** parent emitter. To evaluate the potential of benzothiazole acceptor units for the **4CzIPN (3)** TADF system, DFT calculations were performed using the Turbomole program package. First, geometries of the molecular structures in the ground state in gas phase were optimized employing the BP86 functional and the resolution identity approach (RI). By employing time-dependent DFT (TD-DFT) methods, excitation energies were calculated using the (BP86) optimized structures. For orbital and excited state energies, the B3LYP functional and Def2-SVP basis sets and a m4-grid numerical integration were used. The molecular orbitals were visualized to get initial insights on the influence of replacing one or both nitrile acceptors with benzothiazole groups (Figure 41). Similarly to **4CzIPN (3)** and as expected, the HOMO remains located on the carbazole donor groups. The LUMO however, mainly influenced by the acceptor groups, is predominantly located on the nitrile if present, and thiazole heterocycles, only slightly extending on the annulated arene. Therefore, the introduction of benzothiazoles in silico, which are weaker acceptor groups than nitriles, destabilize the LUMO orbital significantly from −2.45 eV to −2.16 eV when replacing two instead of one nitrile with the benzothiazole acceptor. Interestingly, the double replacement also destabilizes the HOMO from −5.66 eV to −5.46 eV. This leads to an overall increase of the optical band gap for the double replacement from 3.21 eV to 3.30 eV. Both DFT-predicted emitters possess a very low energy gap between excited singlet and triplet state of 0.10 eV and 0.04 eV respectively, allowing efficient RISC while the oscillator strength remains low.

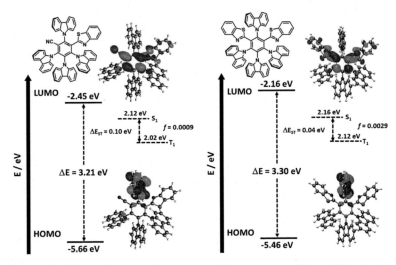

Figure 41. DFT calculations of the benzothiazole-based emitters (**81** and **82**). Results provided by cynora GmbH.

3.1.3.2   Synthesis

In terms of synthetic approach, the literature reports the reaction to be ineffective in case of ortho substitution to the nitriles.[220] Despite the apparent *ortho*-substitution in the case of **4CzIPN** (**3**), the reaction was still evaluated. Unlike reported in literature, ethanol was not found to be a good solvent mainly due to the low solubility of the starting material **4CzIPN** (**3**). Hence, DMSO was used as solvent. As reaction over night at 70 °C following the conditions in literature did not show any conversion, the reaction temperature was increased to 130 °C (Scheme 13).

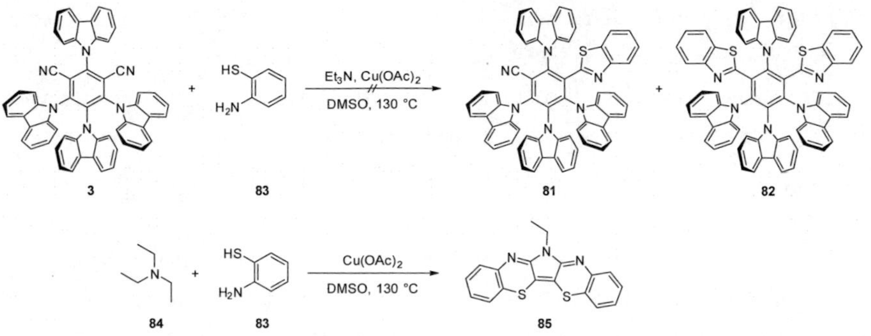

Scheme 13. Expected reaction (top) and observed reaction (bottom) upon reacting **4CzIPN** (3) to benzothiazoles.

Eventually, a strongly emissive product could be isolated. The product showed an intense green fluorescence upon irradiation with UV light at 365 nm in solution. In solid state however it showed a deep red fluorescence. The product was poorly soluble in all common solvents rendering its analysis and purification extraordinarily tedious. Fortunately, single crystals in form of very thin needles suitable for single crystal analysis were obtained upon cooling of a hot, saturated solution of the molecule in dimethylformamide or dimethyl sulfoxide, respectively. Subsequently, the molecular structure was solved and is depicted in Figure 42.

Figure 42. Molecular structure of the formed pyrrolobenzothiazine (**85**). Displacement parameters are drawn at 50% probability level.

The product can be described as two benzothiazines being fused to a centered pyrrole ring which is alkylated at the *N*-terminus with an ethyl group. The centered *N*-ethyl pyrrole is presumably derived from the triethylamine used in the reaction as a base. For now, it is unclear how the carbon atoms in the ethyl chain could be oxidized from -I and -III to +III and +II respectively under the given reaction conditions. It is noteworthy that two of the three ethyl groups of triethylamine were attacked, ultimately fusing them together in a five-membered heterocycle, while leaving one ethyl group absolutely unaffected. The reaction has been repeated without the addition of **4CzIPN (3)** and found to be successful, showing that **4CzIPN (3)** is not participating in this novel reaction.

Similar structures have been reported in literature with different groups attached to the pyrrolic nitrogen. Aside from the NH, phenyl groups, linear alkyl chain[224,225] and branched alkyl chains were reported[226]. The synthesis was first conducted by Dimroth and Reichenender in 1969[227] as seen in Scheme 14 by condensation of 2-aminothiophenol (**83**) and *N*-phenyldichloromaleimide (**86**) in acetic acid. In contrast, the reactivity of the triethylamine and 2-aminothiophenol under copper acetate catalysis in DMSO observed here, is not observed in literature.

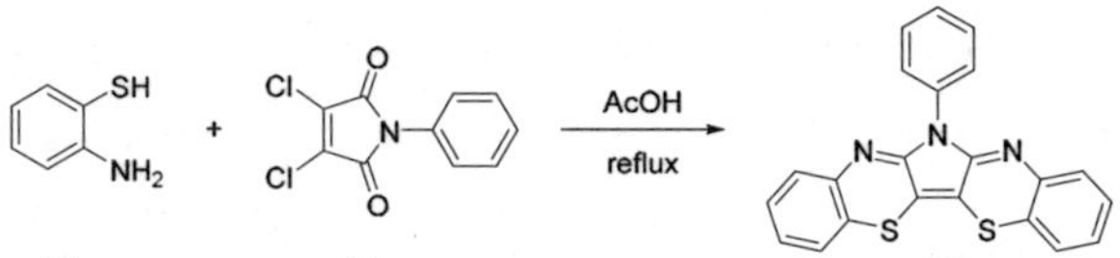

Scheme 14. Synthesis of pyrrolobenzothiazines via condensation of 2-aminothiophenol and *N*-phenyldichloromaleimide.

The respective pyrrolobenzothiazines found application as semiconductors for organic field-effect transistors,[224] (OFET) organic single crystal field-effect transistors (SCFET)[225] and in polymeric photoinitiators.[226]

3.2 TADF Emitters based on the Phthalimide Acceptor System

Aromatic imides such as phthalimides posses excellent photoelectronic properties due to their strong electron withdrawing ability and have therefore found application as electron-transporting semiconductors in organic field-effect transistors (OFET) and in organic solar cells as light-harvesting dyes.[228-230]

The phthalimide-based fluorophores **Fluo-1** (**88**) and **Fluo-2** (**89**) shown in Figure 43 were investigated by Gigmes and co-workers in 2018 and were found to be suitable as green emitter for multilayered OLEDs with an EQE of 3.11% in the optimized solution processed device at an emission maximum of 564 nm.[231]

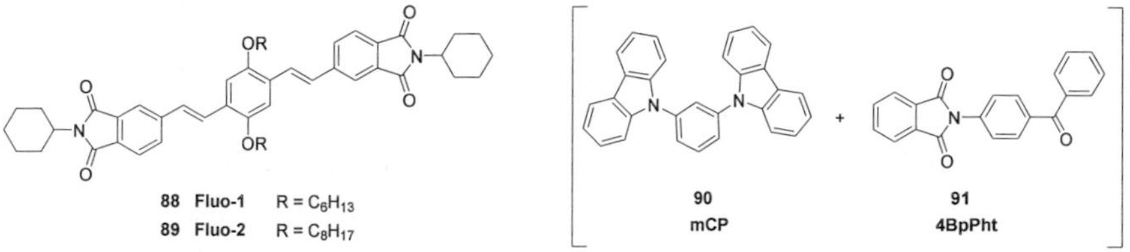

Figure 43. Phthalimide based fluorophores **Fluo-1** (**88**) and **Fluo-2** (**89**) (left) and a TADF exciplex based on mCP and **4BpPht** (**91**) (right).

Grazulevicius and co-workers developed phthalimide-based acceptor units that were combined with carbazole-based donor units as exciplex-forming materials.[232] The high triplet levels and ionization potentials of the phthalimides rendered them not only suitable as different color exciplex components, but also as hole and exciton-blocking materials. By combining different phthalimides with different donors, a large portion of the visible spectrum from sky-blue to red (494–575 nm) was covered. The resulting sky-blue OLED from the combination of mCP as donor and **4BpPht** (**91**) as acceptor as shown in Figure 43 showed an EQE of 2.9%.

In 2015, Adachi and co-workers reported the phthalimide and maleimide based fluorophores **AcPI** (**92**) and **AcMI** (**93**) as depicted in Figure 44.[233] While **AcMI** (**93**) did not show any TADF characteristics, **AcPI** (**92**) showed a much smaller ΔE_{ST} of 0.01 eV accompanied by an intense green TADF emission at 517 nm with a PLQY of 50%. The resulting OLED with its emission maximum at 530 nm showed a high maximum EQE of 11.5%.

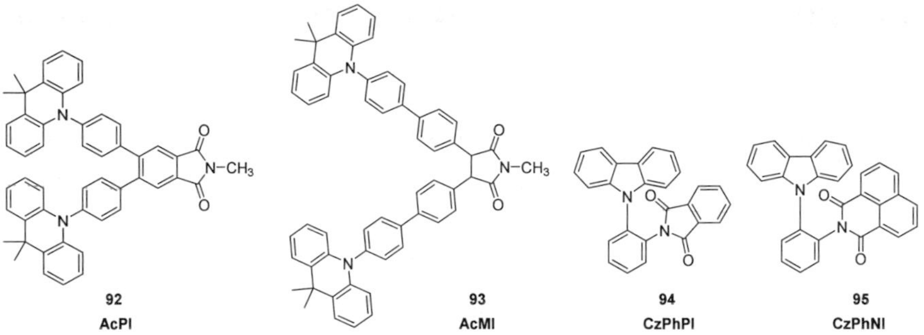

Figure 44. Combination of phthalimide and maleimide as acceptor groups with DMAC as donor, attached via a phenyl spacer and through-space-induced charge transfer-based emitters.

This concept was further investigated by Chen and coworkers in 2017, where they combined the weaker carbazole donor with the phthalimides in a similar arrangement without the phenyl spacers as depicted in Figure 45.[234] The resulting emitters **AI-Cz (96)** and **AI-TBCz (97)** were both found to be efficient TADF materials with small ΔE_{ST} of 0.09 eV and 0.08 eV doped in films. The resulting OLED possessed green electroluminescence for **AI-Cz (96)** at 510 nm and 540 nm for **AI-TBCz (97)** and showed an outstanding performance of 23.2% and 21.1%, respectively.

Later in 2018, the same group modified this emitting system with a chiral unit to obtain both enantiomers (+)-(*S,S*)-**CAI-Cz (98)** and (-)-(*R,R*)-**CAI-Cz (99)** shown in Figure 45 with a small ΔE_{ST} of 0.06 eV and excellent PLQYs of up to 98%.[235] These materials showed intense circularly polarized electroluminescence (CPEL) as well as mirror image CD and CPL activities. The green emission of the OLEDs centered at 520 nm achieved high EQE of up to 19.7% and 19.8%. Hence this group reported on the first CP-OLEDs based on enantiomerically pure TADF materials with high efficiencies and intense CPEL.

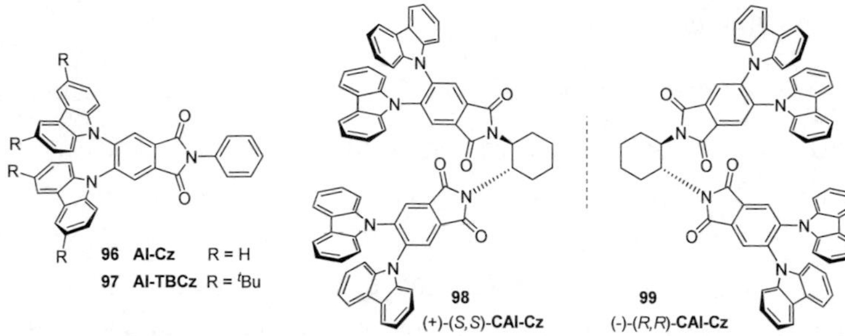

Figure 45. Highly efficient TADF emitters **AI-Cz** (**96**) and **AI-TBCz** (**97**) obtained by combination of the phthalimide acceptor with carbazole donor groups (left) and the attachment of the TADF unit to a chiral unit for emitters showing CPEL (right).

The group of Grazulevicius reported through-space-induced charge transfer-based emitters **CzPhPI** (**94**) and **CzPhNI** (**95**) (Figure 44) containing the phthalimide and naphthalic imide units as acceptor groups, of which **CzPhPI** (**94**) was found to show TADF properties with a small ΔE_{ST} of 0.03 eV.[236] Furthermore, the emitter showed aggregation-induced emission enhancement and a PLQY of up to 20% in neat film at 514 nm. Still, the emitter doped in mCP led to the most efficient device with an EQE of 2.4%.

Most recently this year, isomeric compounds to **CzPhPI** (**94**) with the carbazole unit arranget in *otho*, *meta*, and *para* position to the phthalimide were reported, all showing TADF and room-temperature phosphorescence characteristics.[237]

The following chapters were conducted in collaboration with cynora GmbH. They investigated initial phthalimide-based TADF emitters and reported the results in patents[238–240] and collaborative work.[241] It was found that carbazole substituted phthalimides with only one carbazole in 3-position is the most promising pattern for efficient blue emitters. Furthermore, this compact approach results in lightweight TADF molecules, rendering them particular useful for applications where large and heavy molecules are hindering, e.g. for bioimaging or vacuum-processing. Hence, the 3-carbazolyl phthalimide motif was used as the starting point for the investigations.

3.2.1   Synthesis

3.2.1.1   Acceptormodification

One of the most important and convenient synthetic routes to phthalimides is the dehydrative condensation of phthalic anhydride with primary amines at elevated temperatures.[239,240,242] This approach was used with a multitude of primary alkyl and aryl amines using acetic acid as a solvent at 100 °C to convert the commercially available 3-fluorophthalic anhydride (**100**) into the respective 3-fluorophthalimides (**101a-o**) within 3 to 4 hours of reaction time. As a result, a variety of functional groups and structural motives were attached to obtain the fluorinated phthalimide precursors in fair to excellent yields (Scheme 15). The synthesis was easily applied in multigram scale to yield the pure products after neutralization and filtration over a silica plug. While most of the fluorophthalimides were synthesized in the scale of one to five gram, it was also shown that for e.g. **FPI3BrPh (101k)** 15.2 g of product could be isolated. Also, **FPIMes (101g)** and **FPInDec (101d)** were synthesized in big scale reactions yielding 15.4 g and 8.20 g respectively.

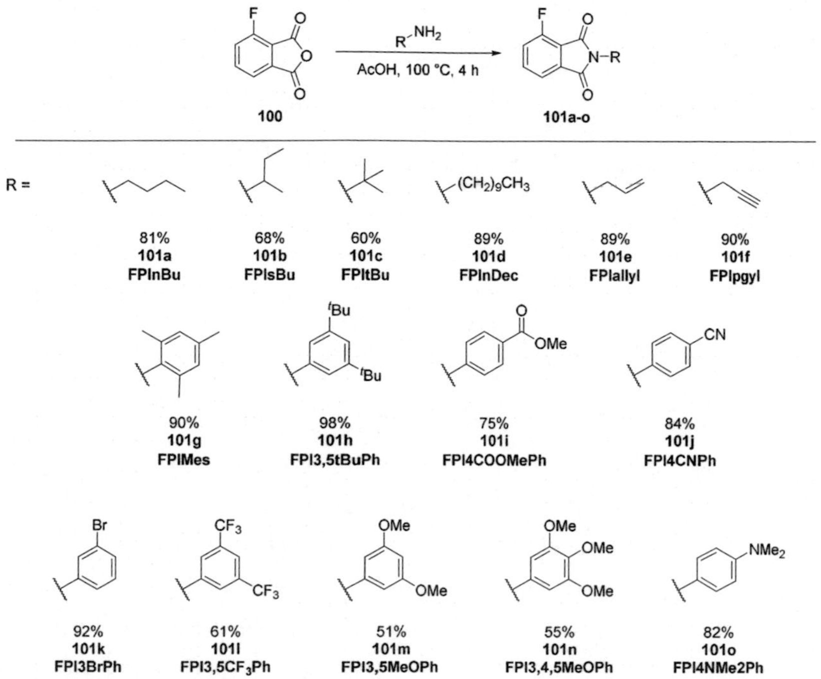

Scheme 15. Substrate scope of the condensation of 3-fluorophthalic anhydride with primary amines.

Single crystals suitable for single crystal analysis were obtained for **FPI3,5MeOPh** (**101m**) and **FPI4NMe₂Ph** (**101o**). The molecular structures are shown in Figure 46 depicting the twist between the substituted arene and the phthalimide due to the steric repulsion of the ortho hydrogens of the arene and the carbonyl oxygens. Although none of the attached arenes to the phthalimide has a sterically demanding group close to its binding site, the dihedral angles differ from 48° for **FPI3,5MeOPh** (**101m**) to 63° **FPI4NMe2Ph** (**101o**). This might be due to intermolecular interactions or interactions with the solvent resulting in different molecular stackings in the crystal structure.

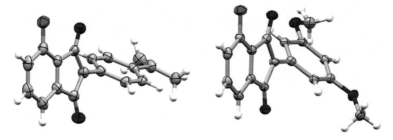

Figure 46. Molecular structure of **FPI4NMe2Ph** (**101m**) and **FPI3,5MeOPh** (**101o**); minor disordered parts omitted for clarity, displacement parameters are drawn at 50% probability level.

Next, the carbazole donor group was attached via nucleophilic aromatic substitution using tripotassium phosphate as base in DMSO at elevated temperatures.[239,240] The reaction was run until completion, typically within 4 to 16 hours monitored by TLC. Here, all synthesized fluorophthalimide precursors were combined with a carbazole moiety to investigate the effect of the *N*-attached group on the optoelectronic properties of the resulting emitters shown in Scheme 16. In most cases, the reaction showed full conversion within 4 to 6 hours and yielded the carbazole-substituted phthalimide emitters in fair to excellent yields.

The resulting materials showed a strong fluorescence in solution and in solid state when irradiated with UV light at 365 nm with considerable solvatochromicity. Furthermore, the color was perceived as sky blue for the phthalimides bearing alkyl groups like **CzPItBu** (**102c**), **CzPInBu** (**102a**) or **CzPInDec** (**102d**), while arene groups showed a bathochromic shift in emission color as the electron density in the arene decreases due to its substituents from **CzPIMes** (**102g**) over **CzPI3,5MeOPh** (**102m**), **CzPI3,5CF₃Ph** (**102l**), to **CzPI4CNPh** (**102j**). The materials **CzPI4NMe2Ph** (**102o**) and **CzPI3,4,5MeOPh** (**102n**) did not show fluorescence upon irradiation giving rise to the assumption that they are either not fluorescent, or the emitted fluorescence is not visible to the human eye. For in-depth details, see 3.2.2 DFT Calculations and 3.2.3 Basic Photophysical Characterization.

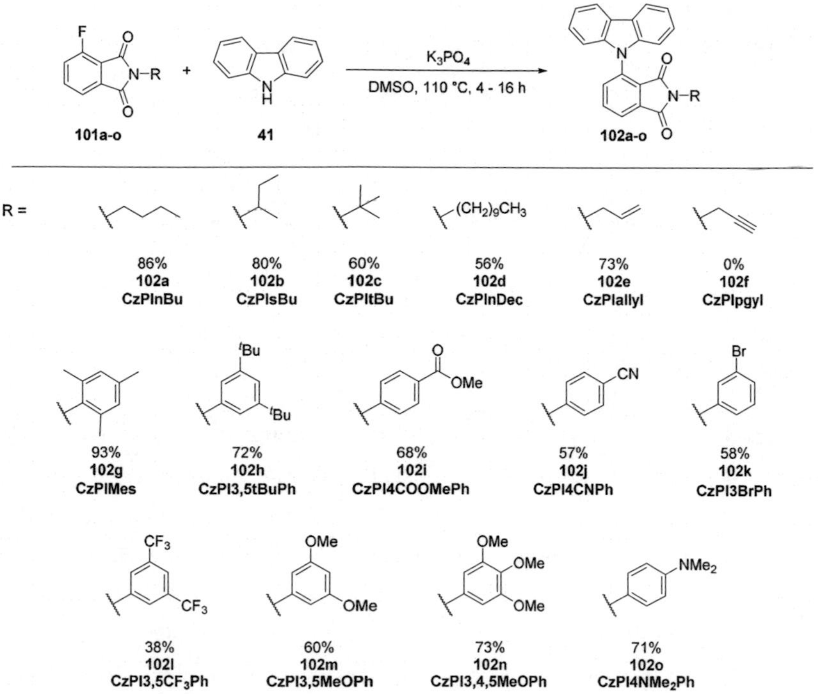

Scheme 16. Substrate scope of the nucleophilic aromatic substitution for the introduction of carbazole groups to fluorophthalimides.

Single crystals suitable for single crystal analysis were obtained for **CzPIMes (102g)** and **CzPI4CNPh (102j)**. The molecular structures are shown in Figure 47. Similar to the fluoro precursors, the *N*-coupled arene substituent to the phthalimide is twisted from the phthalimide plane. While the arene in **CzPI4CNPh (102j)** showed a dihedral angle of 45° similar to the situation in **FPI3,5MeOPh (101m)**, the sterically demanding mesityl group in **CzPIMes (102g)** lead to a drastic increase in the dihedral angle to 71°. As expected, the carbazole substituent is also considerably twisted out of the plane, mainly attributed to the steric repulsion of the carbazole hydrogen in 1-position and the phthalimide carbonyl oxygen adjacent to the carbazole binding carbon. The dihedral angles between the carbazole and the phthalimide plane were measured to be 50° for **CzPI4CNPh (102j)** and 62° for **CzPIMes (102g)** showing a difference of 12° which is in the range of typical experimental variances due to slightly different stackings in the crystal structure.

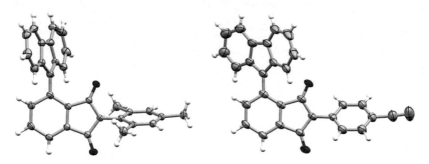

Figure 47. Molecular structure of **CzPIMes (102g)** and **CzPI4CNPh (102j)**; displacement parameters are drawn at 50% probability level.

Unexpectedly, the propargyl-substituted phthalimide **FPIpgyl (101f)** did not yield the desired product. The carbazole group was installed as expected, however, the propargyl group was transformed into an allene (Scheme 17). The transformation of a propargyl substituted phthalimide to the respective allene under the influence of potassium *tert*-butoxide as a strong base is reported in literature,[243] giving rise to the assumption that tripotassium phosphate acts similar under these reaction conditions. Less harsh conditions for the introduction of the carbazole or the protection of the terminal alkyne group may be a suitable choice for the outlook.

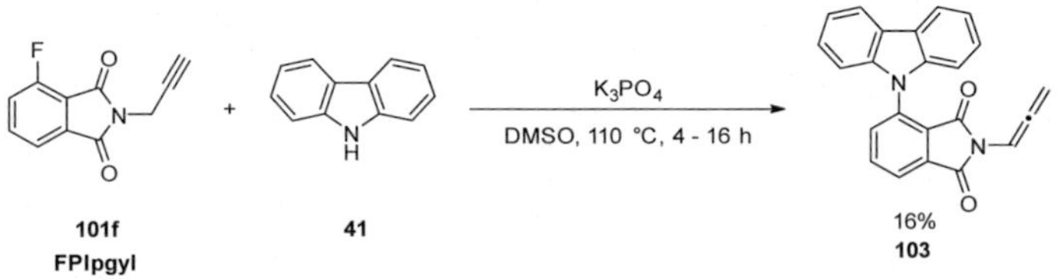

Scheme 17. Unexpected allene formation of the propargylic group upon harsh basic conditions.

The bromide **CzPI2BrPh (102k)** was further reacted in a Suzuki reaction using Pd$_2$(dba)$_3$ as catalyst and SPhos[240] as ligand with 4-toluyl boronic acid to give **CzPI2TolPh (105)** in 31% yield (Scheme 18). Although the yield is poor, this demonstrated that halides at the pending group attached to the phthalimide-*N* offer a straightforward way for follow-up chemistry and further functionalization.

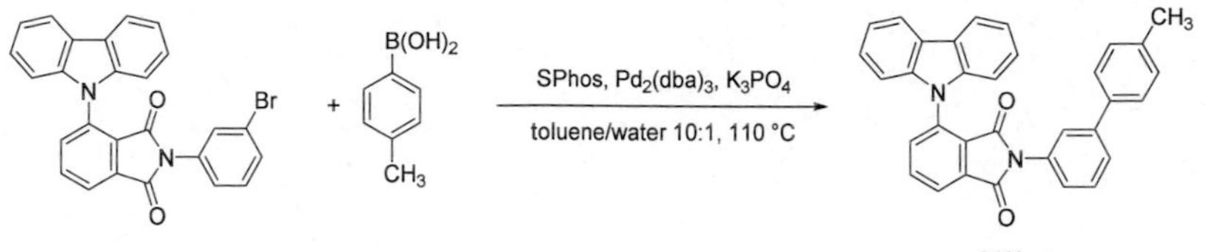

102k **104** **105**

CzPI2BrPh **CzPI2TolPh**

Scheme 18. Protocol for Suzuki reaction at the phthalimide substituted group.

3.2.1.2 Donormodification

As the overall goal is the development of blue emitters, the weak carbazole donor was modified to study the influence of donor strength for this series of molecules. Stronger donor groups would result in an undesired redshift in emission wavelength. In order to assess the effect on the donor modification, the pending group attached to the phthalimide was set to mesityl for the majority of transformations. For some examples, different pending groups were screened as well. The advantage of the mesityl group lies in its stability, solubility and simplified identification via NMR spectroscopy.

The commercially available carbazoles substituted in 3- and 6-position with halides (bromide, iodine) or a nitrile group were attached to the fluoro phthalimide according to the established procedure (Scheme 19). All emitters were obtained in good to excellent yields in gram scale and showed subjectively perceived a blueshift in emission color compared to the parent compound **CzPIMes** (**102g**). In this series, **CNCzPIMes** (**113**) showed the strongest blueshift due to its strongly accepting nitrile group. The present halides can serve as valuable functional groups for further functionalization and as heavy atom centers to facilitate intersystem crossing.

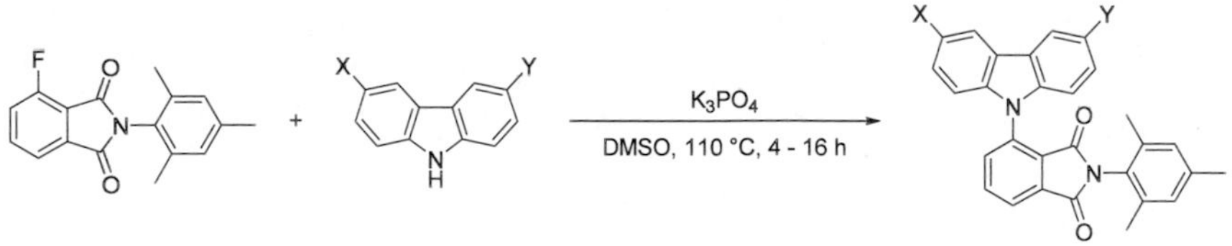

102g	106	X = H	Y = Br	BrCzPIMes	110	96%
FPIMes	107	X = Br	Y = Br	dBrCzPIMes	111	60%
	108	X = H	Y = I	ICzPIMes	112	86%
	109	X = H	Y = CN	CNCzPIMes	113a	84%

Scheme 19. Combination of **FPIMes (101g)** with substituted carbazoles bearing halides and nitriles. **FPItBu (101c)** and **FPInDec (101d)** were reacted as well with **109**, yielding **CNCzPItBu (113b)** and **CNCzPInDec (113c)** in 74% and 71%, respectively.

3.2.1.2.1 Functionalization by an Organometallic Approach

In order to further modify the resulting emitters, **BrCzPIMes (110)** was reacted with organometallic reagents. However, for both *n*-butyllithium and the turbo Grignard reagent *i*PrMgCl·LiCl[244] the attack and resulting ring opening of the phthalimide is preferred over the bromine-metal exchange, rendering this route not feasible as shown in Scheme 20.

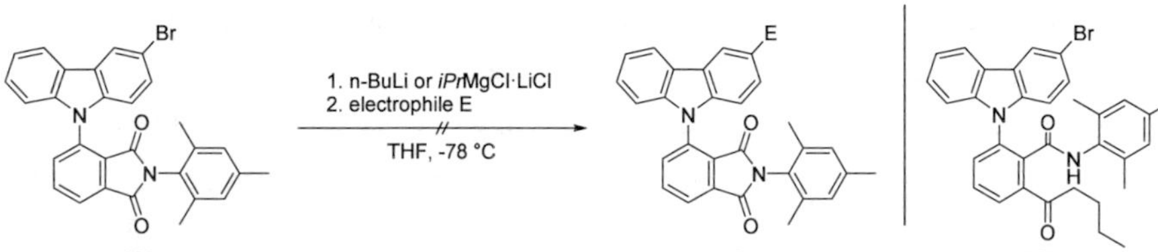

Scheme 20. Reaction of **BrCzPIMes (110)** with organometallic reagents and the expected product **114** and observed product **115** due to an attack on the phthalimide.

Next, the pre-functionalization of the carbazole precursor was tested. Here, the bromide-lithium exchange competes with the deprotonation of the carbazole. 3-Bromocarbazole was treated with one equivalent and with an excess of *n*-butyllithium followed by subsequent quenching with different electrophiles. As electrophiles, water, 1-bromopropane and acetyl chloride were used to attach either a proton, an alkyl chain or an acyl group. The outcome of the reaction was monitored with ASAP-MS. As shown in Table 5, deprotonation of carbazole is favored over the bromide-lithium exchange.

Furthermore, *C*-alkylation/acylation as well as *N*-alkylation/acylation takes place upon addition of the respective electrophiles, resulting in product mixtures and therefore yielding an ineffective route for a selective C-functionalization. Therefore, this approach was discarded.

Table 5. Screening of treatment of 3-bromocarbazole with *n*-butyllithium and subsequent quenching with different electrophiles.

Entry	Electrophile	Equivalents (*n*-BuLi)	Result
1	H₂O	1.0	starting material
2	H₂O	2.5	starting material, little 9*H*-carbazole
3	1-Bromopropane	1.0	starting material
4	1-Bromopropane	2.5	starting material + N-alkylated + C-alkylated carbazole
5	Acetyl chloride	1.0	starting material + N-acetylated carbazole
6	Acetyl chloride	2.5	undefinable

Pre-functionalization of the carbazole starting from 3-cyanocarbazole via transformation of the cyano group into an acyl group upon treatment with methyllithium was feasible. The obtained carbazole was attached to the fluoro precursor as seen in Scheme 21 via the established protocol in good yield. This approach of transforming nitriles with organo lithium reagents by nucleophilic addition into the respective acyl groups can be expanded to other groups such as benzoyl or pivaloyl etc. However, more convenient routes for the installation of acyl groups are available (see 3.2.1.2.2 Functionalization by Friedel-Crafts Reaction).

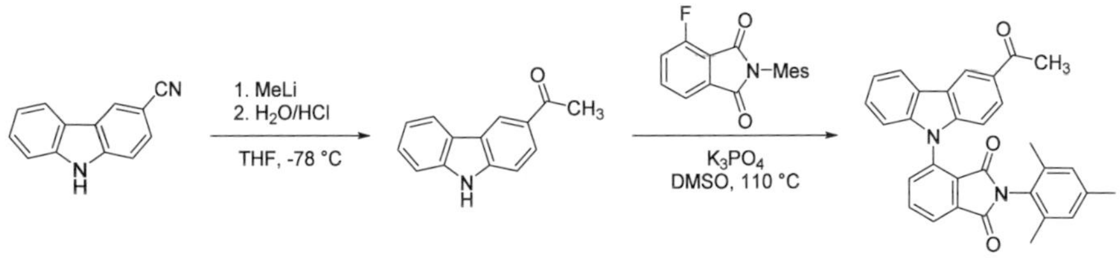

	99%		73%
109	120	101g	121
		FPIMes	COMeCzPIMes

Scheme 21. Transformation of the carbazole nitrile group und subsequent attachment of the functionalized carbazole to yield **COMeCzPIMes (121)**.

3.2.1.2.2   Functionalization by Friedel-Crafts Reaction

Friedel-Crafts acylation allows the introduction of an acyl group via electrophilic aromatic substitution using acyl chlorides or anhydrides. For carbazole, a selectivity of the Friedel-Crafts acylation to the activated 3-position *para* to the nitrogen is reported.[240,245,246] With the carbazole being the electronrichest arene in the TADF emitter structure, it is the favored position for the acylation reaction. Therefore, emitter **CzPIMes (102g)** was reacted in dichlorobenzene or dichloromethane with the respective acyl chloride to yield the acylated products in good to excellent yields of 66% to 97% as shown in Scheme 22. For **122c**, the reaction was successful in a multigram scale yielding 2.71 g of the emitter. The resulting acylated products showed a blueshifted fluorescence compared to the starting material **CzPIMes (102g)** upon irradiation with UV light at 365 nm.

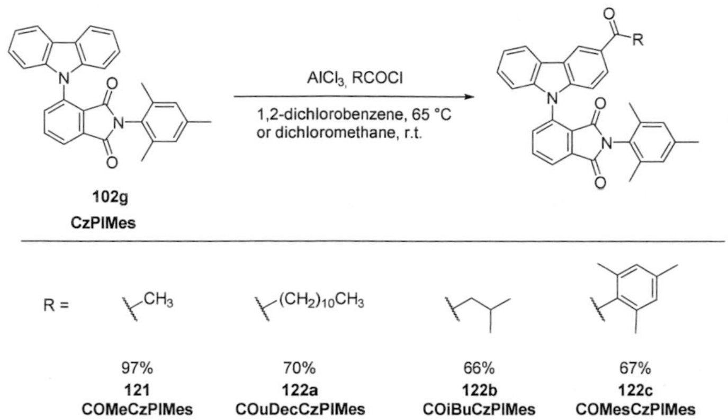

Scheme 22. Friedel-Crafts reaction of **CzPIMes** (**102g**) with different acyl chlorides.

For selected examples, a double acylation was tested. Hence, the equivalents of aluminum chloride and acyl chloride were increased. However, no double acylated product was observed. To circumvent this hinderance, direct Friedel-Crafts acylation of plain 9*H*-carbazole was conducted (see 3.2.1.2.4 Pre-functionalization and other Donor Groups).

In order to proof the selectivity of the acylation at the 3-position of the carbazole, **COMeCzPIMes** (**121**) where the position is clearly known from the starting material (Scheme 21) was compared to the acylated product from the Friedel-Crafts reaction. As seen from the comparison of the proton NMR spectra in Figure 48, both spectra obtained from the prefunctionalized carbazole and the Friedel-Crafts acylated product match perfectly in terms of chemical shifts and integrals of the corresponding signals, verifying that the selectivity of acylation is maintained for both synthetic approaches. The signal at 8.8 ppm, not present in the unacylated molecule, results from the carbazole proton in 4-position being strongly shifted due to the freshly attached acyl group close by.

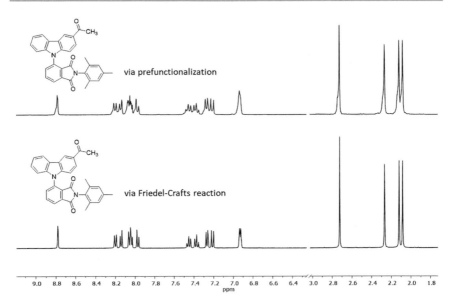

Figure 48. NMR spectra of the acylated emitters **COMeCzPIMes** (**121**) via the pre-functionalization route (top) and via the Friedel-Crafts route (bottom).

In order to install the stronger electron withdrawing trifluoromethyl ketone, a Friedel-Crafts acylation using trifluoroacetic anhydride (TFAA) under similar conditions was executed, yielding the mono acylated product **123**. In contrast to acylation with acyl chloride, the double acylated product (**124**) was readily obtained upon increasing the equivalents of aluminum chloride and anhydride (Scheme 23). Both products showed a drastic blueshift in emission color upon irradiation with UV light at 365 nm compared to the starting material resulting in a bright blue emission.

102g
CzPIMes

123
COCF$_3$CzPIMes

124
dCOCF$_3$CzPIMes

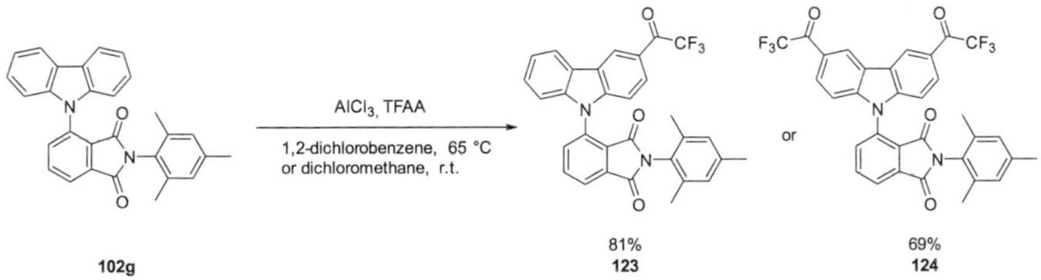

Scheme 23. Friedel-Crafts reaction of **CzPIMes** (**102g**) and TFAA to the mono- and bisacylated products.

Subsequently, **CzPItBu (102c)** was exposed to the same reaction conditions. Although the Friedel-Crafts reaction lead to the bis-acylated carbazole, the *tert*-butyl group at the phthalimide was lost in that process. The expected and observed products are shown in Scheme 24. This is due to the instability of the *tert*-butyl group upon strong acidic conditions, as the *tert*-methylcarbenium ion is stable under acidic conditions.

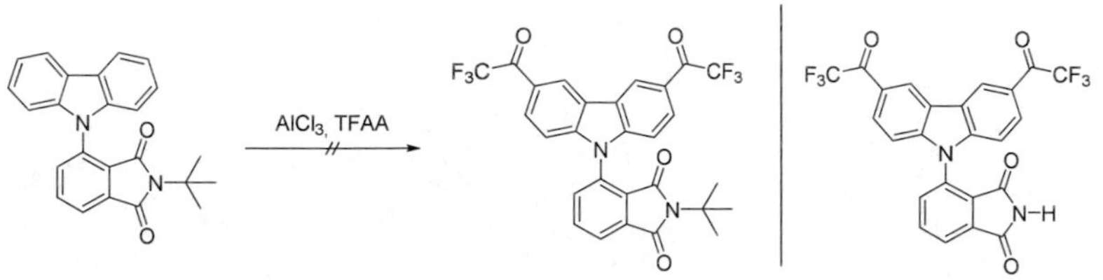

102c
CzPItBu

125
dCOCF$_3$CzPItBu

126
dCOCF$_3$CzPIH

Scheme 24. Friedel-Crafts reaction on **CzPItBu (102c)** and observed product upon loss of the phthalimide substituent.

Furthermore, Friedel-Crafts acylation was conducted with pre-functionalized carbazole systems bearing groups in 3-position, only leaving the 6-position available for acylation. Here, the Friedel-Crafts acylation tolerated the presence of an alky group, bromide and nitrile with the respective products depicted in Scheme 25. The yields were mediocre (38%) to good (71%). Particularly acylation of bromides starting from **BrCzPIMes (110)** performed significantly better when compared to the alkyl group starting from **nBuCzPIMes (132a)**. It is interesting to note that the feasibility of the Friedel-Crafts acylation starting from **CNCzPItBu (113b)** yields **COMeCNCzPItBu (131)** in 57% yield. Here, although aluminum chloride is used as catalyst, the *tert*-butyl group remained stable towards the acidic reaction conditions. The emitters with the strongest blueshift within this series were **COMeCNCzPItBu (131)** and **COMeCNCzPIMes (130)**. This is caused by the strongly deactivated carbazole group through the combination of a nitrile and a methyl ketone.

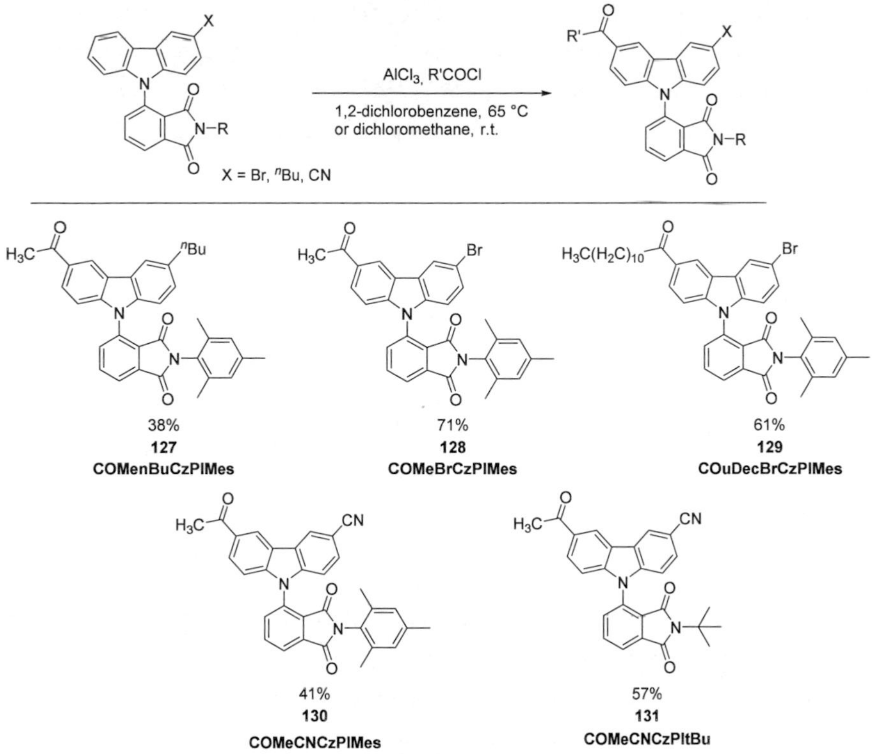

Scheme 25. Friedel-Crafts acylation in the presence of alkyl, halide and nitrile groups.

3.2.1.2.3   Functionalization by Suzuki Cross-Coupling

In addition, a cross-coupling approach via the Suzuki protocol was evaluated in order to install functional groups starting from the halogenated precursors.[240] The goal was to attach groups that allow the enhancement of solubility and to observe the effect on the photophysical properties of the resulting emitters.

As shown in Scheme 26, the bromide of the carbazole substituent of emitter **BrCzPIMes** (**110**) is readily coupled in a Suzuki reaction with aliphatic and electronrich aromatic boronic acids using $Pd_2(dba)_3$ as catalyst and SPhos as ligand, to obtain **nBuCzPIMes** (**132a**) and **TolCzPIMes** (**132b**) in very good yields of 90% and 83%. The limitation of the protocol is observed when a perfluorinated arene is used, where no conversion is observed. **nBuCzPIMes** (**132a**) was subsequently reacted in Friedel-Crafts acylation to achieve an alkylated and acylated emitter in a two-step protocol starting from the brominated carbazole in an overall yield of 34%.

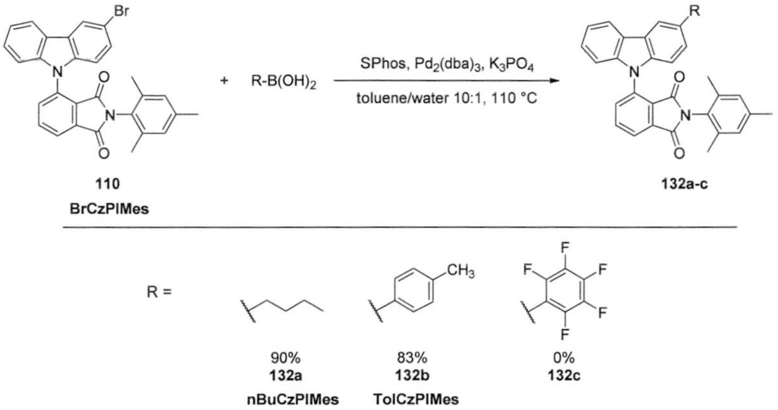

Scheme 26. Protocol for the attachment of aryl- and alkyl groups via Suzuki reaction.

After testing the feasibility of the Suzuki reaction on this system, the acylated and brominated carbazole substituents obtained in section 3.2.1.2.2 (Functionalization by Friedel-Crafts Reaction) were reacted in the Suzuki reaction according to the implemented protocol. The results are shown in Scheme 27. Similar to the prior results, both the alkyl and 4-toluyl boronic acids were effectively coupled to the carbazole group in good to excellent yields of 78% and 93%. The resulting emitter **COuDecnDecPIMes** (**134**) is a prototype example demonstrating the feasibility of this method to implement long alkyl chain and acyl groups with a long alkyl chain. Installing these groups, enhances the solubility of the emitter compared to the initial, unsubstituted system. Furthermore, when Friedel-Crafts acylation is performed prior to Suzuki reaction an overall yield of 48% in two steps is observed, compared to the opposed approach giving an overall yield 34% in two steps.

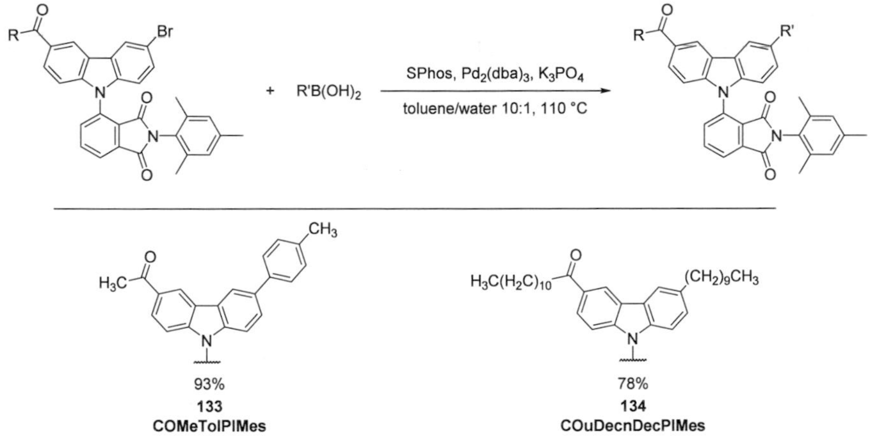

Scheme 27. Suzuki reaction starting from the acylated species.

3.2.1.2.4 Pre-functionalization and other Donor Groups

As the double Friedel-Crafts acylation was not successful for the systems having carbazole already attached to the phthalimide, plain 9H-carbazole was subjected to Friedel-Crafts acylation with acetyl chloride and aluminum chloride in dichloromethane (Scheme 28). The reaction showed full conversion at ambient temperature overnight. Recrystallization with ethanol allowed a facile multigram synthesis, yielding the double acylated carbazole in a yield of 5.22 g (69%).

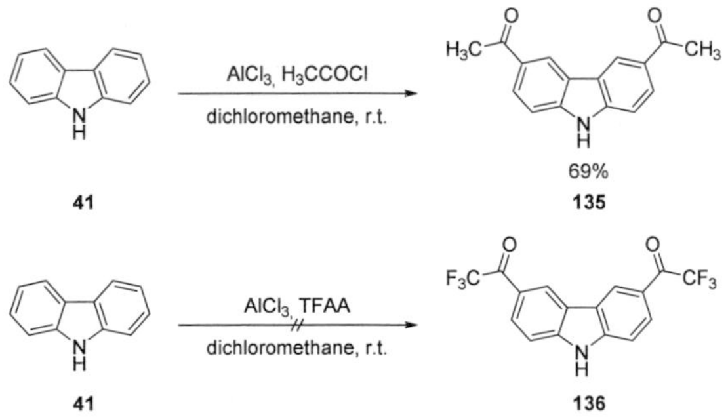

Scheme 28. Friedel-Crafts reaction of carbazole using acetyl chloride and TFAA.

In contrast, acylation using trifluoroacetic anhydride did not succeed with 9H-carbazole, while the same protocol gave the corresponding product when emitter **CzPIMes (102g)** was used. A screening of temperature (r.t. to reflux) and solvent (dichloromethane and dichlorobenzene) did not yield any product even in trace amounts.

The diacylated carbazole **135** was attached to the phthalimide according to the well-established nucleophilic aromatic substitution for two examples. The resulting emitters **dCOMeCzPIMes** and **dCOMeCzPItBu** were obtained in fair yields (Scheme 29).

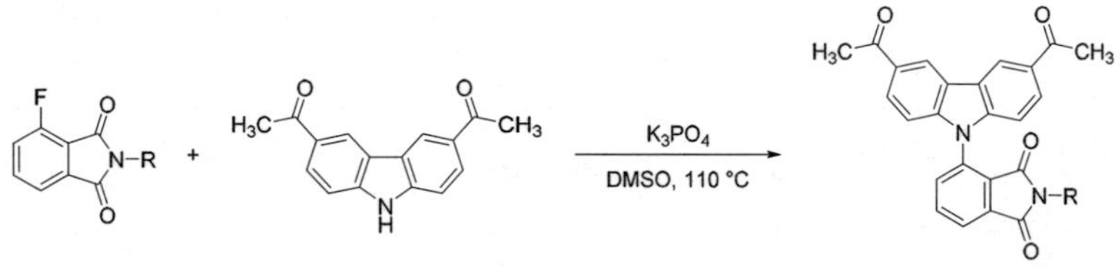

101g **101c** **135** **137** R = Mes 68% **138** R = ^{t}Bu 58%

Scheme 29. Attachment of the acylated carbazole **135** via nucleophilic aromatic substitution.

The molecular structures of both acylated emitters (**dCOMeCzPIMes** (**137**) and **COCF$_3$CzPIMes** (**123**)) were unambiguously determined by single crystal analysis as shown in Figure 49. Similar to **CzPIMes** (**102g**) the mesityl groups are considerably twisted with dihedral angles of 72° for **dCOMeCzPIMes** (**137**) and 82° for **COCF$_3$CzPIMes** (**123**). While **CzPIMes** with the plain carbazole showed a similar dihedral angle as **dCOMeCzPIMes** (**137**) with 71°, the trifluoromethyl ketone at the carbazole in **COCF$_3$CzPIMes** (**123**) further increases the dihedral angle by 10° to an almost perpendicular torsion of 82°. However, the dihedral angle of the carbazole substituent itself decreased to 52° for **dCOMeCzPIMes** (**137**) compared to the initial plain **CzPIMes** (**102g**) system (62°), while for **COCF$_3$CzPIMes** (**123**) the carbazole was twisted similarly at a dihedral angle of 63°.

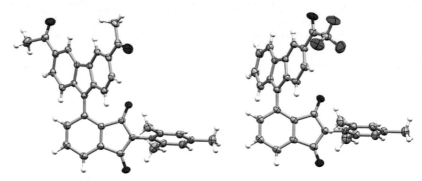

Figure 49. Molecular structure of **dCOMeCzPIMes** (**137**) and **COCF₃CzPIMes** (**123**); displacement parameters are drawn at 50% probability level.

Apart from the main goal of obtaining sky-blue, blue, or deep-blue TADF emitters, where weaker donor groups are of interest due to their effect in lowering the HOMO (see 3.2.2 DFT Calculations), the synthetic scope and limitations of this toolbox approach is of interest. For this, stronger donor groups were also attached to the phthalimide acceptor. The slightly stronger donor 3,6-di-*tert*-butylcarbazole (**42**) was installed to **FPItBu** (**101c**), **FPIMes** (**101g**) and **FPI4CNPh** (**101f**) (Scheme 30).

According to the established procedure for nucleophilic aromatic substitutions, the 3,6-di-*tert*-butylcarbazole was successfully attached to yield **tCzPIMes** (**139**) and **tCzPItBu** (**140**) in good (74% and 79%) yields, while for **tCzPI4CNPh** (**141**) only a moderate yield of 44% was achieved. The pending groups at the phthalimide acceptor were chosen for different reasons. The mesityl group in **tCzPIMes** (**139**) acts as the reference core allowing a competitive evaluation of the donor effects on the optoelectronic properties. *Tert*-butyl groups in general are known to greatly enhance solubility. As already true for **tCzPIMes** (**139**), the combination with the already solubilizing *tert*-butyl group at the phthalimide in **tCzPItBu** (**140**) lead to an emitter with greatly enhanced solubility. Lastly, the combination of the stronger donor group with the electron withdrawing p-benzonitrile group at the phthalimide, shows the strongest redshift in emission color amongst the tested acceptor modifications (3.2.3 Basic Photophysical Characterization) leading to an extension of the obtainable emission wavelengths for this emitter family also towards bathochromic shifts.

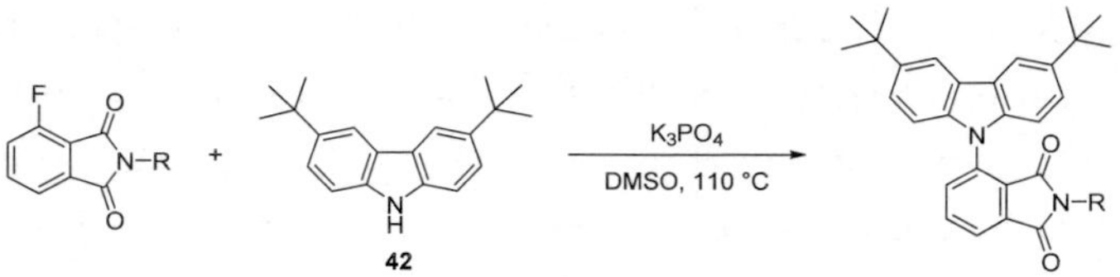

101g	**FPIMes**	R = Mes	**tCzPIMes**	**139**	74%
101c	**FPItBu**	R = ^{t}Bu	**tCzPItBu**	**140**	79%
101j	**FPI4CNPh**	R = 4-CNC_6H_4	**tCzPI4CNPh**	**141**	44%

Scheme 30. Nucleophilic aromatic substitution of 3,6-di-*tert*-butylcarbazole with to a set of phthalimides.

To extend the redshift in emission color further, the strong donor 9,9-dimethyl-9,10-dihydroacridine (DMAC, **142**) was installed for one example (Scheme 31). **AcPIMes** (**143**) was isolated in 62% yield as an intense orange solid showing almost no fluorescence when irradiated with UV light at 365 nm. Also, when doped in PMMA film, barely any fluorescence was detectable for **AcPIMes** (**143**). There was a slight raise around 580 nm but not intense enough and very broad.

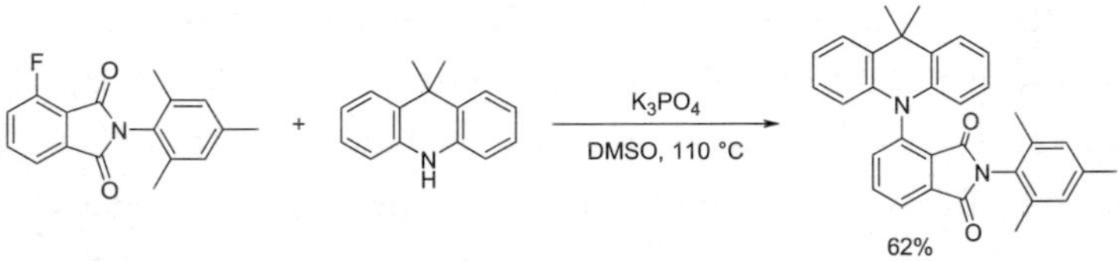

102g **142** 62% **143** **AcPIMes**

Scheme 31. Nucleophilic aromatic substitution reaction using DMAC.

3.2.2 DFT Calculations

DFT calculations were performed using the Turbomole program package. First, geometries of the molecular structures in the ground state in gas phase were optimized employing the BP86 functional and the resolution identity approach (RI). By employing time-dependent DFT (TD-DFT) methods, excitation energies were calculated using the (BP86) optimized structures. For orbital and excited state energies, the B3LYP functional and Def2-SVP basis sets and a m4-grid numerical integration were used. The molecular orbitals were visualized to get initial insights on the HOMO LUMO distribution and the influences of the acceptor and donor modifications.

The energy levels and frontier orbitals of **CzPIMes (102g)** and **CzPI4CNPh (102j)** are shown in Figure 50. As expected, the HOMO is located predominantly on the carbazole donor group with very little contribution on the phthalimide. The LUMO is solely located on the phthalimide group with equal distribution over both cycles of the phthalimide as well as both carbonyl oxygen atoms. The phthalimide nitrogen however, as well as the pending group at the phthalimide do not locate any shares of the LUMO or HOMO. Nevertheless, variation of the pending group has significant influence on the energy levels. Although the pending groups are directly bound to the phthalimide motif which locates and therefore influences the LUMO orbital, both the LUMO and HOMO levels are influenced by variations of the *N*-coupled pending groups. As expected, electron withdrawing groups like the *p*-cyanophenyl substituent in **CzPI4CNPh (102j)** lower the orbital energies compared to **CzPIMes (102g)** with a slight decrease in the HOMO from –5.70 eV to –5.91 eV and a more distinct decrease of the directly affected LUMO from –2.52 eV to –2.92 eV resulting in an overall decreased optical band gap for **CzPI4CNPh (102j)**. Thus, in general, a redshift in emission color is observed for electron withdrawing groups pending at the phthalimide nitrogen. Similar effects were also observed for the electron withdrawing **CzPI3,5CF₃Ph (102l)** (Figure S 12). Interestingly, the electron donating alkyl chains in **CzPInBu (102a)**, **CzPIsBu (102b)**, **CzPItBu (102c)** and **CzPInDec (102d)** barely influence the energy levels compared to **CzPIMes (102g)** resulting in similar optical band gaps with minute differences in the orbitals energies (Figure S 10 and Figure S 11). Neither the degree of branching, nor the length of the alkyl chain showed significant impact.

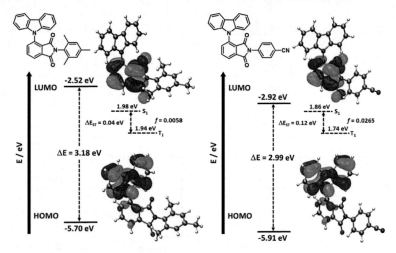

Figure 50. DFT calculations of **CzPIMes (102g)** and **CzPI4CNPh (102j)**. Results provided by cynora GmbH.

Figure 51 visualizes the effects of modifications to the carbazole donor group. As previously observed, both the HOMO and LUMO are influenced by any variations on the system. Presumably, this is due to the very compact TADF system. Still, the impact on the directly attached frontier orbital is typically more distinct such as the decrease in HOMO energy in **COMeCNCzPIMes (130)** from −5.70 eV to −6.34 eV due to the strongly electron withdrawing nitrile and carbonyl group, while the LUMO is decreased only by 0.36 eV. Similarly, the increase in orbital energies in **tCzPI4CNPh (141)** is more distinct for the HOMO. Hence the effects on the optical band gaps are in accordance with the expected increase upon the installation of electron withdrawing groups and decrease by attaching electron donating groups to the carbazole donors. Some modifications such as the nitrile lead to an extension of the LUMO orbital, while the carbonyl group only slightly locates the LUMO orbital, not extending to its substituent (Figure S 13). However, the toluyl group that is directly attached to the carbazole in **COMeTolCzPIMes (133)**, significantly extends the LUMO as shown in Figure S 13.

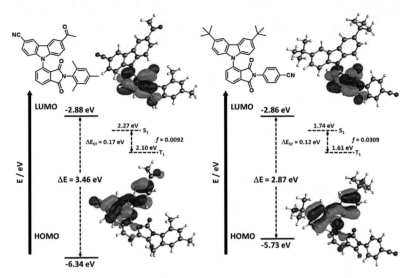

Figure 51. DFT calculations of **COMeCNCzPIMes (130)** and **tCzPI4CNPh (141)**. Results provided by cynora GmbH.

The situation becomes more intriguing when exploring the boundaries of this system as shown in Figure 52. The electron rich trimethoxybenzene substituent in **CzPI3,4,5MeOPh (102n)** changes the localization of the HOMO orbital from the carbazole donor to the trimethoxybenzene. The same effects were observed for the dimethylaniline pending group in **CzPI4NMe₂Ph (102o)** (Figure S 12). Figure 52 also depicts the effects of implementing the strong DMAC donor group in **AcPIMes (143)**, resulting in a significant increase of the HOMO orbital thus reducing the optical band gap. The same effects were observed when attaching phenoxazine, or phenothiazine as donor groups (Figure S 14). For all three strong donor groups, the energy gap ΔE_{ST} is significantly reduced. However, the calculated oscillator strengths are very low, presumably due to the increased steric hindrance and complete separation of the frontier orbitals, which may render these emitters as inefficient.

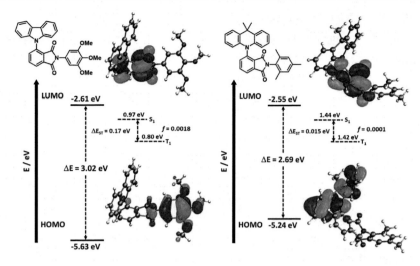

Figure 52. DFT calculations of **CzPI3,4,5MeOPh** (**102n**) and **AcPIMes** (**143**). Results provided by cynora GmbH.

3.2.3   Basic Photophysical Characterization

Basic photophysical characterization was conducted by cynora GmbH. The photoluminescence spectra of the carbazole substituted phthalimides were recorded as 10 wt% dopants in PMMA. Initially, the influence of the *N*-substituted pending group to the phthalimide and their influence on the photophysical properties was investigated (Table 6). A distinct trend on the emission wavelength was observed dependent on the electron donating or withdrawing character of the pending groups. The alkylated phthalimides clearly showed the strongest blueshift in emission wavelength which can be slightly varied between 477 nm and 484 nm dependent on the degree of branching. While for the linear alkyl chains in **CzPInBu** (**102a**) and **CzPInDec** (**102d**) the length barely influences the color, the blueshift increases when the branching of the butyl chain rises from **CzPIsBu** (**102b**) to **CzPItBu** (**102c**). All alkylated phthalimides showed similar photoluminescence quantum yields between 48% and 56% and short lifetimes of the delayed emission component of 7.1 µs to 7.8 µs. The attached allyl group in **CzPIallyl** (**102e**) with its emission maximum at 488 nm bridges the gap to the aryl pending groups attached to the phthalimide nitrogen. The influence of the aryl substituent is in accordance with the results from the DFT calculations. Electron rich arenes such as mesityl in **CzPIMes** (**102g**), 4-toluyl in **CzPI2TolPh** (**105**) and methoxyphenyl in

CzPI3,4,5MeOPh (**102n**) and **CzPI3,5MeOPh** (**102m**) result in emission wavelengths in the range of 491 nm and 495 nm. While the PLQY for **CzPIMes** (**102g**) and **CzPI2TolPh** (**105**) with 45% were found to be similar to the alkylated derivatives, the methoxy groups seem to have a negative influence on the PLQY. While **CzPI3,5MeOPh** (**102m**) still showed a mediocre PLQY of 36%, this significantly decreased to poor 3% in the case of **CzPI3,4,5MeOPh** (**102n**). Presumably, this is attributed to the change in localization of the HOMO orbital from the carbazole donor group to the strongly activated aryl substituent as visualized by DFT calculations earlier. The same effect was observed for **CzPI4NMe2Ph** (**102o**) where no emission and PLQY could be recorded. Similarly, the electron rich N,N-dimethylamine pending group localized the HOMO instead of the carbazole group, as is needed for TADF design.

The electron withdrawing arenes in **CzPI3BrPh** (**102k**), **CzPI3,5CF3Ph** (**102l**) and **CzPI4CNPh** (**102j**) led to the expected redshift in emission color from 498 nm to 505 nm, which is in accordance with the DFT calculations. The strong electron withdrawing nitrile group in **CzPI4CNPh** (**102j**) causes the most distinct redshift in emission color, while still showing a similar PLQY of 40% to the other systems. For all measured emitters, the emission lifetimes of the delayed component were found to be quite short in the range of 5.9 µs to 7.8 µs. Besides the reported influence towards shorter lifetimes induced by halogen atoms,[60] which is also seen for **CzPI3BrPh** (**102k**), there was no observable influence on the lifetimes induced by the nonhalogenated substituents. Also, all emission spectra show similar broad emission bands between 0.45 eV and 0.52 eV, which is typical for this class of emitters.

The emission spectra of selected acceptor-modified emitters are depicted in Figure 53 concluding the feasibility of adjusting the emission wavelength by acceptor substitution. As expected, electron donating groups lead to the expected blueshift, with branched alkyl groups showing the strongest effect, while electron withdrawing groups lead to the respective redshift in emission color. The boundaries of this modification are shown for two examples, where very electron rich arenes lead to a redistribution of the frontier orbitals resulting in non-efficient emitters. The covered range of emission wavelength solely attributed to acceptor modification was found to be 28 nm.

Table 6. Photospectroscopic data in film of the acceptor-modified phthalimide based emitters.

Entry	Compound	#	$\lambda_{max,\,PL}$ [nm]	PLQY [%]	FWHM [eV]	Excited state lifetime τ [µs]
1	**CzPItBu**	**102c**	477	52	0.47	7.0
2	**CzPIsBu**	**102b**	479	56	0.48	7.8
3	**CzPInDec**	**102d**	483	48	0.45	7.5
4	**CzPInBu**	**102a**	484	50	0.48	7.1
5	**CzPIallyl**	**102e**	488	45	0.48	6.4
6	**CzPIMes**	**102g**	491	45	0.47	6.8
7	**CzPI2TolPh**	**105**	493	45	0.48	6.8
8	CzPI3,4,5MeOPh	102n	494	3	0.52	-
9	CzPI3,5MeOPh	102m	495	36	0.48	7.3
10	CzPI4NMe₂Ph	102o	-	-	-	-
11	CzPI3BrPh	102k	498	45	0.49	5.9
12	CzPI3,5CF₃Ph	102l	503	39	0.48	7.7
13	CzPI4CNPh	102j	505	40	0.48	7.2

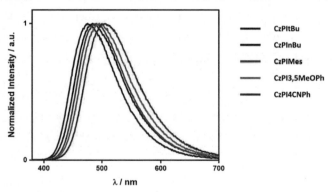

Figure 53. Photoluminescence spectra of acceptor-modified phthalimide based emitters as dopants in PMMA host films.

Next, the influence of modification at the carbazole donor group was investigated (Table 7). The mesityl pending group attached to the phthalimide acceptor was kept fixed to solely assess the influences of the changes on the donor. As expected, and in accordance with DFT calculations, electron withdrawing groups at the carbazole (entries 1–11) lead to a blueshift in emission color in comparison to the initial **CzPIMes** (**102g**) which emits at 491 nm. The strongest effect was shown by the combination of one nitrile and one acyl function in **COMeCNCzPIMes** (**130**) with a blueshift of 39 nm when comparing to the reference emitter, emitting at 452 nm, followed by only one nitrile group in **CNCzPIMes**[240] (**113**) with 462 nm, the trifluoromethyl ketone in **COCF₃CzPIMes** at 464 nm and the diacylated carbazole in **dCOMeCzPIMes** (**137**) with the emission maximum at 466 nm. Out of this series, the nitrile functionalized carbazole donor showed the strongest effect in destabilizing the LUMO and increasing the optical band gap. However, both nitrile substituted emitters **COMeCNCzPIMes** (**130**) and in **CNCzPIMes** (**113**) showed a decrease in PLQY to 24% and 36% respectively, while **dCOMeCzPIMes** (**137**) was more efficient with 44%. Interestingly, the mono-acylated emitters (entries 5, 6, 7, 8) showed similar emission maxima between 471 nm and 478 nm and reasonably good PLQYs between 48% and 55%. The halogenated emitters (entries 9, 10, 11 and 14) acted as precursor molecules for further synthesis but were nevertheless examined. While the iodine in **ICzPIMes** (**112**) did not change the emission wavelength compared to the initial system, the PLQY decreased by 18% to 27%. Bromine substitution (entries 6, 9 and 11) barely had an influence on the photophysical properties when combined with another modification like in **COMeBrCzPIMes** (**128**) and **COuDecBrCzPIMes** (**129**), unless solely the bromine is attached to the carbazole group like in **BrCzPIMes** (**110**), which lead to a blueshift of emission color of 9 nm. As expected, the emitters obtained via Suzuki coupling **nBuCzPIMes** (**132a**) and **TolCzPIMes** (**132b**) show a redshift of emission color to 503 nm and 512 nm, respectively. This tendency is more pronounced upon arylation than alkylation. The *tert*-butyl groups in **tCzPIMes** (**139**) resulted in an emission at 513 nm and a PLQY of 33%. The combined acylation and alkylation in **COuDecnDecPIMes** (**134**) lead to an efficient emitter with a PLQY of 50% emitting at 483 nm.

Table 7. Spectroscopic data of the donor modified phthalimide based emitters.

Entry	Compound	#	λ$_{max, PL}$ [nm]	PLQY [%]	FWHM [eV]	Excited state lifetime τ [μs]
1	**COMeCNCzPIMes**	**130**	452	24	0.47	6.7
2	**CNCzPIMes**	**113**	462	36	0.48	6.6
3	**COCF₃CzPIMes**	**123**	464	32	0.51	71
4	**dCOMeCzPIMes**	**137**	466	44	0.49	7.6
5	**COMesCzPIMes**	**122c**	471	48	0.48	6.3

Entry	Compound	#	$\lambda_{max, PL}$ [nm]	PLQY [%]	FWHM [eV]	Excited state lifetime τ [µs]
6	COuDecBrCzPIMes	129	473	44	0.49	5.7
7	COuDecCzPIMes	122a	475	51	0.48	6.4
8	COMeCzPIMes	121	475	50	0.49	6.4
9	COMeBrCzPIMes	128	475	44	0.49	5.1
10	COiBuCzPIMes	122b	478	55	0.48	6.6
11	BrCzPIMes	110	482	50	0.48	4.0
12	COuDecnDecPIMes	134	483	50	0.48	6.2
13	CzPIMes	102g	491	45	0.47	6.8
14	ICzPIMes	112	491	27	0.47	3.7
15	COMeTolCzPIMes	133	500	30	0.52	3.8
16	nBuCzPIMes	132a	503	36	0.48	5.3
17	TolCzPIMes	132b	512	26	0.50	3.9
18	tCzPIMes	139	513	33	0.48	6.1

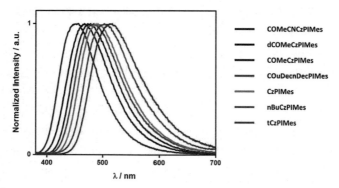

Figure 54. Photoluminescence spectra of donor modified phthalimides as dopants in PMMA host films.

The emission spectra of selected donor modified emitters are depicted in Figure 54 concluding the feasibility of adjusting the emission wavelength by donor substitution. Similar to the acceptor modification series, all emitters showed short excited state lifetimes ranging from 3.7 µs to 7.6 µs and broad emission bands in the range of 0.47 eV to 0.50 eV. Acylation as well as the installation of nitrile groups were efficient ways to significantly blueshift the emission wavelength at the tradeoff of photoluminescence quantum yield beyond an emission wavelength of 465 nm. Similar observations were made upon redshift of emission color by arylation or alkylation. In total, it is demonstrated that the emission wavelength could be varied between 452 nm and 513 nm, covering 61 nm, solely by variation of the carbazole donor.

Finally, the most promising groups from the acceptor and donor screening were combined to explore the limits of this TADF motif. The results are shown in Table 8 and Figure 55. As the *tert*-butyl group in **CzPItBu (102c)** was found to be the acceptor substituent with the strongest blueshift, it was combined with the cyano substituted and bis acylated carbazole, resulting in **dCOMeCzPItBu (138)** and **CNCzPItBu (113b)** with their emission maxima at 448 nm and 450 nm. As already shown for the donor modification, when approaching the deep blue color region, the PLQYs are drastically decreased. This tendency is also observed here with low PLQYs of 26% and 27% for both emitters. Similar results were obtained for **CNCzPInDec (113c)** with the emission maximum at 456 nm and a PLQY of 34%. Lastly, as the strongest blueshift on the carbazole group was found by combining one nitrile with one acyl group, this donor function was combined with the *tert*-butyl group at the phthalimide. The resulting emitter **COMeCNCzPItBu (131)** showed an emission maximum at 437 nm. The PLQY however dropped significantly to 12%, which is even too low to determine the corresponding excited state lifetime. On the other side of the spectral range, the benzonitrile group was combined with the di-*tert*-butyl carbazole o extend the redshift up to 522 nm with a PLQY of 20% for **tCzPI4CNPh (141)**. Apart from the adjustment of the emission color, the solubilizing effects of *tert*-butyl groups at the carbazole and at the phthalimide were combined in emitter **tCzPItBu (140)** to obtain a subjectively perceived highly soluble emitter with a good PLQY of 53% at an emission wavelength of 495 nm. In summary, upon exploring the boundaries for the carbazole substituted phthalimides a significant decrease in photoluminescence quantum yields in the deep blue as well as yellow/orange emission colors were observed. The combined donor and acceptor approach allowed the variation of the emission color between 437 nm and 522 nm covering 85 nm in total.

Table 8. Photoluminescence data of the combination of selected acceptor and donor motifs.

Entry	Compound	#	$\lambda_{max, PL}$ [nm]	PLQY [%]	FWHM [eV]	Excited state lifetime τ [µs]
1	**COMeCNCzPItBu**	**131**	437	12	0.50	-
2	**dCOMeCzPItBu**	**138**	448	26	0.52	9.2
3	**CNCzPItBu**	**113b**	450	27	0.50	7.8
4	**CNCzPInDec**	**113c**	456	34	0.49	7.2
5	**tCzPItBu**	**140**	495	53	0.47	6.5
6	**tCzPI4CNPh**	**141**	522	20	0.48	5.7

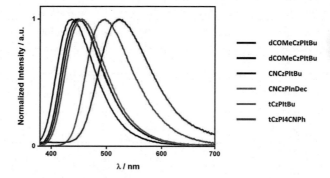

Figure 55. Photoluminescence spectra of the combined acceptor and donor approach modified-phthalimides as dopants in PMMA host films.

3.3   Tristriazolotriazine (TTT) as Acceptor Core

Triazoles,[192] as well as triazines[54,55] have been widely applied as acceptor motifs in TADF emitters due to their electron withdrawing character. The combination of both is investigated for its suitability as accepting unit in TADF emitters. The structural motive of the tris[1,2,4] triazolo[1,3,5] triazine is built by annulation of three triazole heterocycles to a central triazine. There are two possible arrangements shown in Scheme 32 of the triazole rings with their substituents bent regarding the central triazine cycle (**B-TTT**), or linear oriented (**L-TTT**). As the **B-TTT** is mainly investigated in this thesis and for simplicity, the **B-TTT** is subsequently labeled TTT.

First reported in 1911 by Hofmann and Ehrhart,[247,248] Wystrach and co-workers investigated the structure of the triamine of the tris[1,2,4] triazolo[1,3,5]triazine (TTT) core in 1953.[249] The group of Huisgen however was the first to synthesize the first TTT core substituted with phenyl groups in 1961 upon reacting cyanuric chloride with 5-phenyl-tetrazole (Scheme 32).[250]

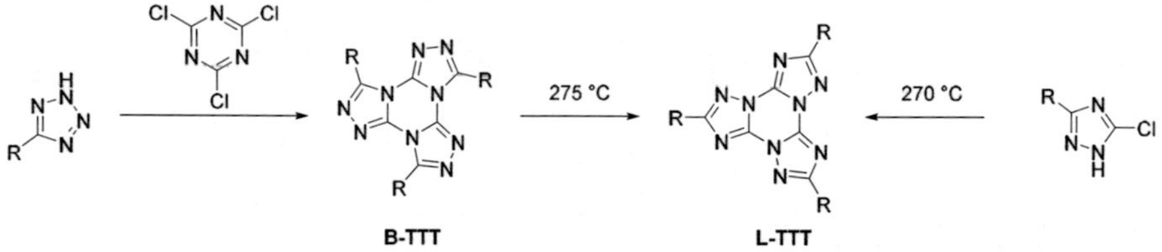

Scheme 32. Synthesis of the TTT core by reacting a tetrazole with cyanuric chloride. Subsequent thermal isomerization and independent formation of the **L-TTT** core.

The proposed mechanism for this reaction is shown in Scheme 33. First, the tetrazole attacks the heterocycle in a nucleophilic aromatic substitution, replacing the chloro substituent. After the loss of molecular nitrogen, the resulting nitrile imine cyclizes with the core heterocycle to a fused triazole system.

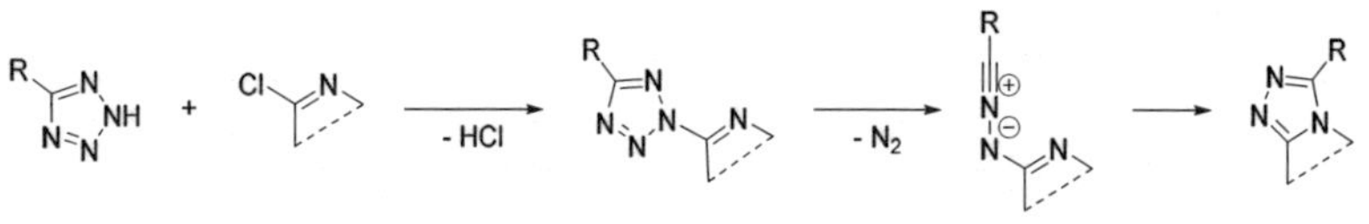

Scheme 33. Proposed mechanism for the formation of the TTT core.

Not considered for any application for many decades, Longo and co-workers investigated the TTT core substituted with peripheral flexible alkyl chains for its application as liquid crystals and advanced materials in 2008.[251] In fact, the disc-like core combined with the aliphatic side chains resulted in luminescent and charge-transporting materials that are also known to self-assemble into columnar superstructures driven by π-stacking, are very suitable for liquid crystal display applications. [251-254] The thermotropic, optical and electrochemical properties were thoroughly investigated, still with the goal of employing the TTT core in liquid crystal applications.[255-257]

Although the TTT core was thoroughly investigated, only two articles report the possible isomerization of the TTT core as shown in (Scheme 32).[254,258] Strelenko and co-workers reported an independent approach to the isomerized L-TTT core by thermocyclization of three 5-chloro-3-phenyl-1,2,4-triazoles (Scheme 32). The thermal isomerization of B-TTT was investigated as well but found to only be partially successful as they found a mixture of partially isomerized triazoles and the fully isomerized product L-TTT upon heating to 350 °C.[258]

Deters and coworkers found similar results when heating a B-TTT derivative to 350 °C, resulting in mixtures with varying composition of non-symmetric isomers and decomposition products. Reducing the temperature and using an inert solvent drastically reduced the decomposition pathway. Time-resolved ^{1}H NMR spectroscopy of the reaction at 235 °C over six days in de-aerated octadecane revealed the stepwise process of the isomerization. Although the starting material was fully converted within 14 hours isomerizing one tetrazole, it took 4 days to complete the full rearrangement to the L-TTT.[254]

The mechanism for the thermal isomerization is believed to follow the Dimroth rearrangement which has been reported to similar structures and is shown in Scheme 34.[259] First, one of the triazine C-N bonds is cleaved heterolytically, allowing the rotation of one ring of the diionic species around the triazole-triazole bond and subsequent recombination. As the positive charge cannot be well distributed and stabilized when compared to the anionic part and taking the high reaction temperatures into account, the radically initiated rotation mechanism starting from a homolytic cleavage of the C-N bond yielding a biradical as the intermediate cannot be ruled out as well.[258]

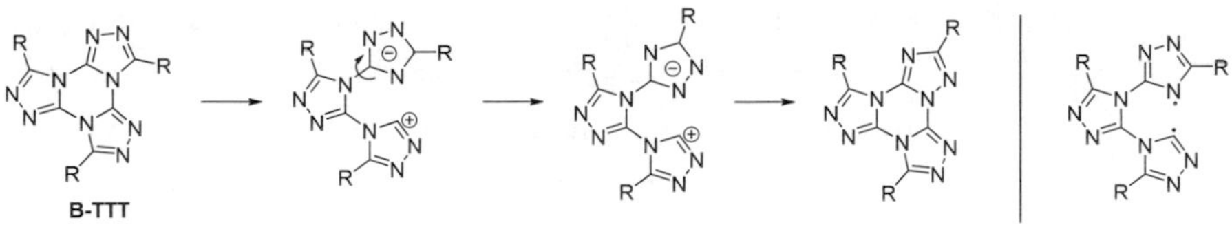

Scheme 34. Proposed mechanism of the isomerization of the **B-TTT** core.

3.3.1   Synthesis

3.3.1.1   Variation on the Donor Type, Count and Arrangement

Starting from the commercially available fluorobenzonitriles, the donor groups were attached via nucleophilic aromatic substitution using tripotassium phosphate as a base in DMSO at elevated temperatures of 110 °C until completion (usually 16 hours). Initially the count and type of donor groups were varied to see the influence of donor strength and arrangement on the optoelectronic properties. As donor groups, diphenylamine, carbazole, 3,6-di-*tert*-butylcarbazole, and 9,9-dimethyl-9,10-dihydroacridine (DMAC) were tested. The rationale behind the choice of donor groups is attributed to their strength as donor groups from the weak diphenylamine, to carbazole, to the slightly stronger di-*tert*-butylcarbazole, to the strong DMAC donor.[103,260–262] The *tert*-butylgroups attached to the carbazole also greatly enhanced the solubility which was necessary as the plain carbazole derivatives showed severe solubility problems at later stages, caused by the planarity and the resulting π-stacking of the TTT cores. Due to the low solubility, di-*tert*-butylcarbazole was used as donor for varying the count and arrangement of TTT-based TADF emitters as well. Here, two and three donor groups were attached to a benzonitrile precursor and for two carbazole groups the arrangement both in bis-*ortho* and bis-*meta* position to the nitrile was tested. To avoid isomers, an asymmetrical attachment of donor groups, e.g. one donor in *meta* and *ortho* position to the nitrile, or two or three donor groups dissimilarly linked, was not pursued.

All donor groups were efficiently linked to the respective fluorobenzonitriles in fair to excellent yields shown in Scheme 35. This could also be performed on multigram scale. Attaching one or two carbazoles in *para*- or *ortho* position resulted in the respective products in excellent yields such as **CzBN (149)** (90%), **tCzBN (150)** (98%) and **2,6-2tCzBN (153)** (94%). The diphenylamine and DMAC groups in **DPABN (148)** (49%) and **AcBN (151)** (67%) were attached less efficiently in fair to good yields. When pursuing the nucleophilic aromatic substitution to the less activated meta position to the nitrile group, the reaction was also not found to be as efficient as for *para* or *ortho* substitutions. Despite this, **3,5-2tCzBN (152)** and **3,4,5-3tCzBN (154)** were obtained in 61% and 50% yield respectively.

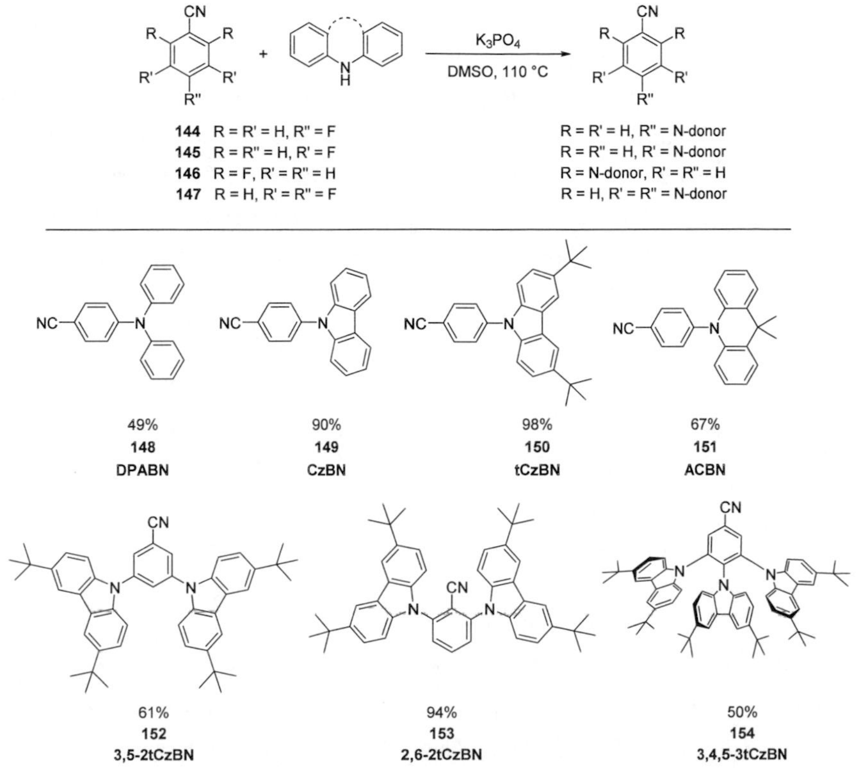

144 R = R' = H, R" = F	R = R' = H, R" = N-donor
145 R = R" = H, R' = F	R = R" = H, R' = N-donor
146 R = F, R' = R" = H	R = N-donor, R' = R" = H
147 R = H, R' = R" = F	R = H, R' = R" = N-donor

49%	90%	98%	67%
148	**149**	**150**	**151**
DPABN	**CzBN**	**tCzBN**	**ACBN**

61%	94%	50%
152	**153**	**154**
3,5-2tCzBN	**2,6-2tCzBN**	**3,4,5-3tCzBN**

Scheme 35. Attachment of different donor groups to the fluorinated precursors via nucleophilic aromatic substitution.

3.3.1.2   Variations on the Phenylene Spacer

Apart from the variation of the donor units, the influence of the phenyl spacer was also subject to investigations. Due to the enhanced solubilizing properties di-*tert*-butylcarbazole was kept as the donor group for this series.

In 2019, the Zysman-Colman group investigated the effects on replacing the phenyl groups of the earlier reported efficient blue TADF material **pDTCz-DPS (155)**[66] by pyridyl groups. Both reported emitters differ in the nitrogen position of the pyridine ring (*ortho* to the sulfone and *ortho* to the donor). One example, **pDTCz-3DPS (156)** alongside with the earlier reported parent compound are shown in Figure 56. While the OLED device fabricated from the parent compound **pDTCz-DPS (155)** showed an EQE of 9.9% with CIE color coordinates of (0.15, 0.07), analogues OLED stacks refabricated using the reported emitter and the pyridyl derivative showed an emission at 428 nm and CIE coordinates of (0.15, 0.08) for **pDTCz-DPS (155)** and 452 nm for **pDTCz-3DPS (156)** with CIE of (0.15, 0.13). However, the performance of the OLED device of **pDTCz-3DPS (156)** showed a significant increase of the maximum EQE to 13.4% compared to 4.6% for the parent compound. The pyridyl groups led to a more planar conformation and emphasized, that small changes in the structure can play a crucial role on the overall effectiveness of an emitter and performance of an OLED device.

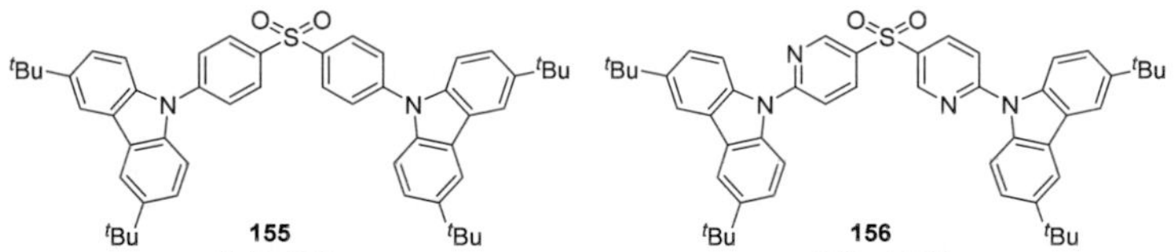

Figure 56. Sulfone-carbazole based TADF emitters and the variation on the bridging unit.

Next, the dihedral angle between the donor and acceptor motive in a molecule is crucial for TADF characteristics as well. The simple combination of the triazine acceptor with different donor groups via a phenyl spacer often serves as a model system to assess the steric demand and resulting dihedral angles of different donor units (Table 9). The combination of plain carbazole[263,264] and the slightly modified emitter **Cz-TRZ1 (157)**[92] was found to be insufficient for the TADF turn-on as the dihedral angle of 49.8° and the weakness of the donor unit results in an energy gap ΔE_{ST} of 0.43 eV caused by significant spatial overlap of the HOMO and LUMO orbitals, which is too high for efficient RISC. By addition of methyl groups, either in 1- and 8-position on the carbazole as in **Cz-TRZ2 (158)** or in *ortho* position to the carbazole on the phenyl spacer as in **Cz-TRZ4 (159)**, the dihedral angles were significantly increased to 86.7° and 82.3°, respectively. This also led to a significant reduction of the energy gaps to 0.08 eV and 0.15 eV. The TADF characteristics were also ultimately proven by highly efficient devices with maximum EQEs of 22.0% and 18.3% compared to 7.2% for the fluorescent emitter **Cz-TRZ1**

(157).[92] The dihedral angle was also greatly increased upon addition of other donor groups like phenoxazines[265] and acridines.[116] Due to the centered six-membered ring compared to the five-membered ring in carbazole, the *ortho*-hydrogen atoms have a stronger steric repulsion from the respective hydrogens at the phenyl spacer ring. The resulting emitter **DMAC-TRZ (160)** with an almost orthogonal dihedral angle of 88° possesses a very small ΔE$_{ST}$ of 0.05 eV, resulting in a highly efficient OLED device with an EQE of 26.5%. The use of stronger donor units like DMAC however also lead to a significant redshift in emission color.

Table 9. Effects of the dihedral angle between donor and acceptor group on the photophysical data and device performance.

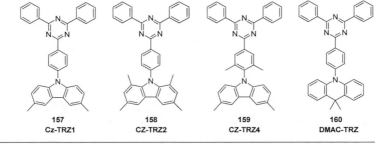

Entry	Compound	#	Dihedral angle	ΔE$_{ST}$ [eV]	EQE$_{max}$ [%]	λ$_{EL}$ [nm]
1	**CZ-TRZ1**	**157**	49.8°	0.43	7.20	450
2	**CZ-TRZ2**	**158**	86.7°	0.08	22.0	485
3	**CZ-TRZ4**	**159**	82.3°	0.15	18.3	448
4	**DMAC-TRZ**	**160**	88°	0.05	26.5	498

As the phenyl group in the final TTT systems between the donor group and the TTT core is mostly attributed to the TTT, hence acceptor unit, the pyridyl approach was transferred to these systems as well. Here, according to the findings in literature, pyridyl spacers were implemented differing in the regiochemistry of the pyridyl nitrogen in 2- or 3-position starting from 5-fluoropicolinonitrile and 6-fluoronicotinonitrile as depicted in Scheme 36. Both pyridyl nitriles, **tCz2PyN (164)** and **tCz3PyN (165)** were isolated in good to fair yields of 77% and 59% (Scheme 36).

As the dihedral angle between donor and acceptor unit in a TADF emitter is crucial for its effectiveness, the installation of methyl groups in *ortho*-position to the attached donor was used as a measure to increase the dihedral angle of the donor substituent in this TTT project. Therefore, 4-fluoro-3,5-dimethylbenzonitrile was subject to the nucleophilic aromatic substitution to obtain **tCz3,5dMeN (166)** in excellent yield of 90% (Scheme 36).

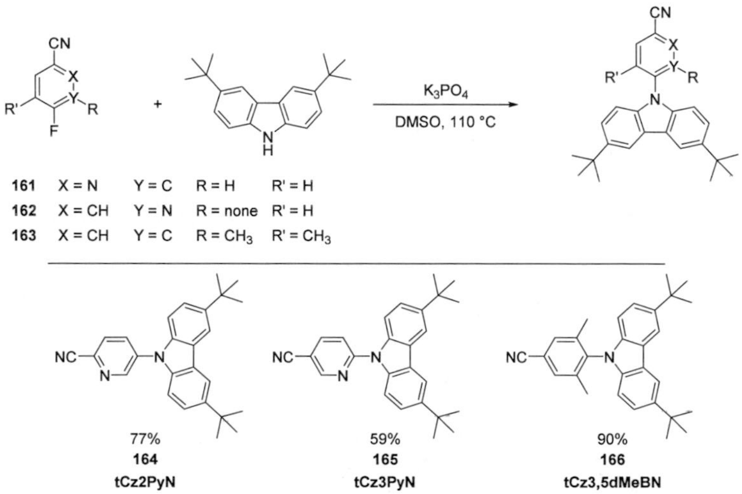

161	X = N	Y = C	R = H	R' = H
162	X = CH	Y = N	R = none	R' = H
163	X = CH	Y = C	R = CH$_3$	R' = CH$_3$

77%	59%	90%
164	165	166
tCz2PyN	tCz3PyN	tCz3,5dMeBN

Scheme 36. Nucleophilic aromatic substitution with the variation at the phenyl spacer.

Another approach in increasing the dihedral angle to greatly affect the TADF properties is using a carbazole derivative methylated in 1- and 8-position. Apart from the alkylation in 1- and 8-position, the tetra-methylated 1,3,6,8-tetramethyl-carbazole was used due to the enhanced electrochemical stability of the carbazole ring.[266,267] It was synthesized as shown in Scheme 37 in a two-step protocol starting from the commercially available methylated bromo-aniline **167** and the dimethyliodobenzene **168**.[268] After Buchwald-Hartwig coupling using Pd$_2$(dba)$_3$ as catalyst and dppf as ligand, the resulting methylated bromo-diphenylamine was oxidatively cyclized in a subsequent palladium-catalyzed reaction to form the pyrrole ring resulting in the final tetramethyl-carbazole donor **170** at an overall yield of 77% over two steps on a multigram scale.

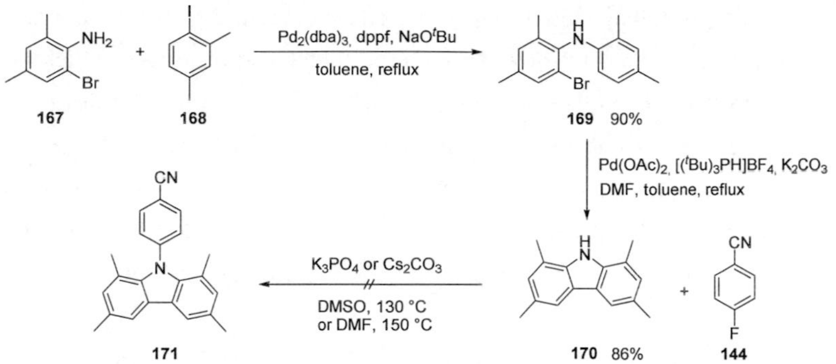

Scheme 37. Buchwald-Hartwig coupling and subsequent cyclization towards the tetramethylcarbazole **170** and subsequent nucleophilic substitution attempts.

Unfortunately, the subsequent otherwise reliable and well working procedure for nucleophilic aromatic substitutions did not yield any product. Even the adjustment to more harsh conditions reported to be working for the tetramethyl-carbazole, was not successful here.[92]

Eventually, the Buchwald-Hartwig coupling with 4-bromobenzonitrile using palladium acetate as catalyst and tri-*tert*-butylphosphonium tetrafluoroborate as ligand lead to the carbazole-substituted benzonitrile **171** in a fair yield of 45% (Scheme 38).[269]

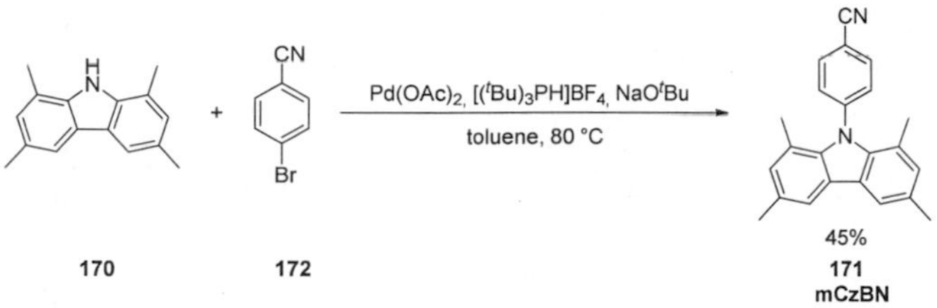

Scheme 38. Successful Buchwald-Hartwig coupling of the tetramethylcarbazole **170**.

3.3.1.3   Synthesis of the Tetrazole Precursors

An efficient and simple procedure for transforming aromatic nitriles into tetrazoles by cyclization using sodium azide and ammonium chloride in DMF at elevated temperatures similar to the tetrazole synthesis in chapter 3.1.1 Tetrazole Derivatives was used to convert the obtained donor substituted benzonitriles into the respective tetrazoles (Scheme 39).[194] The ease of this procedure lies in the isolation of the tetrazoles, which can be directly precipitated after full conversion by pouring the reaction mixture into diluted hydrochloric acid. This allowed the synthesis of the tetrazoles on multigram scale in excellent to nearly quantitative yields without the need of further purification steps. Solely the reaction starting with multiple donor attached benzonitriles (**2,6-2tCzBN (153), 3,5-2tCzBN (152),** and **3,4,5-3tCzBN (154)**) did not show full conversion. Hence, the resulting products **2,6-2tCzBTl (182), 3,5-2tCzBTl (181), 3,4,5-3tCzBTl (183)** were purified by silica column chromatography, still yielding good to excellent.

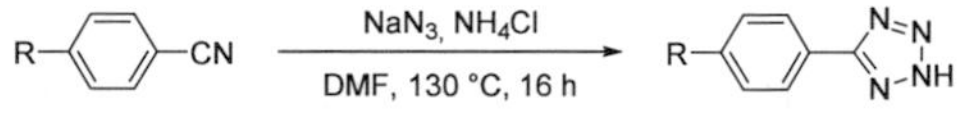

DPABN	148	R = diphenylamine	DPABTl	173	91%
CzBN	149	R = carbazole	CzBTl	174	98%
tCzBN	150	R = 3,6-di-*tert*-butyl-carbazole	tCzBTl	175	99%
AcBN	151	R = 9,9-dimethyl-9,10-dihydroacridine	AcBTl	176	98%
mCzBN	171	R = 1,3,6,8-tetramethyl-carbazole	mCzBTl	177	99%

tCz2PyCN	164	X = N	Y = C	R = H	R' = H	tCz2PyTl	178	97%
tCz3PyCN	165	X = CH	Y = N	R = none	R' = H	tCz3PyTl	179	100%
tCz3,5dMeBCN	166	X = CH	Y = C	R = CH$_3$	R' = CH$_3$	tCz3,5dMeBTl	180	99%

3,5-2tCzBN	152	R = R'' = H, R' = 3,6-di-*tert*-butyl-carbazole	3,5-2tCzBTl	181	70%
2,6-2tCzBN	153	R = R' = H, R'' = 3,6-di-*tert*-butyl-carbazole	2,6-2tCzBTl	182	93%
3,4,5-3tCzBN	154	R = R' = 3,6-di-*tert*-butyl-carbazole, R'' = H	3,4,5-3tCzBTl	183	87%

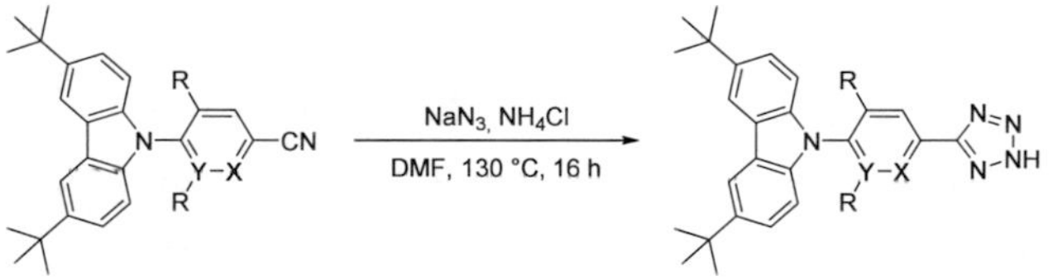

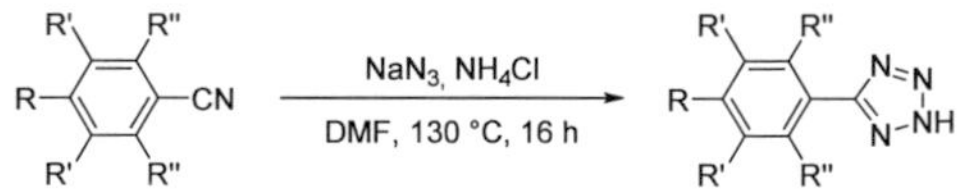

Scheme 39. Cyclization of aromatic nitriles to tetrazoles with sodium azide at different sets of structures: variation of the donor type (top), variation of the spacer (middle), variation of the donor count and arrangement (bottom).

3.3.1.4   Synthesis of the TTT Emitters

A slightly modified procedure from literature[250,257] was used for the synthesis of the final TTT molecules, using toluene as solvent and lutidine as additive as is shown in Scheme

40. Typically, the reaction was completed within a few hours yielding the desired TTT molecules in fair to very good yields ranging from 67% to 89%. Unexpectedly, for the 2-pyridyl derivative **tCz2PyTTT (189)** the yield was poor with only 23%.

Reacting **2,6-2tCzBTl (193)** did not yield any TTT-based product under these reaction conditions. Presumably, the *ortho*-substituted carbazole units inhibit the reaction due to their steric hinderance.

Nevertheless, the reaction worked smoothly for the majority of tested derivatives, usually yielding no observable side products and therefore giving the desired TTT molecules as very pure compounds after filtration over silica or after column chromatography. All resulting compounds show an intense luminescence when irradiated with UV light at 365 nm. The subjectively perceived color ranged from deep blue for **DPATTT (185)**, **tCzTTT (186)** and **tCz3,5dMeBTTT (191)** over sky-blue for **tCz2PyTTT (189)**, **tCz3PyTTT (190)** and **3,5-2tCzTTT (192)** to greenish-blue for **AcTTT (187)** and **3,4,5-3tCzTTT (194)**. As expected, there is a trend to a redshift in emission color as either the donor strength increases from diphenylamine to DMAC, or upon increase in donor count, as this increases the overall donor strength of the system.

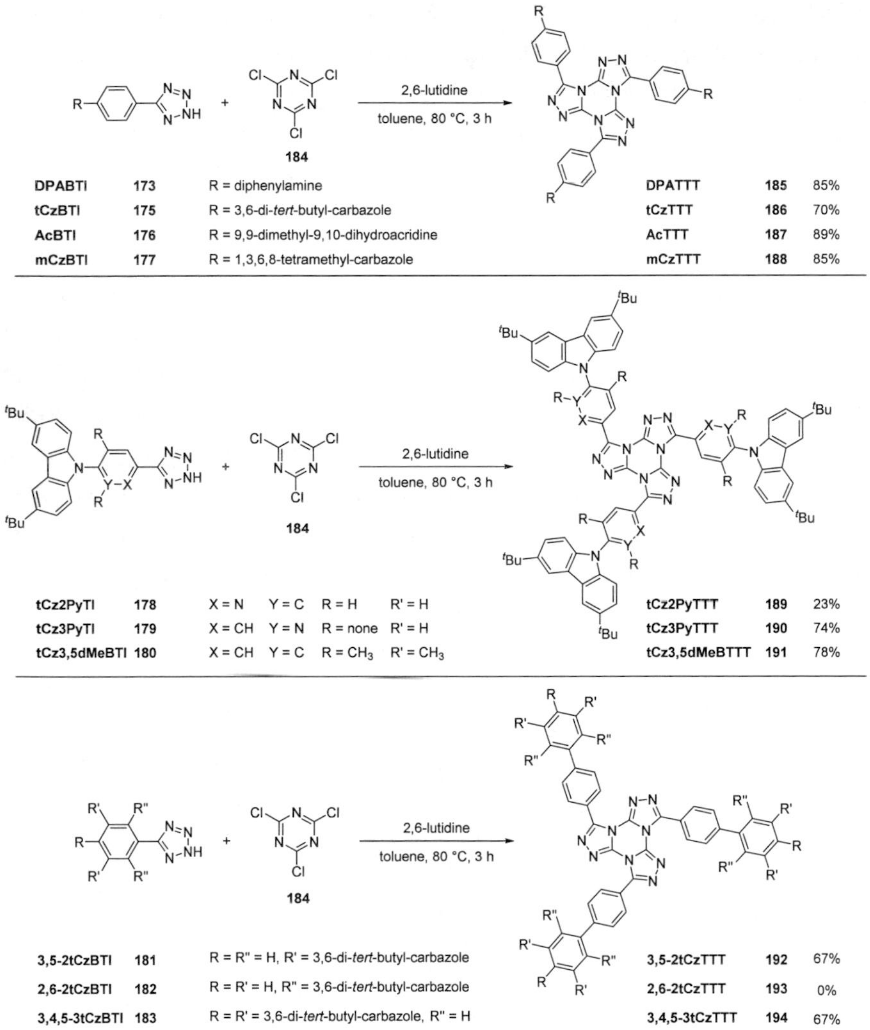
2,6-lutidine
toluene, 80 °C, 3 h
184
DPABTI 173 R = diphenylamine DPATTT 185 85%
tCzBTI 175 R = 3,6-di-tert-butyl-carbazole tCzTTT 186 70%
AcBTI 176 R = 9,9-dimethyl-9,10-dihydroacridine AcTTT 187 89%
mCzBTI 177 R = 1,3,6,8-tetramethyl-carbazole mCzTTT 188 85%
tCz2PyTI 178 X = N Y = C R = H R' = H tCz2PyTTT 189 23%
tCz3PyTI 179 X = CH Y = N R = none R' = H tCz3PyTTT 190 74%
tCz3,5dMeBTI 180 X = CH Y = C R = CH3 R' = CH3 tCz3,5dMeBTTT 191 78%
3,5-2tCzBTI 181 R = R" = H, R' = 3,6-di-tert-butyl-carbazole 3,5-2tCzTTT 192 67%
2,6-2tCzBTI 182 R = R' = H, R" = 3,6-di-tert-butyl-carbazole 2,6-2tCzTTT 193 0%
3,4,5-3tCzBTI 183 R = R' = 3,6-di-tert-butyl-carbazole, R" = H 3,4,5-3tCzTTT 194 67%

Scheme 40. Formation of the TTT core upon reaction of the tetrazoles with cyanuric chloride with different sets of structures: variation of the donor type (top), variation of the spacer (middle), variation of the donor count and arrangement (bottom).

3.3.1.5 Isomerization of the TTT Core

The triazole isomerization was tested for **tCzTTT (186)** using different protocols. The starting material and expected product are shown in Scheme 41. First the molecule was heated similar to a protocol from literature for 72 h at 260 °C in hexadecane.[254] As a result, mostly starting material was isolated, some minor fractions did not yield sufficient compound for a suitable characterization. Next, **tCzTTT (186)** was heated as a neat compound at 360 °C for 1.5 hours. Here, the decomposition of the material started slowly shown by brownish discolorations that were insoluble in any solvent. The crude product was filtered through a short silica plug and analyzed by NMR. As a result, solely the starting material was detected. Also, a quick heating protocol from 300 °C to 400 °C within 10 minutes solely yielded the starting material. Lastly, the material was kept at 390 °C to 400 °C for one hour resulting in intense brown to black discolorations, thus loss of most material. The residue was extracted with dichloromethane and analyzed by NMR showing different pattern and shifts in the proton spectrum that may match the desired product.

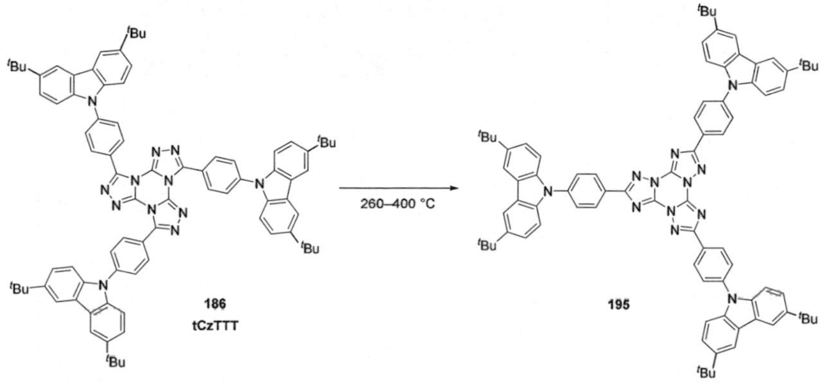

Scheme 41. Isomerization and expected product of the TTT core.

Despite possibly yielding the desired product in traces, no efficient protocol was found to isomerize **tCzTTT (186)** without significant decomposition and loss of starting material so far. Therefore, upscaling to produce sufficient material for optoelectrical characterization seems not feasible for now.

3.3.1.6   Expanding the Concept to MTT and BTT

Aside from the synthesized TTT core starting from cyanuric chloride, both, 2-chloro-4,6-diphenyl-1,3,5-triazine **(196)** and 2,4-dichloro-6-phenyl-1,3,5-triazine **(199)**, were subject to this kind of ring forming reaction. In fact, using the same reaction conditions, once reacted with **tCzBTl (175)** and once with **tCz3,5dMeBTl (180)** led to the desired monotriazolotriazines (MTT) and bistriazolotriazines (BTT) respectively as shown in Scheme 42. All resulting emitters showed intense luminescence both in solution and in solid state when irradiated with UV light at 365 nm. Interestingly, both for the MTT and BTT class of compounds, a strong solvatochromic effect was observed, changing the emission color from deep blue in cyclohexane, over sky blue in toluene to orange in dichloromethane (Figure 57).

Figure 57. Solvatochromism of **tCzMTT** irradiated at 365 nm in different solvents from left to right: cyclohexane, toluene, chloroform, dichloromethane.

Following the same reaction conditions as used for the formation of the TTT emitters, the reaction forming the MTT derivatives was not effective in terms of isolated yields, yielding poorly for both reactions with only 12% and 19%. The synthesis of the BTT derivatives resulted in good yields of 71% and 86%, respectively.

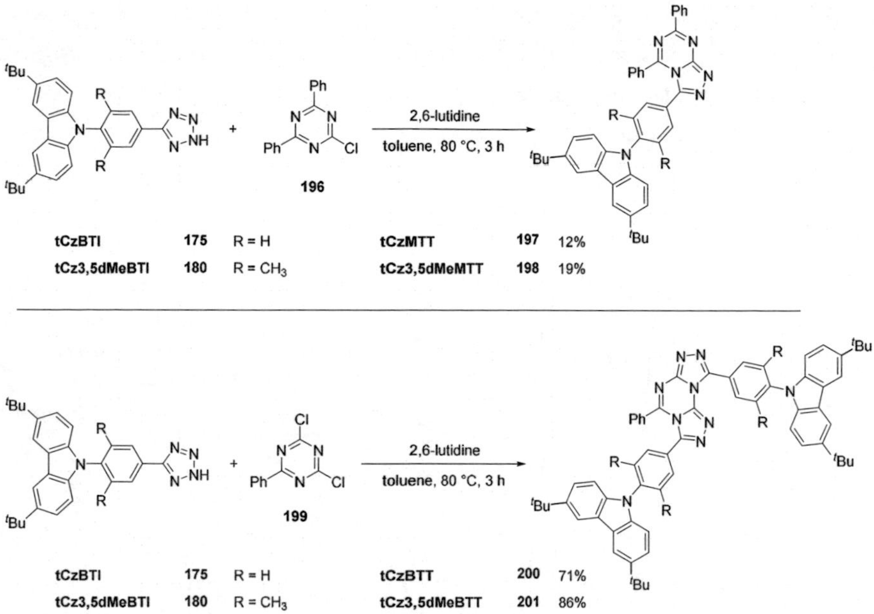

tCzBTI   175   R = H   tCzMTT   197   12%
tCz3,5dMeBTI   180   R = CH₃   tCz3,5dMeMTT 198   19%

tCzBTI   175   R = H   tCzBTT   200   71%
tCz3,5dMeBTI   180   R = CH₃   tCz3,5dMeBTT 201   86%

Scheme 42. Reaction of tetrazoles similar to the TTT formation with mono-chloro and di-chloro analogues resulting in the MTT and BTT motive.

3.3.2   Physical Properties

In order to assess the thermal stability and possible thermal transitions, simultaneous thermal analysis (STA) was conducted with a set of TTT emitters and **tCzMTT (197)**. Thermal gravimetric analysis (TGA) and differential scanning calorimetry (DSC) were measured to investigate the decomposition temperature and processes and the transitions such as melting and to observe the possible isomerization.

All tested emitters showed excellent thermal stability with decomposition temperatures beyond 400 °C. The respective TGA and DSC curves are shown in Scheme 43. The onset temperatures of the decompositions were found to be at 475 °C for **tCzTTT (186)**, 428 °C for **AcTTT (187)**, 484 °C for **3,5-2tCzTTT (192)** and 445 °C for the MTT derivative **tCzMTT (197)**. There were no distinguished decomposition processes for any emitter as the decompositions were observed with one single significant loss in mass, loosing at least 50% of mass for all emitters presumably due to the total decomposition of the molecular structure. Interestingly, the DSC measurement showed either exothermic and endothermic processes for the different emitters above 300 °C. **tCzTTT (186)** and **AcTTT**

(**187**) showed an endothermic process at 313 °C and 303 °C, while **tCzMTT** (**197**) showed an endothermic process at 316 °C. As a separate melting point measurement did not show a melting point for these emitters up to 400 °C, it can be assumed that these processes are attributed to the isomerization of the triazole rings. No rationale was found for one isomerization being endothermic and the other exothermic so far.

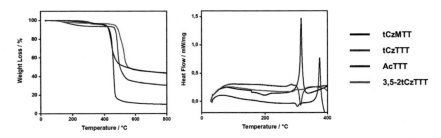

Scheme 43. TGA (left) and DSC (right) measurements for **tCzMTT** (**197**), **tCzTTT** (**186**), **AcTTT** (**187**) and **3,5-2tCzTTT** (**192**). Results provided by Dr. Dominique Moock.

3.3.3 DFT Calculations

DFT calculations were performed using the Gaussian 09 revision D.018 suite.[270] First, geometries of the molecular structures in the ground state in gas phase were optimized employing the PBE0[270] functional with the standard People 6-31G(d,p) basis set.[271] By employing time-dependent DFT (TD-DFT) methods using the Tamm-Dancoff approximation (TDA)[63] excitation energies were calculated using the optimized structures. The molecular orbitals were visualized using GaussView software to get initial insights on the influence of the type and count of donor groups combined with the MTT, BTT and TTT acceptor core.

The energy levels and frontier orbitals of **tCzTTT** (**186**) and its isomer are shown in Figure 58. As expected, the HOMO in **tCzTTT** (**186**) is located on the donor groups, in this case the *tert*-butylcarbazole groups. Due to the presence of several donor groups, many donor-located molecular orbitals are formed partially delocalized over several carbazole moieties (HOMO, HOMO-1, HOMO-2, etc.). The LUMO is located mainly on the TTT core, also occupying the phenyl spacer between the donor and the TTT core. Hence, the overlap between HOMO and LUMO is at the interface of donor and TTT core plus the respective spacer. When comparing both isomers of the TTT core, there are slight differences. Upon isomerization, the optical band gap is increased slightly from 3.73 eV to 3.77 eV, mainly attributed to a destabilization of the LUMO. This also results in an increase of the excited singlet state, hence an increase of ΔE_{ST} from 0.29 eV to 0.34 eV. The energy gaps for both emitters are borderline to the limits to be considered efficient TADF materials.

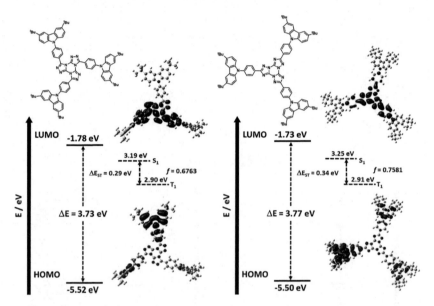

Figure 58. DFT calculations of **tCzTTT** (**186**) (left) and the product of isomerization **195** (right). Results provided by Ettore Crovini.

Figure 59 visualizes the effects of replacing the *tert*-butylcarbazole donor group by the weaker diphenylamine (left) and the stronger DMAC (right). As expected, diphenylamine in **DPATTT** (**185**), being the weaker donor greatly destabilizes the LUMO orbital compared to **tCzTTT** (**186**) by 0.39 eV to −1.39 eV. The HOMO however gets stabilized by 0.23 eV resulting in an overall increase in the optical band gap of 0.17 eV. Also, the energy gap ΔE_{ST} in **DPATTT** (**185**) of 0.56 eV is greatly increased compared to 0.29 eV for **tCzTTT** (**186**), also resulting in an increase in oscillator strength (1.114). This is also visualized by the frontier orbitals, where the HOMO is not only located on the diphenylamine, but in similar shares on the phenyl spacer partially overlapping in space with the LUMO. The resulting overlap in frontier orbitals causes an increased exchange interaction, rendering the molecule as less efficient TADF material. Opposed trends are observed for **AcTTT** (**187**) where the strong DMAC donor stabilizes the LUMO from –1.78 eV for **tCzTTT** (**186**) to –1.93 eV, while also stabilizing the HOMO to –5.39 eV. This results in an overall band gap of 3.46 eV for **AcTTT** (**187**) which amounts to 0.27 eV less than for **tCzTTT** (**186**). The ΔE_{ST} is also significantly reduced to 0.01 eV combined with a low oscillator strength of 0.0014. The situation for the frontier orbitals is similar to **tCzTTT** (**187**), localizing the HOMO on the DMAC donor groups, while the LUMO is mainly located on the TTT core and the respective phenyl spacer groups.

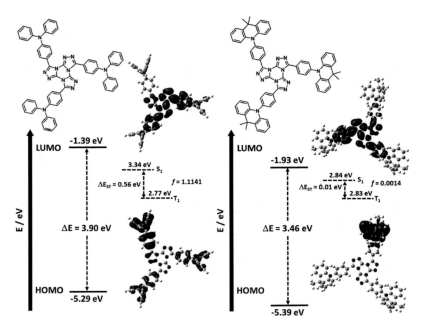

Figure 59. DFT calculations of **DPATTT** (**185**) (left) and **AcTTT** (**187**) (right). Results provided by Ettore Crovini.

Figure 60 visualizes the effects on replacing the phenyl spacer by pyridyl groups. For both, the 2-pyridyl and 3-pyridyl spacer the effects on the optical band gap are marginal resulting in 3.72 eV and 3.78 eV compared to 3.73 eV for the parent phenyl spacer. This is mainly attributed to the slight changes in the LUMO and HOMO. Due to the lack of one proton on the pyridyl rings, the steric repulsion between donor or acceptor group is lower resulting in a better overlap of the frontier orbitals which is shown in an increase of ΔE_{ST}

from 0.29 eV to 0.39 eV for **tCz3PyTTT** (**190**) and to 0.36 eV for **tCz2PyTTT** (**189**). This also results in an increase in oscillator strength especially for **tCz3PyTTT** (**190**) to 0.9573. These effects are more distinct for **tCz3PyTTT** (**189**) as the torsion between the donor group and the spacer which is partially part of the acceptor is crucial for a low ΔE_{ST}. As **tCzTTT** (**186**) with an ΔE_{ST} of 0.29 eV is questionably high to claim sufficient TADF characteristics, it is summarized that replacing the phenyl spacer by the pyridyl group for the TTT motif has in sum a negative influence on the TADF performance judging from the DFT data.

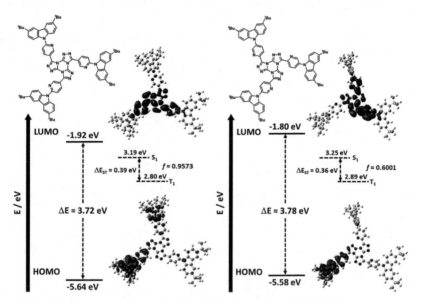

Figure 60. DFT calculations of **tCz3PyTTT** (**190**) (left) and **tCz2PyTTT** (**189**) (right). Results provided by Ettore Crovini.

The effects of the *ortho* methyl groups in **tCz3,5dMeBTTT** (**191**) are depicted in Figure 61. Compared to the parent emitter **tCzTTT** (**186**), the changes of the energy levels are minute. While the HOMO is barely affected, the LUMO is slightly destabilized from −1.78 eV to −1.73 eV. The slight increase on the LUMO level is presumably caused by the electron donating effect of the methyl group to the phenyl spacer which is partially part of the acceptor moiety. Hence, there is also a slight increase in the optical band gap leading to a blueshift for **tCz3,5dMeBTTT** (**191**). More distinct however is the influence on the energy gap ΔE_{ST} which is significantly reduced from 0.29 eV to 0.10 eV, reaching a value which is small enough for efficient RISC, hence the presence of TADF characteristics. As expected, this also results in a decrease of the oscillator strength from 0.676 to 0.136 which is still sufficient for the communication of acceptor and donor groups.

A similar measure in decreasing the energy gap is observed for the sterically more demanding tetramethylcarbazole in **mCzTTT (188)** (Figure 61), resulting in a small ΔE_{ST} of 0.06 and an oscillator strength of 0.0854. However, the electron donating methyl groups also result in an increase of the HOMO level and decrease of the LUMO level, thus a significant smaller optical band gap of 3.41 eV, 0.32 eV smaller than the parent emitter **tCzTTT (186)**.

The *ortho* methyl groups and the tetramethylcarbazole present a powerful measure to significantly lower the energy gap ΔE_{ST}, thus possibly turning on the TADF characteristics of the mono carbazole TTT systems due to an increased dihedral angle.

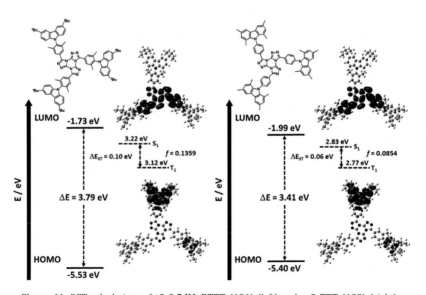

Figure 61. DFT calculations of **tCz3,5dMeBTTT (191)** (left) and **mCzTTT (188)** (right). Results provided by Ettore Crovini.

Figure 62 shows the results of the DFT calculations for the MTT and BTT systems **tCzMTT (197)** and **tCzBTT (200)**. The frontier orbitals are localized similarly to the respective TTT motive with the HOMO at the carbazole group and the LUMO at the MTT and BTT core. The LUMO is partially extended to the residual phenyl groups at the triazine core, while the phenyl spacer is partially occupied by the HOMO, with slight contribution on the directly bound triazole cycle. Surprisingly, the optical band gap first decreases from 3.27 eV for **tCzMTT (197)** to 3.11 eV for **tCzBTT (200)**, while for **tCzTTT (185)** with the TTT system, a drastic increase to 3.73 eV is observed, mainly attributed to a destabilization of the LUMO. As the HOMOs are quite similar, and the donor count increases when going from the MTT to the BTT and to the TTT core, the TTT core seems to be a weaker acceptor compared to the other two. This might be attributed to the complete surrounding of the triazine center by triazole cycles and the accompanying resonance effects. Still, both MTT and BTT systems possess a reasonably low ΔE_{ST} of 0.28 eV each. Interestingly, a significant difference of the oscillator strengths of 0.0187 for **tCzMTT (197)** and 0.2467 for **tCzBTT (200)** is observed. These rather high values render both the MTT and BTT system as potential TADF motifs.

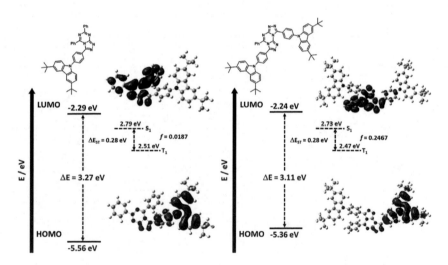

Figure 62. DFT calculations of **tCzMTT (197)** (left) and **tCzBTT (200)** (right). Results provided by Ettore Crovini.

Surprisingly, the methyl groups in *ortho*-position to the carbazole donors does not have the same effect for the MTT and BTT as for the TTT as shown in Figure S 15. For **tCz3,5dMeBMTT (198)** both, the LUMO and HOMO levels are nearly unaffected compared to **tCzMTT (197)**, while ΔE_{ST} is slightly increased from 0.28 to 0.32 eV. Similar effects are observed for **tCz3,5dMeBBTT (201)**, where the energy levels of the frontier orbitals are marginally affected. However, the energy gap ΔE_{ST} is decreased from 0.28 to 0.12 eV by attachment of the *ortho* methyl groups.

3.3.4 Structural Investigations

As no crystals suitable for single crystal analysis of any TTT molecule could be obtained, the DFT-optimized geometries in the ground state were consulted for structural investigations. Still, for two nitrile precursor molecules **tCzBN (149)** and **3,5-2tCzBN (152)**, suitable crystals for x-ray analysis were obtained. The molecular structures are shown in Figure 63. The dihedral angles of the carbazole donors to the benzonitrile group were measured at 47.4° for **tCzBN (149)** and slightly more twisted with 51.3° for both carbazoles in **3,5-2tCzBN (152)**. This is in good agreement with the respective dihedral angles of the optimized TTT structures from the calculation demonstrating their suitability for structural discussions.

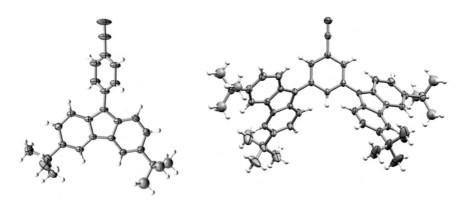

Figure 63. Molecular structure of **tCzBN** (**149**, left), **3,5-2tCzBN** (**152**, right); minor disordered part at the *tert*butyl-groups omitted for clarity, displacement parameters are drawn at 50% probability level.

The dihedral angles of the TTT core to the spacer, as well as the dihedral angles of the spacer to the donor units are shown in Table 10 with the respective structures in Figure 64. Typically, there are only minute differences in the dihedral angles between the three branches of the structures. The dihedral angles between the TTT core and the respective spacer molecule are typically between 29.7° to 36.4° with the exception of **tCz2PyTTT (189)**, which reports slightly higher angles. Hence, the major differences occur from the dihedral angles of the donor units to the spacer units which are typically part of the LUMO orbital. The dihedral angles for one carbazole donor like in **tCzTTT (186)** was calculated at 48.0° to 48.8°. Very similar angles were observed for **tCz2PyTTT (189)**, while the isomer **tCz3PyTTT (190)**, with its nitrogen atom in *ortho*-position to the carbazole donor lowered the dihedral angle by more than 15° due to the comparatively decreased steric repulsion. In contrast, the methyl groups in close proximity to the carbazole donor in **tCz3,5dMeBTTT (191)** led to a significant increase of the dihedral angle to roughly 75°. A more drastic increase in dihedral angle was observed when employing DMAC as donor group like in **AcTTT (197)** resulting in an almost perpendicular orientation of the donor group to the spacer of up to 88.5°. The structural changes in the dihedral angles are directly linked to the calculated singlet-triplet energy gaps and oscillator strengths as a large dihedral angle results in a decreased ΔE_{ST} and a decreased oscillator strength due to a less efficient overlap of frontier orbitals and less degrees of freedom. Due to the lack of structural information from single crystal analysis, no distinct assessment concerning the planarity of the TTT core is made.

Table 10. Dihedral angles of TTT cores to the spacer and spacer to the donor groups of selected TTT-class molecules.

Entry	Compound	#	TTT-spacer1	Donor-spacer1	TTT-spacer2	Donor-spacer2	TTT-spacer3	Donor-spacer3
1	tCzTTT	186	29.7°	48.5°	32.1°	48.8°	32.8°	48.0°
2	AcTTT	187	33.2°	88.5°	34.8°	86.9°	36.4°	87.6°
3	tCz3PyTTT	190	31.8°	33.3°	28.3°	34.1°	32.0°	34.1°
4	tCz2PyTTT	189	43.8°	48.6°	34.5°	48.1°	38.2°	48.7°
5	tCz3,5dMeBTTT	191	32.7°	74.9°	33.9°	75.2°	35.5°	74.9°

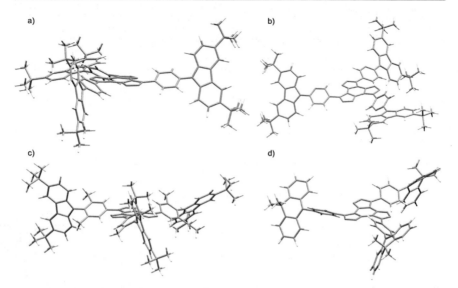

Figure 64. Optimized structures of a) **tCzTTT** (**186**), b) **tCz3PyTTT** (**190**), c) **tCz3,5dMeBTTT** (**191**), and d) **AcTTT** (**187**).

3.3.5 Photophysical Characterization

The photophysical characterization were partially performed during a research stay in the Zysman-Colman group at the University of St. Andrews under guidance of Dr. Nidhi Sharma and beyond that in collaboration with Ettore Crovini.

First, a host screening was performed for the TTT class of molecules to find the optimal host to determine photoluminescence quantum yields. As vacuum processed devices require the emitting materials to be sublimable, which is typically realized for molecular masses below 1000 u none of the synthesized TTT molecules was further assessed for vacuum processing. Hence, in the case of device fabrication, solution processing is the technique of choice. In this regard, solubility properties are crucial. The **3,4,5-3tCzTTT** (**194**) showed excellent solubility in common solvents due to the *tert*-butyl groups. Hence, it was used as the emitter for the host screen. As hosts, the carbazole based mCP, commonly used for blue emitting materials was tested.[272–274] The phosphine oxides PPT[275–277] and DPEPO,[278–280] also known for their application as host materials for blue and green emitters were tested as well. Besides the host itself, the doping concentration of the emitter was varied ranging from 3 wt% to 20 wt%. The impact on the photoluminescence quantum yield with regards to the embedded host and the doping concentration is shown in Table 11.

Table 11. Screening of different host materials for optimal photoluminescence quantum yields using **3,4,5-3tCzTTT (194)** as emitter.

Entry	Host material	Doping concentration	$\Phi_{PL}^{thin\ film}$ air/ %	$\Phi_{PL}^{thin\ film}$ nitrogen/ %
1		3 wt%	30.8	34
2		5 wt%	40.6	44
3	mCP	10 wt%	40.0	43
4		15 wt%	51.1	56
5		20 wt%	43.2	50
6		3 wt%	6.6	8.4
7		5 wt%	10	13.8
8	PPT	10 wt%	14	17.2
9		15 wt%	12.8	15.6
10		20 wt%	20.4	25
11		3 wt%	5	6.7
12		5 wt%	7.6	9.7
13	DPEPO	10 wt%	20	26.5
14		15 wt%	18.7	23.9
15		20 wt%	16	20.6

As expected, all films showed an increase of PLQY when measuring under nitrogen atmosphere instead of air of few percent. mCP was clearly found to be the optimal host with PLQY of up to 56% at a doping concentration of 15 wt%, while PPT and DPEPO gave maximal PLQY values of 25% and 26.5% respectively. For all hosts, higher doping concentrations in the range of 10 wt% to 15 wt% were more effective, while for mCP and DPEPO the PLQY decreased again when increasing the doping concentration further to 20 wt%. For photophysical characterization in thin-film of the other emitters, mCP was chosen as host as well.

Photoluminescence and absorbance spectra of emitter **tCzTTT (186)** in neat film and as dopant in the unpolar hosts PMMA and mCP were recorded (Figure 65). In addition, the photoluminescence quantum yields and photoluminescence lifetimes were measured and recorded as shown in Table 12. The emission maximum in neat film at 462 nm was significantly blueshifted by embedding the emitter in the unpolar hosts PMMA and mCP with emission maxima at 426 nm and 418 nm, respectively, resulting in a deep blue emission. Also, the PLQY greatly increased from 28.2% in neat film to 59.2% in PMMA and 53% in mCP. For the mCP film, 5 wt% was found to give similar PLQY as higher doping concentrations, thus the low concentration was chosen. The emission decay curve

(Figure 65) with a multiexponential fitting showed a prompt decay time of 3.15 ns in mCP (5 wt%) and up to 4.26 ns in the neat film. Unfortunately, no delayed component was found for the emission, giving rise to the assumption that **tCzTTT (186)** shows no TADF characteristics neither in neat film, nor as a dopant.

Table 12. Photophysical characterization of **tCzTTT (186)** in neat thin films and doped in host materials.

Entry	Host	λ_{PL} thin film / nm	Φ_{PL} thin film air / %	Φ_{PL} thin film nitrogen / %	τ_P / ns	τ_D
1	neat	462	25.4	28.2	4.26	-
2	PMMA	426	57.8	59.2	4.14	-
3	mCP (5wt%)	418	52.5	53	3.15	-

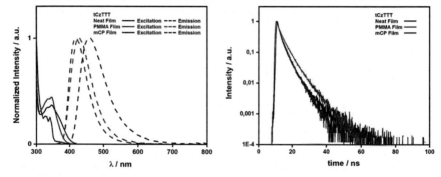

Figure 65. Photoluminescence spectra (left) and photoluminescence decay curves (right) of **tCzTTT (186)**.

As expected for **AcTTT (187)**, the emission wavelength is red-shifted in neat film as well when comparing as dopant in the host materials PMMA and mCP, showing the emission maxima at 491 nm, 461 nm and 469 nm, respectively (Figure 66). An overall redshift of 51 nm in the unpolar OLED host mCP is observed when varying the donor from *tert*-butylcarbazole to DMAC. This is in accordance with the DFT calculations, which showed a decrease of the optical band gap of 0.27 eV due to the stronger DMAC donor. The photoluminescence quantum yield of the **AcTTT (187)** ranged from 21.8% in PMMA to up to 37.5% in mCP at a doping concentration of 20 wt%, which was found to be most efficient for this emitter (Table 13). The observed lower PLQY is reflected by the calculated low oscillator strength. In contrast to **tCzTTT (186)**, the emission decay curve for **AcTTT (187)** showed a delayed component which was found to be at 561 µs with a multiexponential fit. Although this is comparatively long, the presence of the delayed decay component demonstrates the presence of TADF characteristics.

Table 13. Photophysical characterization of **AcTTT (187)** in neat thin films and doped in host materials.

Entry	Host	$\lambda_{PL}^{\text{thin film}}$ / nm	$\Phi_{PL}^{\text{thin film}}$ air / %	$\Phi_{PL}^{\text{thin film}}$ nitrogen / %	τ_P / ns	τ_D / µs
1	neat	491	23.2	30.1	12.3	73.5
2	PMMA	461	11.4	21.8	16.5	762
3	mCP (20wt%)	469	33.6	37.5	12.3	452

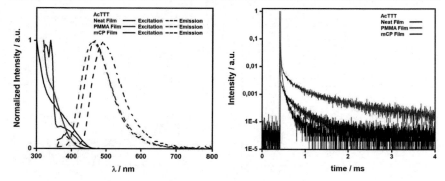

Figure 66. Photoluminescence spectra (left) and photoluminescence decay curves (right) of **AcTTT (187)**.

Similar host effects concerning the emission wavelength were observed for **3,5-2tCzTTT (192)** as well. The addition of a second donor group led to a redshift in emission wavelength by 16 nm to 434 nm in mCP (15 wt%) when compared to **tCzTTT (186)**. Surprisingly, the shift is minute with only 4 nm in PMMA and a blueshift of 16 nm at 446 nm in the neat film. The redshift in emission wavelength is in accordance with the calculated reduced optical band gap. However, compared to **AcTTT (187)**, it is not as significant as predicted by the DFT calculations. Unfortunately, the photoluminescence quantum yields drastically decreased for **3,5-2tCzTTT (192)** to 16.5% in the neat film, 19.9% in PMMA and even 9.4% as 15 wt% dopant in mCP which was found to be the most efficient doping concentration. The low PLQY can be caused by the *meta* position of the donor groups to the TTT core which is known to be less efficient when it comes to electronical communication compared to the ortho or para position. Nevertheless, a prompt component in emission in the range of 5.81 ns to 8.0 ns and delayed components for the neat film and PMMA film of 25.0 µs and 36.6 ms were observed, respectively. While 25.0 µs are in the range of typical TADF emitters, it is unclear why the delayed emission in the PMMA film shows a thousand-fold increase to 36.6 ms.

Table 14. Photophysical characterization of **3,5-2tCzTTT (192)** in neat thin films and doped in host materials.

Entry	Host	$\lambda_{PL,}$ thin film / nm	Φ_{PL}thin film air / %	Φ_{PL}thin film nitrogen / %	τ_P / ns	τ_D
1	neat	446	6.9	16.5	7.55	25.0 µs
2	PMMA	430	6.9	19.9	5.81	36.6 ms
3	mCP (15wt%)	434	9	9.4	8.0	none

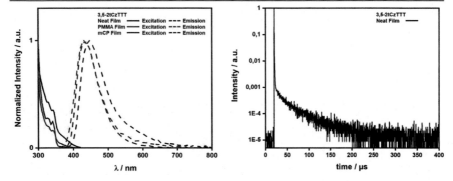

Figure 67. Photoluminescence spectra (left) and photoluminescence decay curve (right) of **3,5-2tCzTTT (192)**.

Further addition of a donor group resulted in a stronger redshift in emission wavelength for **3,4,5-3tCzTTT (194)** (Table 15 and Figure 68). Similar to **AcTTT (187)**, emission maxima at 490 nm in neat film, at 466 nm doped in PMMA and at 462 nm doped at 15 wt% in mCP were observed, which was determined to be the most efficient doping concentration in the previous screening. In this case, the donor strength of one DMAC group is comparable to the donor strength of the three *tert*-butylcarbazole donors. For **3,4,5-3tCzTTT (194)** however, a photoluminescence quantum yield of up to 56% in mCP was observed, which is similar to the PLQY recorded for **tCzTTT (186)**. Also, the prompt component in the range of 10 ns is accompanied by delayed components for the neat, PMMA, and mCP films of 284 µs, 2.90 ms and 454 µs, respectively.

Table 15. Photophysical characterization of **3,4,5-3tCzTTT (194)** in neat thin films and doped in host materials.

Entry	Host	$\lambda_{PL}^{\text{thin film}}$ / nm	$\Phi_{PL}^{\text{thin film}}$ air / %	$\Phi_{PL}^{\text{thin film}}$ nitrogen / %	τ_P / ns	τ_D / ms
1	neat	490	24.4	36.7	10.0	0.284
2	PMMA	466	35.7	54.7	11.9	2.90
3	mCP (20wt%)	462	51.1	56.0	9.78	0.454

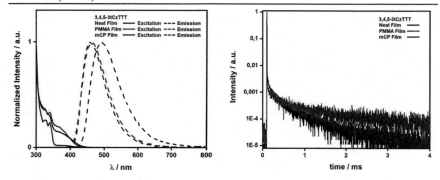

Figure 68. Photoluminescence spectra (left) and photoluminescence decay curves (right) of **3,4,5-3tCzTTT (194)**.

The photoluminescence and absorbance spectra of **tCz2PyTTT (189)** and **DPATTT (185)** are shown in Figure 69. The solubility of the isomer **tCz3PyTTT (190)** was insufficient for solution processed films and their investigations. **DPATTT (185)**, despite showing good PLQYs, especially when embedded in the mCP host of up to 61.5%, no delayed emission component was found, hence no TADF character is attributed to this emitter (Table 17). This is in accordance with the calculated ΔE_{ST} of 0.56 eV which is too high for TADF to efficiently occur at room temperature. The emission maxima for **DPATTT (185)** were dependent on the molecular environment with 453 nm in the neat film and 438 nm doped in mCP. The decay time of the prompt component was fitted between 2.04 ns and 7.76 ns. Similar observations concerning the TADF character were made for **tCz2PyTTT (189)** with no delayed emission component (Table 16). The absence of TADF properties is attributed to the large calculated singlet-triplet gap of 0.36 eV. The highest PLQY of **tCz2PyTTT (189)** was found when doped in mCP at a concentration of 2 wt% with 44.1% at an emission maximum of 438 nm.

Table 16. Photophysical characterization of **tCz2PyTTT** (**189**) in neat thin films and doped in host materials.

Entry	Host	$\lambda_{PL}^{\text{thin film}}$ / nm	$\Phi_{PL}^{\text{thin film}}$ air / %	$\Phi_{PL}^{\text{thin film}}$ nitrogen / %	τ_P / ns	τ_D / µs
1	neat	470	12.4	12.2	9.15	none
2	PMMA	444	27.7	29.1	9.24	none
3	mCP (2wt%)	438	44.4	44.1	6.53	none

Table 17. Photophysical characterization of **DPATTT** (**185**) in neat thin films and doped in host materials.

Entry	Host	$\lambda_{PL}^{\text{thin film}}$ / nm	$\Phi_{PL}^{\text{thin film}}$ air / %	$\Phi_{PL}^{\text{thin film}}$ nitrogen / %	τ_P / ns	τ_D / µs
1	neat	453	33.7	33.6	2.04	none
2	PMMA	429	51.7	52.9	6.92	none
3	mCP (5wt%)	438	61.3	61.5	7.76	none

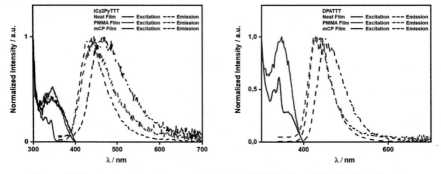

Figure 69. Photoluminescence spectra of **tCz2PyTTT** (**189**) (left) and **DPATTT** (**185**) (right).

The photophysical characterization of **tCz3,5dMeBTTT** (**191**) is summarized in Table 18 and Figure 70. There is a slight blueshift in emission color for **tCz3,5dMeBTTT** (**191**) when compared to the parent emitter **tCzTTT** (**186**) by up to 28 nm which is in accordance with DFT calculations. Especially as 15 wt% dopant in the common OLED host mCP, **tCz3,5dMeBTTT** (**191**) shows a deep blue emission at 409 nm with a PLQY of 29.9%. Surprisingly, the delayed component of emission is significantly reduced to 5.57 µs when compared to any other investigated TTT emitters. It seems, that the RISC process is enhanced by the installation of the *ortho* methyl groups to the carbazole donors rendering this modification crucial for this specific system in order to achieve efficient TADF properties. **mCzTTT** (**188**) showed emission between 448 nm to 464 nm with poor PLQY up to 10%, thus no suitable analyses of the delayed emission component was conducted.

Table 18. Photophysical characterization of **tCz3,5dMeBTTT (191)** in neat thin films and doped in host materials.

Entry	Host	λ_{PL} thin film / nm	Φ_{PL} thin film air / %	Φ_{PL} thin film nitrogen / %	τ_P / ns	τ_D / μs
1	neat	444	17.2	18.9	6.47	none
2	PMMA	398	23.3	25.8	6.67	none
3	mCP (15wt%)	409	29.4	29.9	6.47	5.57

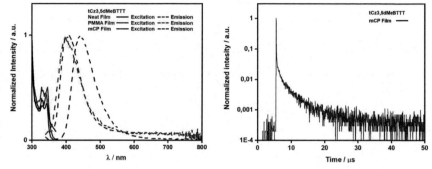

Figure 70. Photoluminescence spectra (left) and photoluminescence decay curves (right) of **tCz3,5dMeBTTT (191)**.

Photoluminescence and absorbance spectra of the MTT- and BTT-based emitters **tCzMTT (197)** and **tCzBTT (200)** were recorded in neat film and as dopant in the unpolar hosts PMMA and mCP (Figure 71). The photoluminescence quantum yields and photoluminescence lifetimes were measured and recorded as well (Table 19 and Table 20). Surprisingly, despite showing a strong solvatochromic effect in solution, the emission wavelengths of **tCzMTT (197)** in neat film or doped in PMMA and mCP (5 wt%) barely show a difference with the emission maxima ranging between 487 nm and 489 nm. The redshift in emission color when compared to **tCzTTT (186)** is in accordance to the difference in calculated optical band gap. The photoluminescence quantum yields increased from the neat film (13.6%) to 33% as a 5 wt% dopant in mCP. Apart from the prompt component of 29.3 ns, the emission decay curves showed a long-lived delayed component for the emission of 423 μs which is similar to the findings in **3,4,5-3tCzTTT (194)**. The BTT analogue **tCzBTT (200)** shows a distinct difference in emission color in the different molecular environment from 541 nm in neat film to 486 when doped in mCP. The PLQY as a 5 wt% dopant in mCP is up to 54.5%. The absence of any delayed emission component is in accordance with the rather high energy splitting ΔE_{ST} of 0.28 eV from the DFT calculations.

Table 19. Photophysical characterization of **tCzMTT** (**197**) in neat thin films and doped in host materials.

Entry	Host	$\lambda_{PL}^{\text{thin film}}$ / nm	$\Phi_{PL}^{\text{thin film}}$ air / %	$\Phi_{PL}^{\text{thin film}}$ nitrogen / %	τ_P / ns	τ_D / μs
1	neat	489	13.2	13.6	11.9	none
2	PMMA	487	24.4	31.7	29.7	532
3	mCP (5wt%)	488	29.3	33	29.3	423

Table 20. Photophysical characterization of **tCzBTT** (**200**) in neat thin films and doped in host materials.

Entry	Host	$\lambda_{PL}^{\text{thin film}}$ / nm	$\Phi_{PL}^{\text{thin film}}$ air / %	$\Phi_{PL}^{\text{thin film}}$ nitrogen / %	τ_P / ns	τ_D / μs
1	neat	541	34	36	24.1	none
2	PMMA	502	31	38	12.2	none
3	mCP (5wt%)	486	54.1	54.5	11.2	none

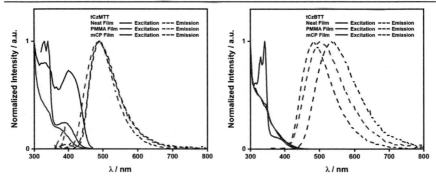

Figure 71. Photoluminescence spectra of **tCzMTT** (**197**) (left) and **tCzBTT** (**200**) (right).

3.3.6   OLED Fabrication

All experiments and analyses concerning the OLED fabrication were conducted via collaboration with Yoshimasa Wada of the Prof. Kaji group at the Kyoto University.

Out of the MTT/BTT/TTT series, **AcTTT** (**187**), **3,4,5-3tCzTTT** (**194**) and **tCz3,5dMeBTTT** (**191**) were the most promising candidates based on their initial photophysical properties. Hence, these three emitters were considered first for OLED fabrication. As optimal host, the silyl-carbazole CzSi, commonly used for blue electrophosphorescence,[281,282] was identified as it showed high PLQYs in the doped films for **AcTTT** (**187**) and **3,4,5-3tCzTTT** (**194**) with 78% and 79%, respectively. Unfortunately, no good PLQYs for **tCz3,5dMeBTTT** (**191**) were found in any tested host so far. OLED devices with the stack architecture shown in Figure S 16 were fabricated for **AcTTT** (**187**) and **3,4,5-3tCzTTT** (**194**). The optimized device for **3,4,5-3tCzTTT** (**194**) emitted in the sky-blue region with 474 nm with an EQE_{MAX} of 5.8% at CIE (0.17, 0.28) (Figure 72). The sky-blue (470 nm) OLED of **AcTTT** (**187**) showed an EQE_{MAX} of 11% at CIE (0.16, 0.23) (Figure 72). The lifetimes of the devices were demonstrated by repeated EQE measurements. They were found to slightly degrade within multiple cycles. (Figure S 17).

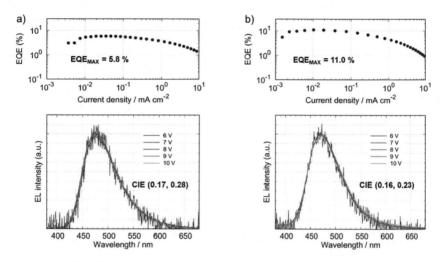

Figure 72. EQE (top) and electroluminescent (bottom) measurements of **3,4,5-3tCzTTT** (**194**) (a) and **AcTTT** (**187**) (b).

4 Summary and Outlook

4.1 Derivatization of the 4CzIPN TADF System

The **4CzIPN** (**3**) system was thoroughly investigated regarding the chemical transformation of the nitrile acceptor groups. Both, substituted tetrazole and oxadiazole derivatives were found as suitable acceptor groups and have been conveniently synthesized via a two-step protocol. Particularly when using oxadiazoles, the optoelectronic properties could be easily manipulated by adjusting the electron donating or withdrawing character of the substituents. Hence, by selective synthesis of either mono- or bis-oxadiazole derivatives the emission wavelength could be varied between 473 nm and 527 nm.

In collaboration with the Colsmann group, selected tetrazole and oxadiazole derivatives were shown to maintain promising TADF characteristics and their suitability in OLED devices was demonstrated with high EQEs of 16% for the mono-oxadiazole **4CzCNOXDtBu (71a)** and 5.9% for the bis-oxadiazole **4CzdOXDtBu (73a)** accompanied by a blueshift when compared to the reference emitter **4CzIPN (3)** (17% EQE).

The newly attached functional groups embody binding sites for further conjugation and follow up chemistry. In collaboration with the Schepers group, the alkyne bearing tetrazole derivative **4CzCNTlpgyl (47e)** has been attached to peptoide biotransporters and was used as fluorescent label for fluorescence imaging experiments *in vitro* (HeLa, SK-Mel 28 and RAW cells) and *in vivo* (zebra fish).

The iodine motif in **4CzCNOXD4IPh (71k)** was used for the attachment of rod-like oligomers in the context of the Collaborative Research Centre 1176 "Molecular Structuring of Soft Matter" to investigate photophysical phenomena such as exciton transport and charge separation in collaboration with the Meier and Lemmer groups. Further investigations are ongoing.

A combined tetrazole/oxadiazole hybrid was successfully synthesized via iterative reaction control, demonstrating orthogonal functionalization possibilities with this model system.

In summary, starting from the reference emitter **4CzIPN (3)**, a multitude of functional groups and structural motives was introduced either via tetrazole or oxadiazole acceptor units resulting in highly efficient tailor-made TADF emitters (Figure 73). Besides succeeding in attaching a TADF emitter to biomolecules and oligomers for various applications, the optoelectronic properties were manipulated mainly with regards to the emission wavelength covering a total range of 438 nm to 541 nm (Figure 73).

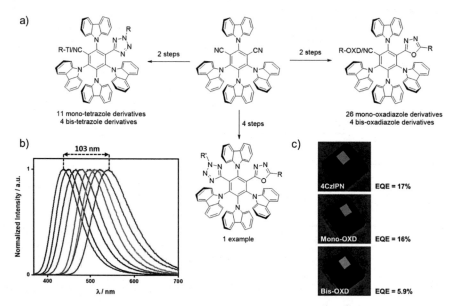

Figure 73. Summary of the **4CzIPN (3)** derivatives: a) Tetrazole, oxadiazole and mixed derivatives. b) Coverage of the emission wavelength depicted with selected emitters. c) OLED performance of selected emitters.

4.1.1 Outlook for 4CzIPN Derivatives

The ongoing investigations particularly in the context of the CRC 1176 can be steered in different directions. As the current TADF-oligomer conjugate serves as an initial model system, the TADF emitter may be in need of change regarding preliminary results. The presented modular reaction control allows just that. Furthermore, orthogonal functional groups for follow up chemistry can be attached which allow the targeted and sequence-controlled attachment of oligomeric building blocks.

The concept of nitrile derivatization can be further applied to other nitrile containing TADF emitters in order to manipulate their physical and optoelectronic properties. Besides the improvement of the solubility, the observed blueshift in emission color is of great interest in order to achieve deep blue TADF emitters.

4.2   TADF   Emitters   based   on   the   Phthalimide   Acceptor System

The phthalimide-based TADF system was thoroughly investigated (in collaboration with cynora GmbH) with modifications to the phthalimide pending group, the carbazole donor and the combination of both (Figure 74). The fine adjustment of the emission color by solely varying the electronic character of the acceptor pending groups for twelve examples was realized covering an emission range between 477 nm and 505 nm with small increments of typically 2 nm. Similarly, the carbazole donor unit was modified with electron donating and withdrawing substituents resulting in the coverage of blue to green emission colors between 452 nm and 513 nm. The groups with the most distinct influences on the emission color were combined in order to assess the limitations of this modular approach. As a result, variation of the phthalimide pending group and the substituents on the carbazole donor, resulted in a tuning of the emission wavelength from 437 nm to 522 nm (Figure 74). Different measures for an increased solubility were conducted like branched and linear alkyl groups. Particularly the introduction of alkyl groups connected via a ketone to the carbazole are of great interest as the electron withdrawing acyl group simultaneously results in a blueshift of the emission color. The limits of this system were demonstrated when very electron rich pending groups at the phthalimide shifted the spatial distribution of the frontier orbitals resulting in non-emissive materials. Furthermore, the replacement of the carbazole donor by the much stronger DMAC resulted in a poorly emissive material, rendering the combination of phthalimide-acceptor and stronger donor groups inefficient.

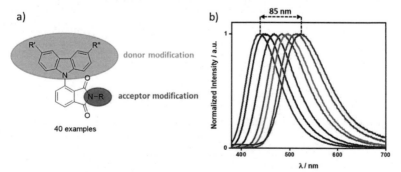

Figure 74. Summary of the phthalimide-based TADF emitters: a) Modification sites on the carbazolyl phthalimide system. b) Coverage of the emission wavelength depicted with selected emitters.

4.2.1   Outlook for Phthalimide-based TADF Emitters

As a multitude of altogether 40 phthalimide-based TADF emitters have been synthesized
and their optoelectronic properties characterized on a basic level, the most promising
candidates will be subjected to advanced photophysical investigations and eventually
OLED fabrication to assess their performance in electroluminescent devices. In addition,
their suitability for both vacuum and solution processed devices has to be determined. As
the structural changes in this series of emitters are minute and a vast plethora of emitters
has been synthesized, this data can be further consulted especially for the investigation of
structure-property relationships. Horizontally oriented emitters are of great interest in
order to further increase the light outcoupling. Thus, orientation studies of this large set
of TADF emitters may give helpful insights for emitter orientation studies.

Chemically, the phthalimide-based TADF emitters can be further modified in order to
assess the effects on the optoelectronic properties. No modifications on the phthalimide
body itself have been reported so far, with the exception of additional donor groups. An
ortho-methyl group to the carbazole donor can be used to adjust the dihedral angle
(Figure 75) and therefore to significantly modulate the TADF characteristics. Fluorine
groups or a strongly withdrawing nitrile group can strengthen the phthalimide acceptor,
rendering it compatible with stronger donor groups as well. Functional groups installed
at the donor or acceptor groups possibly via the nitrile to tetrazole/oxadiazole strategy
will allow the conjugation of the TADF emitter to advanced derived applications such as
biotransporters for bioimaging or in the context of CRC 1176 for the attachment of rod-
like oligomers in order to investigate exciton mobility.

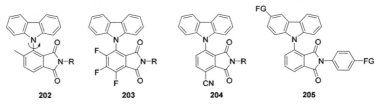

Figure 75. Outlook for phthalimide-based emitters: Variation on the dihedral angle (**202**), the
phthalimide acceptor core (**203** and **204**) and functional groups (FG) for follow-up chemistry
(**205**).

4.3   Tristriazolotriazine (TTT) as Acceptor Core

The tris[1,2,4]triazolo[1,3,5]triazine (TTT) core was successfully employed as acceptor group for TADF emitters. Hence, a three-step protocol was developed for the synthesis of the TTT containing emitters with peripheral donor groups. The donor type and count were varied as well as the connecting phenyl group at the TTT core (Figure 76). Optoelectronic characterization was conducted for all synthesized TTT molecules in collaboration with the Zysman-Colman and Kaji group. The donor count and strength were determined to be the dominant levers to adjust the emission color of the resulting emitter from 398 to 466 nm in PMMA doped films. The dihedral angle between the donor groups and the spacer unit was identified as a critical measure for these systems to show TADF characteristics either by an increased donor count, thus more cramped packing, by installing more space demanding donor groups like DMAC or by *ortho* methyl groups to a single carbazole donor.

The concept was expanded to monotriazolo triazines (MTT) and bistriazolo triazines (BTT) which show promising initial TADF properties.

The most promising candidates, **3,4,5-3tCzTTT (194)** and **AcTTT (187)** were employed as emitters in OLEDs, resulting in sky-blue emitting devices (474 nm and 470 nm) with EQEs of 5.8% and 11% and CIE (0.17, 0.28) and (0.16, 0.23), respectively (Figure 76).

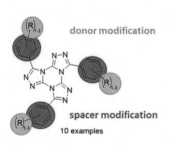
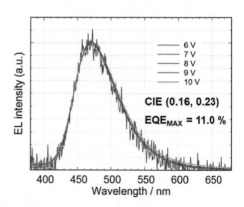

Figure 76. Summary of the TTT-based emitters: a) Scope of TTT derived emitters. b) Electroluminescence spectra of a selected OLED device of **AcTTT (187)**.

4.3.1 Outlook for Tristriazolotriazines

The optoelectronic characterization of the MTT/BTT/TTT class of emitters and therefore their full potential in electroluminescent devices is still under investigation. As the currently reported PLQY are high, further optimization of the device architecture is a promising measure in achieving more efficient devices.

Synthetically, the TTT systems can be further explored by constructing them asymmetrically when starting from a mixture of different tetrazole precursors (Figure 77c). This can facilitate the fine adjustment of optoelectronic properties to further optimize the performance of the resulting emitter. Furthermore, the TTT core with its three branches is perfectly suited for dendrimeric TADF emitter structures for solution processed devices by prolonging the donor branches. The isomerized TTT core (**L-TTT**) and the corresponding emitter can be further investigated. Either, successful conditions for the thermal isomerization from the **B-TTT** to the **L-TTT** core are explored or this core can be built via another route starting from 5-chloro-1,2,4-triazoles (Figure 77a). Once succeeded, the optoelectronic properties of the emitter isomer pair have to be compared in order to assess the structure-property relationships. The concept of annulating triazole rings to the triazine center can be further expanded to other heterocycles. Pyridazines, pyrimidines and even tetrazines may be suitable (Figure 77b).

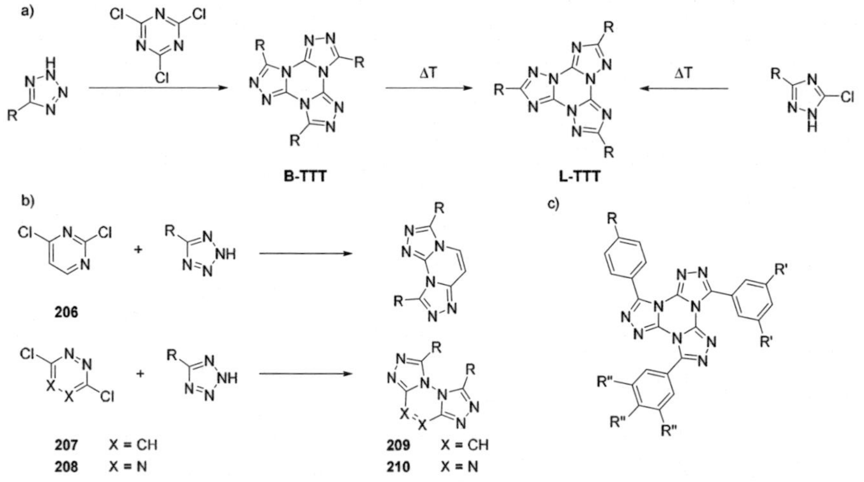

Figure 77. Outlook for the TTT project: a) Thermal isomerization or synthesis to realize the L-TTT core. b) Expanded concept to pyridazines, pyrimidines and tetrazines. c) TTT core with different substituents at its branches.

5   Experimental Section

5.1   General Remarks

5.1.1   Materials and Methods

Nuclear Magnetic Resonance Spectroscopy (NMR)

The NMR spectra of the compounds described herein were recorded on a Bruker Avance 300 NMR instrument at 300 MHz for ^{1}H NMR and 75 MHz for ^{13}C NMR, and a Bruker Avance 400 NMR instrument at 400 MHz for ^{1}H NMR and 101 MHz for ^{13}C NMR.

The NMR spectra were recorded at room temperature in deuterated solvents acquired from Eurisotop. The chemical shift δ is displayed in parts per million [ppm] and the references used were the ^{1}H and ^{13}C peaks of the solvents themselves: d_1-chloroform (CDCl$_3$): 7.26 ppm for ^{1}H and 77.0 ppm for ^{13}C, d_6-dimethyl sulfoxide (DMSO-d_6): 2.50 ppm for ^{1}H and 39.4 ppm for ^{13}C, d_6-acetone (acetone-d_6): 2.05 ppm for ^{1}H and 206.26 ppm for ^{13}C, methylene chloride-d_2 (CD$_2$Cl$_2$): 5.32 ppm for ^{1}H and 54.0 ppm for ^{13}C.

For the characterization of centrosymmetric signals, the signal's median point was chosen, for multiplets the signal range. The following abbreviations were used to describe the proton splitting pattern: d = doublet, t = triplet, m = multiplet, dd = doublet of doublet, ddd = doublet of doublet of doublet, dt = doublet of triplet. Absolute values of the coupling constants "J" are given in Hertz [Hz] in absolute value and decreasing order.

Infrared Spectroscopy (IR)

The infrared spectra were recorded with a Bruker, IFS 88 instrument. Solids were measured by attenuated total reflection (ATR) method. The positions of the respective transmittance bands are given in wave numbers $\tilde{v}$ [cm^{-1}] and was measured in the range from 3600 cm^{-1} to 500 cm^{-1}.

Characterization of the transmittance bands was done in sequence of transmission strength T with following abbreviations: vs (very strong, 0–9% T), s (strong, 10–39% T), m (medium, 40–69% T), w (weak, 70–89% T), vw (very weak, 90–100% T) and br (broad).

Mass Spectrometry (MS)

Fast atom bombardment (FAB) experiments were conducted using a Finnigan, MAT 90 (70 eV) instrument, with 3-nitrobenzyl alcohol (3-NBA) as matrix and reference for high resolution. For the interpretation of the spectra, molecular peaks [M]+, peaks of protonated molecules [M+H]+ and characteristic fragment peaks are indicated with their mass-to-charge ratio (m/z) and in case of EI their intensity in percent, relative to the base peak (100%) is given. In case of high-resolution measurements, the tolerated error is 0.0005 m/z.

APCI and ESI experiments were recorded on a Q-Exactive (Orbitrap) mass spectrometer (Thermo Fisher Scientific, San Jose, CA, USA) equipped with a HESI II probe to record high resolution. The tolerated error is 5 ppm of the molecular mass. Again, the spectra were interpreted by molecular peaks [M]+, peaks of protonated molecules [M+H]+ and characteristic fragment peaks and indicated with their mass-to-charge ratio (m/z).

Elemental Analysis (EA)

Elemental analysis was done on an Elementar vario MICRO instrument. The weight scale used was a Sartorius M2P. Calculated (calc.) and found percentage by mass values for carbon, hydrogen, nitrogen and sulfur are indicated in fractions of 100%.

Thin Layer Chromatography (TLC)

For the analytical thin layer chromatography, TLC silica plates coated with fluorescence indicator, from Merck (silica gel 60 F254, thickness 0.2 mm) were used. UV-active compounds were detected at 254 nm and 366 nm excitation wavelength with a Heraeus UV-lamp, model Fluotest.

Solvents and Chemicals

Solvents of p.a. quality (per analysis) were commercially acquired from Sigma Aldrich, Carl Roth or Acros Fisher Scientific and, unless otherwise stated, used without further purification. Dry solvents were either purchased from Carl Roth, Acros or Sigma Aldrich (< 50 ppm H_2O over molecular sieves). All reagents were commercially acquired from abcr, Acros, Alfa Aesar, Sigma Aldrich, TCI, Chempur, Carbolution or Synchemie, or were available in the group. Unless otherwise stated, all chemicals were used without further purification.

Experimental Procedure

Air- and moisture-sensitive reactions were carried out under argon atmosphere in previously baked out glassware using standard Schlenk techniques. Solid compounds were ground using a mortar and pestle before use, liquid reagents and solvents were injected with plastic syringes and stainless-steel cannula of different sizes, unless otherwise specified.

Reactions at low temperature were cooled using shallow vacuum flasks produced by Isotherm, Karlsruhe, filled with a water/ice mixture for 0 °C, water/ice/sodium chloride for –20 °C or isopropanol/dry ice mixture for –78 °C. For reactions at high temperature, the reaction flask was equipped with a reflux condenser and connected to the argon line. Solvents were evaporated under reduced pressure at 40 °C using a rotary evaporator. Unless otherwise stated, solutions of inorganic salts are saturated aqueous solutions.

Reaction Monitoring

The progress of the reaction in the liquid phase was monitored by TLC. UV active compounds were detected with a UV-lamp at 254 nm and 366 nm excitation wavelength. When required, vanillin solution, potassium permanganate solution or methanolic bromocresol green solution was used as TLC-stain, followed by heating. Additionally, APCI-MS (atmospheric pressure chemical ionization mass spectrometry) was recorded on an Advion expression CMS in positive ion mode with a single quadrupole mass analyzer. The observed molecule ion is interpreted as $[M+H]^+$.

Product Purification

Unless otherwise stated, the crude compounds were purified by column chromatography. For the stationary phase of the column, silica gel, produced by Merck (silica gel 60, 0.040 × 0.063 mm, 260–400 mesh ASTM), and sea sand by Riedel de-Haën (baked out and washed with hydrochloric acid) were used. Solvents used were commercially acquired in HPLC-grade and individually measured volumetrically before mixing.

5.1.2   Advanced Optoelectronic Data

5.1.2.1   Tetrazole Derivatives

Time-resolved and temperature dependent photoluminescence

All photoluminescence measurements for the advanced TADF analysis were measured with an Edinburgh FS5 Fluorometer. For oxygen-free measurements in solution, **4CzCNTlallyl (47d)** was dissolved in anhydrous THF (Sigma-Aldrich) and filled in a sealable cuvette. Everything was prepared in an N_2-filled glovebox. For solid state measurements a 3 g/l solution of mCP:**4CzCNTlallyl** (10 wt%) was spincoated onto a glass substrate (1000 rpm, 45 s). It was mounted onto the cold finger of a liquid nitrogen cryostat and the cryostat was evacuated inside the glovebox to avoid oxygen contact with the sample. Afterwards the cryostat was transferred to the fluorometer in air. All solid-state measurements were carried out in vacuum. For emission decays, the samples were excited with a 340 nm LED with a pulse length of 1 ns. Solid state measurements were conducted in time correlated single photon counting (TCSPC) mode and the long component of **4CzCNTlallyl (47d)** in solution measured with multichannel scaling (MCS). Multiexponential fitting was used to extract the individual lifetimes from the decay curves.

Figure S 1 shows the irradiation dependence of the emission of mCP: **4CzCNTlallyl (47d)** at room temperature. The trend is clearly linear with a slope of 1.06. For TTA one would expect a quadratic behavior as it is a bimolecular recombination process, whereas TADF is a monomolecular process. The intensity was varied by decreasing the slit-width of the monochromator after the Xenon-lamp. The intensity was recorded by the transmission diode of the FS5 after passing through the sample.

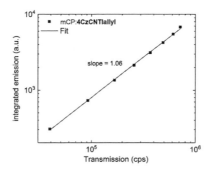

Figure S 1. Irradiation dependence of mCP:**4CzCNTlallyl (47d)** at room temperature.

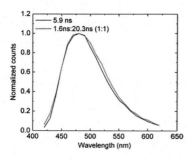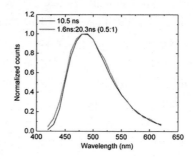

Figure S 2. Time resolved spectra of mCP:**4CzCNTlallyl** (**47d**) at room temperature in the ns-regime can be perfectly modelled by linear combination of the spectra at 1.6 ns and 20.3 ns. We suppose that these two spectra arise from the ^{1}LE and ^{1}CT state, respectively. The ^{1}LE state is excited first and transfers energy to the ^{1}CT state as well as emits fluorescence. With increasing time, the ^{1}LE state is depopulated and after a time where both states show fluorescence only emission occurs from the ^{1}CT state.

OLED Fabrication and Additional Data

The ionization potential of **4CzCNTlallyl** (**47d**) was determined by photo-electron spectroscopy in air (PESA, Riken Keiki AC-2E).

OLED fabrication: All OLEDs were fabricated on indium tin oxide (ITO) coated glass substrates ($R_\square \approx 13\ \Omega$) with an active area of 0.105 cm^2. The substrates were subsequently cleaned with acetone and isopropanol in an ultrasonic bath (10 min). Afterwards the substrates were exposed to oxygen plasma (2 min) in order to remove organic residues and to polarize the ITO surface for better PEDOT:PSS adhesion. After plasma treatment, the samples were transferred to a nitrogen-filled glovebox and kept there for the remaining fabrication process and optoelectronic characterization. PEDOT:PSS (Clevios P VPAI 4083, Heraeus Deutschland GmbH & Co. KG) was diluted in ethanol 1:3 (v/v), spin cast (4000 rpm, 45 s, d = 25 nm) and annealed (120°C, 10 min). mCP (1,3-Bis(N-carbazolyl)benzene; Lumtec) was used as received. The blend of mCP:**4CzCNTlallyl** (100:15 wt/wt) was spin coated (1000 rpm, 30 s) from tetrahydrofuran (THF) solution (2 mg mL^{-1}) onto the PEDOT:PSS layer, followed by sample annealing (60°C, 10 min) to achieve a 30 nm thick layer. Then the samples were transferred to a high vacuum chamber (10^{-7} mbar), where an electron transport layer from BP4mPy (3,3',5,5'-Tetra[(m-pyridyl)-phen-3-yl]biphenyl; Lumtec) (50 nm) as well as a Liq (8-Hydroxyquinolinolato-lithium; Lumtec) (2 nm) /Al (50 nm) cathode were thermally evaporated.

Device characterization: The electrical properties of the OLEDs were recorded on a source measurement unit (Keithley 2400), the luminous flux was measured using an integrating sphere coupled a photometer (Gigahertz Optics VL-1101) and a spectrometer (Instrument Systems CAS140-151). The luminance was calculated assuming Lambertian emission. The setup was calibrated with a halogen standard and spectral mismatch as well as integrating sphere response were corrected for each OLED using an auxiliary lamp. The layer thicknesses were measured on a tactile profiler (DektakXT, Bruker). The OLEDs showed a very strong degradation already during the first measurement (Figure S 3). This is attributed to a degradation of the emitter as the J-V curve doesn't change much with repeated measurements and as a red shift in emission occurs after multiple measurements (Figure S 4).

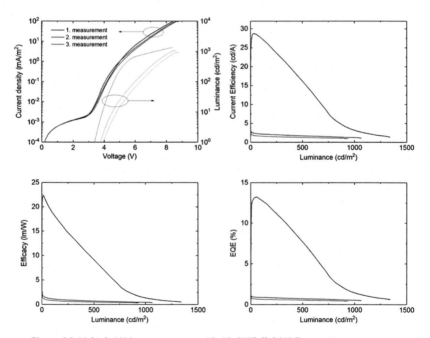

Figure S 3. Multiple J-V-L measurements with **4CzCNTlallyl** (**47d**) as emitter.

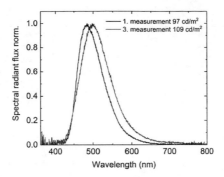

Figure S 4. Redshift of emission spectrum of **4CzCNTlallyl**-OLED during three measurements up to 100 mA/cm^2.

5.1.2.2   Oxadiazole Derivatives

Excerpts of this chapter have been already reported in a peer reviewed article of the author.[180]

Solid samples were prepared by spincoating a 3 g/l solution of host:guest (10 wt%) in THF onto quartz samples, resulting in a layer thickness of 50 nm. Measurements were performed with a Edinburgh Instruments FS5 fluorometer. Temperature and time-dependent measurements on solid samples were done in an Oxford Instruments liquid nitrogen cryostat under vacuum (<5E-5 mbar) using TCSPC as measurement technique. Room temperature measurements were also performed in the cryostat under vacuum. Excitation wavelength was 340 nm (1 ns pulsewidth). PLQY was measured in a calibrated integrating sphere (SC30, Edinburgh Instruments) under nitrogen flushing.

a)   b)

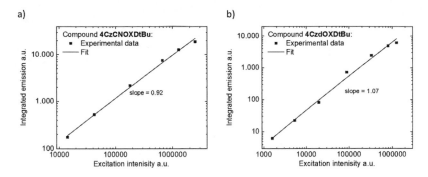

Figure S 5. Irradiation dependence of the PL-emission at room temperature of compound **4CzCNOXDtBu (71a)** (a) and **4CzdOXDtBu (73a)** (b) in mCP (10 wt%). Both show a slope close to 1 indicating that TADF is the dominant mechanism for delayed fluorescence and not triplet-triplet annihilation where we would expect a slope of 2 for bimolecular recombination. The intensity was varied by opening the slit of the spectrometer (Edinburgh Instruments FS5) and the intensity was measured with the calibrated transmission photodiode. Therefore, the values are in arbitrary units.

a)   b)

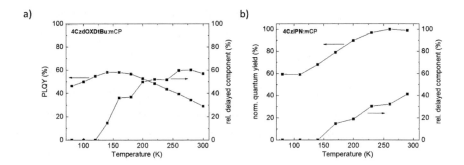

Figure S 6. Temperature dependent PL intensity and portion of the delayed fluorescent component of (a) **4CzdOXDtBu**:mCP and (b) **4CzIPN**:mCP. The PL intensity was measured in a Cryostat and PLQY was calculated by integrating of the emission spectra and referencing with the PLQY measured at room temperature in the integrating sphere. During measurement the transmission of the sample was monitored to rule out that the absorption of the sample changes with temperature.

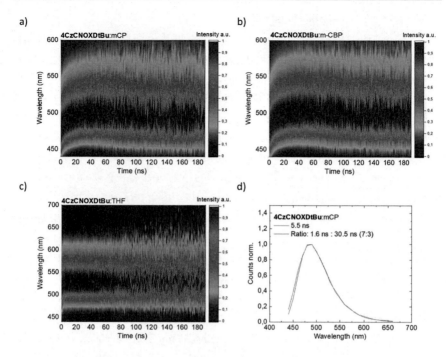

Figure S 7. TRES map of (a) **4CzCNOXDtBu**:mCP, (b) **4CzCNOXDtBu**:m-CBP and (c) **4CzCNOXDtBu**:THF. (d) Fit of time resolved spectrum of **4CzCNOXDtBu**:mCP at 5.5 ns by a linear combination of the spectra at 1.6 ns and 30.5 ns.

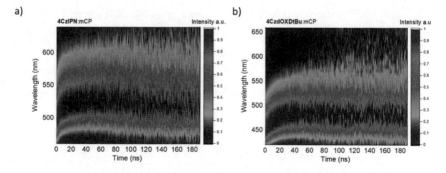

Figure S 8. TRES map of (a) **4CzIPN**:mCP, (b) **4CzdOXDtBu**:mCP.

OLED Fabrication and Additional Data

mCP (1,3-bis(N-carbazolyl)benzene), m-CBP (3,3-di(9*H*-carbazol-9-yl)-1,1'-biphenyl), BP4mPy (3,3',5,5'-tetra[(m-pyridyl)-phen-3-yl]biphenyl) and Liq ((8-Hydroxyquinolinato)lithium) were purchased from Lumtec and used as received. OLED devices were fabricated onto glass substrates covered with prepatterned ITO with an active area of 10.5 mm^2. Substrates were cleaned by sonification in acetone and 2-Propanol (10 min each). To improve the wettability the substrates were treated with oxygen plasma for 2 min (Diener Alto). Immediately after the plasma treatment, the substrates were transferred to a nitrogen filled glovebox and the PEDOT:PSS was spincoated (Hereaus VPAI 4083 diluted in Ethanol (1:3); 4000 rpm for 45 s) yielding a layer thickness of 25 nm. After annealing for 10 min at 120 °C the emissive layer was spincoated from 3 g/l THF solution with mCP as host (3000 rpm for 45 s, 40 nm). Emitter concentration for optimized devices was 10 wt% for **4CzCNOXDtBu** (**71a**) and **4CzIPN** (**3**) and 20 wt% for **4CzdOXDtBu** (**73a**). To remove any residual solvents the substrates were annealed for 10 min at 100 °C. Lower annealing temperatures yielded higher efficiencies but lower stability of the devices. Electron transport materials (BP4mPy, 60 nm and Liq, 2 nm) and the aluminum cathode (50 nm) were evaporated in a multi-chamber evaporation system (Philips self-construction) at a pressure of <1E-6 mbar and rates of 1 Å/s, 0.1 Å/s and 1 Å/s, respectively.

The OLEDs were measured in an integrating sphere (Gigahertz Optics, 19 cm diameter) in 2π configuration. The substrate edges were covered, so only photons emitted by the surface can enter the sphere and not photons emitted by the edges. The luminous flux was measured with a photometer (VL-1101-2, Gigahertz Optics). Spectral mismatch of the photometer was corrected with a spectrometer (CAS140 CT-151, Instrument Systems) coupled to the same integrating sphere. Self-absorption of the OLED was corrected by a halogen aiding lamp. It is important to consider that a device under test with a different reflectivity also changes the spectral sensitivity of the photometer in the sphere and therefor the spectral mismatch correction factor. The integrating sphere was calibrated with a calibrated halogen lamp (Technoteam LN-15) and checked with a calibrated white LED (Technoteam LN-3). Luminances were calculated from the luminous flux assuming Lambertian emission. The current-voltage-characteristics were measured with a Keithley 2400 source-measure-unit. OLED lifetimes were measured with the same setup at constant current.

The lifetime of the fabricated OLEDs is displayed in Figure S 9. OLEDs containing **4CzCNOXDtBu (71a)** and **4CzdOXDtBu (73a)** have a lower lifetime than **4CzIPN (3)** at an initial luminance of 300 cd m^{-2}. This might be due to the higher current density in the device because the OLEDs containing **4CzCNOXDtBu (71a)** and **4CzdOXDtBu (73a)** have a lower EQE. The current density in OLEDs with **4CzdOXDtBu (73a)** as emitter is about three times higher than in the reference device to achieve the same luminance.

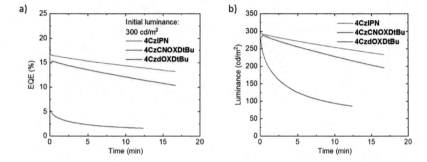

Figure S 9. Lifetime of the fabricated OLEDs containing **4CzIPN (3)**, **4CzCNOXDtBu (71a)** and **4CzdOXDtBu (73a)** as emitting layer. a) Drop in EQE at constant current over time. The initial luminance was ~300 cd m^{-2}. b) Luminance degradation at constant current with time.

5.1.2.3   Phthalimide-based TADF Emitter Project

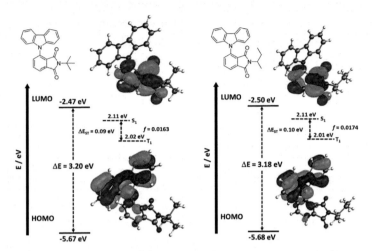

Figure S 10. DFT calculations of **CzPItBu** (**102c**) and **CzPIsBu** (**102b**). Results provided by cynora GmbH.

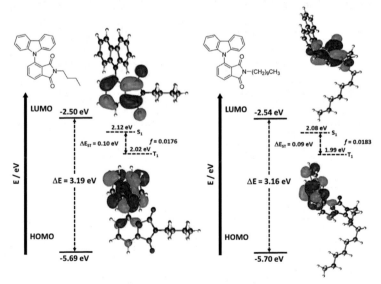

Figure S 11. DFT calculations of **CzPInBu** (**102a**) and **CzPInDec** (**102d**). Results provided by cynora GmbH.

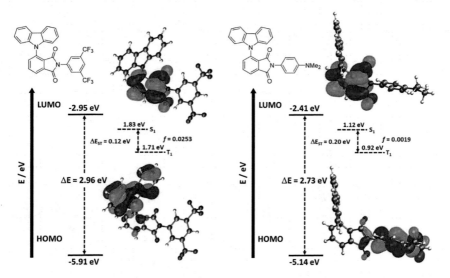

Figure S 12. DFT calculations of **CzPI3,5CF₃Ph (102l)** and **CzPI4NMe₂Ph (102o)**. Results provided by cynora GmbH.

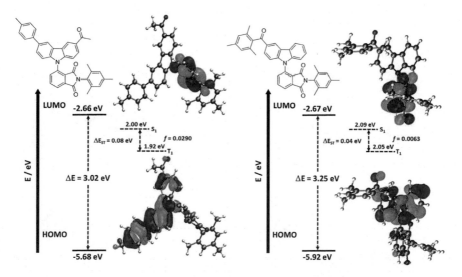

Figure S 13. DFT calculations of **COMeTolCzPIMes (133)** and **COMesCzPIMes (122c)**. Results provided by cynora GmbH.

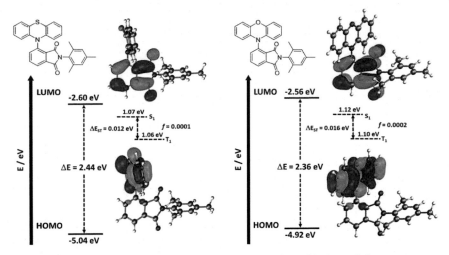

Figure S 14. DFT calculations of two emitters employing phenothiazine and phenoxazine as donor groups. Results provided by cynora GmbH.

5.1.2.4   Tristriazolotriazine (TTT) Project

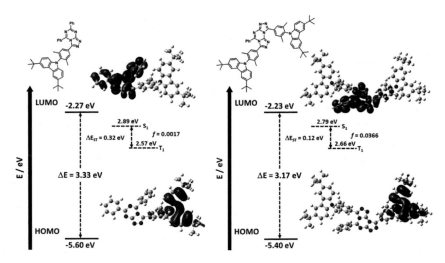

Figure S 15. DFT calculations of **tCz3,5dMeBMTT** (**198**) (left) and **tCz3,5dMeBBTT** (**201**) (right). Results provided by Ettore Crovini.

OLED Fabrication and Additional Data

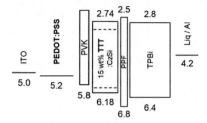

	Procedure	Condition
PEDOT:PSS	spin-coating	500 rpm, 1 s → 4000 rpm, 10 s → 500 rpm, 1 s
	baking	150 °C, 10 min baked at Air
PVK	spin-coating	2000 rpm, 30 s
	drying	120 °C, 10 min baked at Air
EML	spin-coating	2200 rpm, 30 s
	drying	100 °C, 10 min baked at Air
PPF	vap. dep.	5 nm, 0.1-0.2 nm s^{-1}, < ~ 10^{-5} Pa
TPBi	vap. dep.	50 nm, 0.1-0.2 nm s^{-1}, < ~ 10^{-5} Pa
Liq	vap. dep.	1 nm, < 0.1 nm s^{-1}, < ~ 10^{-5} Pa
Al	vap. dep.	80 nm, < 0.15 nm s^{-1}, < ~ 10^{-4} Pa

Figure S 16. Device architecture and processing conditions.

a)

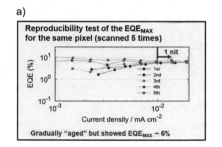

b)

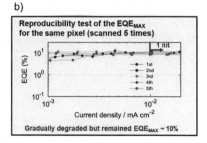

Figure S 17. Reproducibility test for **3,4,5-3tCzTTT** (**194**) (a) and **AcTTT** (**187**) (b).

5.1.3   Reaction Procedures and Analytical Data

5.1.3.1   Derivatization of the 4CzIPN TADF System Project

General Procedure for the Alkylation of Tetrazoles (GP 1)[144]

In a sealable vial, a tetrazole derivative (1.00 equiv.), sodium carbonate (1.50 equiv.) and an alkyl halide (1.50 equiv.) were added to *N,N*-dimethylformamide. The mixture was stirred at ambient temperature until full conversion (usually 3–4 h, monitored with TLC). The reaction mixture was added to an excess of water and extracted with dichloromethane three times. The combined organic layers were washed with brine, dried over sodium sulfate, reduced in vacuum and purified by silica column chromatography.

General Procedure for the Arylation of Tetrazoles (GP 2)[147]

In a sealable vial, a tetrazole derivative (1.00 equiv.), potassium carbonate (1.50 equiv.), [Cu(OH)(TMEDA)]$_2$Cl$_2$ (12 mol%) and an arylboronic acid (2.0–2.4 equiv.) were evacuated and backfilled with oxygen. Dichloromethane was added and the mixture was stirred at ambient temperature until full conversion (usually 14 h, monitored with TLC). The reaction mixture was added to an excess of water and extracted with dichloromethane three times. The combined organic layers were washed with brine, dried over sodium sulfate, reduced in vacuum and purified by silica column chromatography.

General Procedure for Oxadiazole Synthesis with Acyl Chlorides (GP 3)[200]

In a sealable vial, a tetrazole derivative (1.00 equiv.) and an acyl chloride (2.00 equiv.) were added to trichloromethane. The mixture was heated and stirred at 100 °C for 14 hours. After cooling to room temperature, the reaction mixture was poured into a saturated aqueous sodium hydrogen carbonate solution and stirred for 15 minutes. The mixture was extracted with dichloromethane three times. The combined organic layers were washed with brine, dried over sodium sulfate, reduced in vacuum and purified by silica column chromatography.

General Procedure for Oxadiazole Synthesis with Carboxylic Acids (GP 4)[283]

In a sealable vial, a tetrazole derivative (1.00 equiv.), a carboxylic acid (4.00 equiv.) and *N,N'*-diisopropylcarbodiimide (4.00 equiv.) were added to trichloromethane and heated and stirred at 100 °C until full conversion. After cooling to room temperature, the reaction mixture was poured into a saturated aqueous sodium hydrogen carbonate solution and stirred for 15 minutes. The mixture was extracted with dichloromethane three times. The combined organic layers were washed with brine, dried over sodium sulfate, reduced in vacuum and purified by silica column chromatography.

2,3,4,6-Tetrakis(*N*-carbazol-9-yl)isophthalonitrile (4CzIPN)[51]

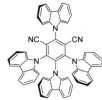

To a solution of carbazole (37.6 g, 225 mmol, 4.50 equiv.) in dry THF (260 mL) under argon, sodium hydride (10.0 g, 50.0 mmol, 5.00 equiv.) was added at 0 °C and the resulting mixture was stirred for 30 minutes at room temperature. After addition of tetrafluoroisophthalonitrile (10.0 g, 50.0 mmol, 1.00 equiv.), the mixture was stirred at room temperature overnight. The reaction progress was monitored via TLC. Afterwards the mixture was quenched with water and extracted with dichloromethane. The organic layers were neutralized with aqueous sodium bicarbonate solution, washed with brine and dried over sodium sulfate and reduced in vacuum. The crude product was purified by recrystallization in THF to yield 36.4 g of the title compound (46.2 mmol, 92%) as a yellow solid.

^{1}H NMR (400 MHz, CDCl$_3$, δ): 8.23 (d, J = 7.8 Hz, 2H), 7.74 – 7.68 (m, 8H), 7.49 (ddd, J = 8.0, 6.4, 1.8 Hz, 2H), 7.33 (d, J = 7.6 Hz, 2H), 7.24 (m, 2H), 7.22 (d, J = 2.2 Hz, 2H) 7.12 – 7.06 (m, 8H), 6.86 – 6.81 (m, 4H), 6.64 (ddd, J = 8.4, 7.2, 1.2 Hz, 2H).

^{13}C NMR (101 MHz, CDCl$_3$, δ) 145.3, 144.8, 140.1, 138.3, 137.1, 134.9, 127.1, 125.9, 125.1, 124.9, 124.7, 124.0, 122.5, 122.1, 121.5, 121.1, 120.6, 119.8, 116.5, 111.8, 110.1, 109.6, 109.6.

IR (ATR) ṽ [cm^{-1}] = 3046 (vw), 2981 (vw), 1719 (w), 1597 (w), 1542 (w), 1490 (w), 1477 (w), 1445 (m), 1371 (w), 1333 (w), 1308 (w), 1252 (w), 1220 (m), 1151 (w), 1117 (w), 1043 (w), 766 (vw), 743 (m), 721 (m), 617 (w), 5556 (w), 527 (w), 473 (w), 421 (w).

HRMS (APCI, C$_{56}$H$_{33}$N$_6$) calc. 789.2761 [M+H]$^+$; found 789.2736 [M+H]$^+$.

The analytical data is consistent with literature.

2,4,5,6-Tetrakis(3,6-di-*tert*-butyl-9*H*-carbazol-9-yl)isophthalonitrile[138]

To a solution of 3,6-di-*tert*-butyl-9*H*-carbazole (9.43 g, 33.8 mmol, 4.50 equiv.) in dry THF (100 mL) under argon, sodium hydride (1.65 g, 41.3 mmol, 5.00 equiv.) was added at 0 °C and the resulting mixture was stirred for 30 minutes at room temperature. After addition of tetrafluoroisophthalonitrile (1.50 g, 7.50 mmol, 1.00 equiv.),

the mixture was stirred at room temperature overnight. The reaction progress was monitored via TLC. Afterwards the mixture was quenched with water and extracted with dichloromethane (3 × 50 mL). The organic layers were neutralized with aqueous sodium bicarbonate solution, washed with brine and dried over sodium sulfate and reduced in vacuum. The crude product was purified by column chromatography over SiO_2 (dichloromethane/pentane 1:1) to yield 8.37 g of the title compound (6.76 mmol, 90%) as a yellow solid.

R_f (dichloromethane/pentane 1:1) = 0.43

^{1}H NMR (400 MHz, Chloroform-*d*) δ 8.23 (t, *J* = 1.7 Hz, 2H), 7.77 (dt, *J* = 8.6, 1.8 Hz, 2H), 7.66 – 7.58 (m, 6H), 7.20 (s, 2H), 7.05 (d, *J* = 2.0 Hz, 8H), 6.52 (d, *J* = 8.6 Hz, 2H), 6.46 (d, *J* = 8.6 Hz, 2H), 1.55 (s, 18H), 1.31 (s, 36H), 1.24 (s, 18H).

^{13}C NMR (101 MHz, CDCl$_3$) δ 146.0, 145.1, 144.7, 144.5, 143.5, 138.6, 137.2, 135.6, 134.1, 125.1, 124.6, 124.2, 123.4, 122.2, 117.6, 116.2, 115.5, 115.1, 112.4, 109.9, 109.1, 109.0, 35.1, 34.7, 34.4, 32.2, 31.9, 31.9.

IR (ATR) $\tilde{v}$ [cm^{-1}] = 2957 (w), 2233 (vw), 1469 (m), 1362 (w), 1297 (m), 1035 (w), 888 (w), 809 (w), 737 (w), 611 (w), 468 (w).

HRMS (ESI, C$_{88}$H$_{96}$N$_6$) calc. 1237.7769 [M+H]$^+$; found 1237.7718 [M+H]$^+$.

The analytical data is consistent with literature.

2,3,4,6-Tetrakis(9H-carbazol-9-yl)-5-(1H-tetrazol-5-yl)benzonitrile[180]

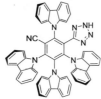

In a vial, 2,3,4,6-tetrakis(N-carbazolyl)isophthalonitrile (2.00 g, 2.54 mmol, 1.00 equiv.), sodium azide (330 mg, 5.07 mmol, 2.00 equiv.) and copper sulfate (40.0 mg, 0.254 mmol, 0.10 equiv.) were sealed and flushed with argon three times. Subsequently, dimethyl sulfoxide (15 mL) was added. The mixture was stirred at 140 °C for 16 h. Afterwards the solution was added to water, carefully acidified with HCl and extracted with dichloromethane (2 × 100 mL). The combined organic layers were washed with brine, dried over sodium sulfate and concentrated under reduced pressure. The crude product was purified by column chromatography on silica gel (ethyl acetate/dichloromethane 0:1 to 1:0). The crude product was washed with cold ethyl acetate to obtain the product as a yellow solid in 1.53 g (69%) yield.

R_f (cyclohexane/ethyl acetate) = 0.50

^{1}H NMR (400 MHz, CDCl$_3$) δ 8.16 (d, J = 7.7 Hz, 2H), 7.89 (d, J = 6.4 Hz, 2H), 7.84 – 7.77 (m, 4H), 7.63 (dd, J = 16.1, 7.7 Hz, 4H), 7.55 (s, 2H), 7.48 (t, J = 7.5 Hz, 2H), 7.41 (d, J = 7.6 Hz, 2H), 7.29 (t, J = 7.4 Hz, 2H), 7.08 (dt, J = 27.2, 7.5 Hz, 4H), 6.92 (t, J = 9.8 Hz, 4H), 6.72 (dt, J = 36.7, 7.4 Hz, 4H).

^{13}C NMR (101 MHz, CDCl$_3$) δ 140.9, 139.1, 138.3, 137.5, 126.0, 125.1, 124.5, 123.8, 122.8, 122.8, 122.6, 122.5, 120.7, 120.4, 120.2, 120.0, 119.9, 119.9, 119.4, 119.1, 117.4, 112.8, 111.5, 111.5, 111.2, 111.0.

IR (ATR) ṽ [cm^{-1}] = 3652 (vw), 3047 (vw), 2771 (vw), 1708 (w), 1597 (w), 1490 (w), 1478 (w), 1446 (m), 1373 (w), 1333 (w), 1309 (w), 1222 (m), 1151 (w), 1118 (w), 1022 (w), 744 (m), 720 (m), 618 (w), 597 (w), 577 (w), 550 (w), 527 (w), 485 (vw), 441 (w), 422 (w), 382 (vw).

HRMS (APCI, C$_{56}$H$_{34}$N$_9$) calc. 832.2932 [M+H]$^+$; found 832.2910 [M+H]$^+$.

2,3,4,6-Tetrakis(3,6-di-*tert*-butyl-9*H*-carbazol-9-yl)-5-(2*H*-tetrazol-5-yl)benzonitrile

In a vial, 2,4,5,6-tetrakis(3,6-di-*tert*-butyl-9*H*-carbazol-9-yl)isophthalonitrile (2.00 g, 1.62 mmol, 1.00 equiv.), sodium azide (210 mg, 3.23 mmol, 2.00 equiv.) and copper sulfate (26.0 mg, 0.162 mmol, 0.10 equiv.) were sealed and flushed with argon three times. Subsequently, dimethyl sulfoxide (15 mL) was added. The mixture was stirred at 140 °C for 16 h. Afterwards the solution was added to water, carefully acidified with HCl and extracted with dichloromethane (2 × 100 mL). The combined organic layers were washed with brine, dried over sodium sulfate and concentrated under reduced pressure. The crude product was purified by column chromatography over SiO2 (dichloromethane/ethyl acetate 95:5) to yield 1.47 g of the title compound (1.11 mmol, 69%) as a yellow solid.

R$_f$ (dichloromethane/ethyl acetate 95:5) = 0.81

¹H NMR (400 MHz, Chloroform-*d*) δ 8.10 (d, *J* = 1.8 Hz, 2H), 7.62 – 7.59 (m, 2H), 7.58 – 7.52 (m, 4H), 7.33 (d, *J* = 8.6 Hz, 2H), 7.21 (d, *J* = 1.8 Hz, 2H), 7.03 – 6.92 (m, 8H), 6.60 (d, *J* = 8.6 Hz, 2H), 6.54 (dd, *J* = 8.6, 1.9 Hz, 2H), 1.47 (s, 18H), 1.31 (s, 18H), 1.25 (s, 18H), 1.23 (s, 18H).

¹³C NMR (101 MHz, CDCl3) δ 144.3, 143.3, 143.2, 139.6, 137.7, 137.5, 136.4, 124.8, 124.5, 124.4, 124.3, 124.2, 123.4, 123.2, 122.1, 117.4, 116.1, 115.0, 109.8, 109.5, 108.7, 35.0, 34.7, 34.6, 34.4, 32.1, 32.1, 31.9, 31.9, 31.9, 31.9.

IR (ATR) $\tilde{v}$ [cm⁻¹] = 3295 (vw), 3046 (vw), 2953 (w), 2236 (vw), 1584 (vw), 1470 (w), 1363 (w), 1295 (w), 1035 (vw), 876 (w), 804 (w), 609 (w), 421 (vw).

HRMS (APCI, C88H97N9) calc. 1280.7940 [M+H]⁺; found 1280.7881 [M+H]⁺.

9,9',9'',9'''-(4,6-Di(2H-tetrazol-5-yl)benzene-1,2,3,5-tetrayl)tetrakis(9H-carbazole)[180]

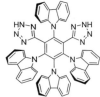

In a vial, 2,3,4,6-tetrakis(N-carbazolyl)isophthalonitrile (2.00 g, 2.54 mmol, 1.00 equiv.), sodium azide (660 mg, 10.1 mmol, 4.00 equiv.) and copper sulfate (40.0 mg, 0.254 mmol, 0.10 equiv.) were sealed and flushed with argon three times. Subsequently, dimethyl sulfoxide (15 mL) was added. The mixture was stirred at 140 °C for 16 h. Afterwards the solution was added to water, carefully acidified with HCl and extracted with dichloromethane (2 × 100 mL). The combined organic layers were washed with brine, dried over sodium sulfate and concentrated under reduced pressure. The crude product was purified by column chromatography on silica gel (ethyl acetate/dichloromethane 1:1 to 1:0 to methanol/ethyl acetate 1:9) to obtain the product as a white solid in 680 mg (31%) yield.

R_f (ethyl acetate/methanol 10:1) = 0.37

^{1}H NMR (400 MHz, DMSO-d_6) δ 7.88 (d, J = 7.7 Hz, 2H), 7.54 (d, J = 7.6 Hz, 4H), 7.41 – 7.28 (m, 8H), 7.28 – 7.17 (m, 4H), 7.03 (t, J = 7.4 Hz, 2H), 6.92 – 6.85 (m, 4H), 6.81 (t, J = 7.4 Hz, 4H), 6.69 – 6.57 (m, 4H).

^{13}C NMR (101 MHz, DMSO) δ 141.3, 139.8, 138.8, 137.5, 135.9, 124.9, 124.2, 123.6, 122.3, 122.2, 122.2, 119.3, 119.1, 119.0, 118.9, 118.8, 118.5, 111.5, 111.2, 110.8.

IR (ATR) ṽ = 3047 (vw), 2985 (vw), 2930 (vw), 2789 (vw), 1724 (w), 1689 (w), 1597 (w), 1448 (m), 1373 (w), 1334 (w), 1312 (w), 1269 (w), 1226 (m), 1153 (w), 1121 (w), 1038 (w), 846 (w), 744 (s), 723 (m), 620 (w), 583 (w), 548 (w), 528 (w), 470 (w), 443 (w), 424 (w).

HRMS (APCI, $C_{56}H_{35}N_{12}$) calc. 875.3102 [M+H]$^+$, found 875.3077 [M+H]$^+$.

9,9',9'',9'''-(4,6-di(2*H*-tetrazol-5-yl)benzene-1,2,3,5-tetrayl)tetrakis(3,6-di-*tert*-butyl-9*H*-carbazole)

In a vial, 2,4,5,6-tetrakis(3,6-di-*tert*-butyl-9*H*-carbazol-9-yl)isophthalonitrile (1.50 g, 1.21 mmol, 1.00 equiv.), sodium azide (315 mg, 4.85 mmol, 4.00 equiv.) and copper sulfate (39.0 mg, 0.242 mmol, 0.200 equiv.) were sealed and flushed with argon three times. Subsequently, dimethyl sulfoxide (15 mL) was added. The mixture was stirred at 140 °C for 48 h. Afterwards the solution was added to water, carefully acidified with HCl and extracted with dichloromethane (2 × 100 mL). The combined organic layers were washed with brine, dried over sodium sulfate and concentrated under reduced pressure. The crude product was purified by column chromatography over SiO_2 (dichloromethane/ethyl acetate 98:2) to yield 1.22 g of the title compound (0.920 mmol, 76%) as a yellow solid.

**R*f* (dichloromethane/ethyl acetate 98:2) = 0.44

¹H NMR (400 MHz, Chloroform-*d*) δ 11.61 (s, 2H), 7.98 (d, *J* = 1.8 Hz, 2H), 7.55 (d, *J* = 1.7 Hz, 4H), 7.39 (dd, *J* = 8.5, 1.9 Hz, 2H), 7.24 (d, *J* = 1.9 Hz, 2H), 7.13 (d, *J* = 8.5 Hz, 2H), 6.98 – 6.88 (m, 8H), 6.76 (d, *J* = 8.6 Hz, 2H), 6.58 (dd, *J* = 8.6, 1.9 Hz, 2H), 1.40 (s, 18H), 1.27 (s, 36H), 1.22 (s, 18H).

¹³C NMR (101 MHz, CDCl₃) δ 143.8, 143.6, 143.0, 141.4, 139.2, 138.3, 138.1, 137.7, 137.2, 124.6, 124.4, 124.1, 123.5, 123.3, 122.2, 117.1, 116.0, 115.0, 109.8, 109.1, 108.1, 34.9, 34.6, 34.4, 32.0, 31.9, 31.9.

IR (ATR) ṽ [cm⁻¹] = 3277 (vw), 3044 (vw), 2954 (w), 1585 (vw), 1472 (m), 1362 (w), 1295 (w), 1034 (w), 875 (w), 805 (m), 607 (w), 421 (w).

HRMS (ESI, C₈₈H₉₈N₁₂) calc. 1323.8110 [M+H]⁺; found 1323.8059 [M+H]⁺.

Methyl 2-(5-(-2,3,4,6-tetra(carbazol-9-yl)-5-cyanophenyl)-*1H*-tetrazol-2-yl)acetate[142]

According to the general procedure (GP 1), 2,3,4,6-tetra(carbazol-9-yl)-5-(*1H*-tetrazol-5-yl)benzonitrile (500 mg, 0.601 mmol, 1.00 equiv.), methyl bromoacetate (137 mg, 0.903 mmol, 1.50 equiv.) and sodium carbonate (144 mg, 0.903 mmol, 1.50 equiv.) were reacted in *N,N*-dimethylformamide (8 mL). The reaction mixture was extracted and purified by column chromatography over SiO_2 (dichloromethane/cyclohexane 1:3 to 3:1) to yield 451 mg of the title compound (83%) as a yellow solid.

R_f (cyclohexane/ethyl acetate 1:1) = 0.69

¹H NMR (500 MHz, CDCl₃) δ 8.17 (d, *J* = 7.9 Hz, 2H), 7.93 (d, *J* = 8.3 Hz, 2H), 7.81 (dd, *J* = 13.9, 7.9 Hz, 4H), 7.66 (d, *J* = 7.5 Hz, 2H), 7.59 (dd, *J* = 18.6, 8.1 Hz, 4H), 7.50 (t, *J* = 7.7 Hz, 2H), 7.41 (d, *J* = 7.7 Hz, 2H), 7.30 (t, *J* = 7.5 Hz, 2H), 7.08 (dt, *J* = 25.3, 7.5 Hz, 4H), 6.99 – 6.88 (m, 4H), 6.76 (t, *J* = 7.4 Hz, 2H), 6.68 (t, *J* = 7.6 Hz, 2H), 4.92 (s, 2H), 3.18 (s, 3H).

¹³C NMR (126 MHz, CDCl₃) δ 164.5, 157.7, 143.1, 143.0, 142.0, 141.0, 139.1, 139.0, 138.2, 137.7, 132.4, 126.1, 125.2, 124.7, 123.9, 122.9, 122.9, 122.7, 122.5, 120.8, 120.6, 120.3, 120.2, 120.0, 119.9, 119.5, 119.2, 117.6, 112.6, 111.5, 111.6, 111.1, 110.9, 52.5, 52.4.

IR (ATR) $\tilde{v}$ [cm⁻¹] = 2949 (vw), 1742 (w), 1596 (vw), 1478 (vw), 1446 (w), 1333 (w), 1307 (w), 1271 (vw), 1222 (w), 1151 (vw), 1119 (vw), 1008 (vw), 919 (vw), 789 (vw), 740 (w), 723 (w), 618 (vw), 592 (vw), 549 (vw), 528 (vw), 446 (vw), 424 (vw).

HRMS (FAB-MS, 3-NBA, C₅₉H₃₈O₂N₉) calc. 904.3143, found 904.3141.

3-(1-(2-Bromoethyl)-1H-tetrazol-5-yl)-2,4,5,6-tetra(carbazol-9-yl)benzonitrile[142]

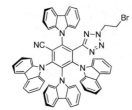

According to the general procedure (GP 1), 2,3,4,6-tetra(carbazol-9-yl)-5-(1H-tetrazol-5-yl)benzonitrile (300 mg, 0.361 mmol, 1.00 equiv.), dibromoethane (102 mg, 0.541 mmol, 1.50 equiv.) and sodium carbonate (86.0 mg, 0.541 mmol, 1.50 equiv.) were reacted in N,N-dimethylformamide (10 mL). The reaction mixture was extracted and purified by column chromatography over SiO2 (dichloromethane/cyclohexane 1:1 to dichloromethane) to yield 263 mg of the title compound (78%) as a yellow solid.

R_f (cyclohexane/ethyl acetate 3:1) = 0.26

1H NMR (400 MHz, DMSO-d_6) δ 8.17 (d, J = 7.8 Hz, 2H), 7.94 (d, J = 8.2 Hz, 2H), 7.83 (dt, J = 7.8, 1.5 Hz, 4H), 7.70 – 7.63 (m, 4H), 7.59 (dd, J = 7.8, 1.2 Hz, 2H), 7.50 (ddd, J = 8.3, 7.2, 1.2 Hz, 2H), 7.45 – 7.40 (m, 2H), 7.35 – 7.25 (m, 2H), 7.16 – 7.03 (m, 4H), 6.94 (td, J = 20.0, 7.3, 1.2 Hz, 4H), 6.80 – 6.67 (m, 4H), 4.26 (t, J = 5.7 Hz, 2H), 2.85 (t, J = 5.7 Hz, 2H).

13C NMR (101 MHz, DMSO) δ 157.5, 143.0, 142.9, 141.8, 141.0, 139.2, 139.0, 138.3, 137.8, 132.5, 126.1, 125.1, 124.7, 123.9, 123.0, 122.9, 122.7, 122.5, 120.7, 120.6, 120.3, 120.2, 120.0, 119.9, 119.5, 119.2, 117.7, 112.7, 111.5, 111.5, 111.1, 110.8, 54.9, 53.4, 28.2.

IR (ATR) $\tilde{v}$ [cm^{-1}] = 2923 (vw), 1597 (vw), 1490 (vw), 1478 (w), 1448 (m), 1333 (w), 1309 (w), 1223 (w), 1150 (w), 1118 (vw), 1025 (vw), 925 (vw), 743 (m), 721 (w), 618 (vw), 550 (vw), 528 (vw), 421 (w).

HRMS (FAB-MS, 3-NBA, $C_{58}H_{37}N_9{}^{79}Br$) calc. 938.2350, found 938.2351.

3-(1-allyl-1*H*-tetrazol-5-yl)-2,4,5,6-Tetra(carbazol-9-yl)benzonitrile[142]

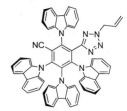

According to the general procedure (GP 1), 2,3,4,6-tetra(-carbazol-9-yl)-5-(1*H*-tetrazol-5-yl)benzonitrile (200 mg, 0.241 mmol, 1.00 equiv.), allyl bromide (43.7 mg, 0.361 mmol, 1.50 equiv.) and sodium carbonate (58 mg, 0.361 mmol, 1.50 equiv.) were reacted in *N,N*-dimethylformamide (8 mL).

The reaction mixture was extracted and purified by column chromatography over SiO$_2$ (dichloromethane/cyclohexane 3:1 to dichloromethane) to yield 392 mg of the title compound (75%) as a yellow solid.

R$_f$ (cyclohexane/ethyl acetate 1:5) = 0.28

^{1}H NMR (400 MHz, DMSO-d_6) δ 8.17 (dt, *J* = 7.8, 1.0 Hz, 2H), 7.89 (d, *J* = 8.3 Hz, 2H), 7.83 (ddd, *J* = 7.7, 4.9, 1.3 Hz, 4H), 7.71 – 7.60 (m, 4H), 7.58 – 7.52 (m, 2H), 7.51 – 7.40 (m, 4H), 7.32 – 7.26 (m, 2H), 7.15 – 7.03 (m, 4H), 6.93 (m, 4H), 6.81 – 6.65 (m, 4H), 4.89 (ddt, *J* = 17.2, 10.5, 5.3 Hz, 1H), 4.66 (dd, *J* = 10.4, 1.1 Hz, 1H), 4.48 (d, *J* = 5.4 Hz, 1H), 3.93 (dd, *J* = 17.1, 1.1 Hz, 1H).

^{13}C NMR (101 MHz, DMSO) δ 157.5, 142.9, 142.9, 141.8, 140.8, 139.2, 139.1, 138.3, 137.6, 132.4, 129.5, 126.1, 125.2, 124.7, 123.9, 122.9, 122.9, 122.8, 122.6, 120.8, 120.6, 120.3, 120.2, 120.0, 119.9, 119.5, 119.2, 118.1, 117.5, 111.5, 111.1, 110.8, 54.9, 53.8.

IR (ATR) $\tilde{v}$ [cm^{-1}] = 1668 (vw), 1597 (w), 1477 (w), 1446 (m), 1333 (w), 1308 (w), 1222 (m), 1150 (w), 1118 (w), 1025 (w), 923 (w), 743 (m), 720 (m), 617 (w), 548 (w), 527 (w), 486 (vw), 441 (w), 421 (w).

HRMS (FAB-MS, 3-NBA, C$_{59}$H$_{38}$N$_9$) calc. 872.3245, found 872.3244.

2,3,4,6-Tetra(carbazol-9-yl)-5-(1-(prop-2-yn-1-yl)-1*H*-tetrazol-5-yl)benzonitrile[142]

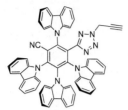

According to the general procedure (GP 1), 2,3,4,6-tetra(carbazol-9-yl)-5-(*1H*-tetrazol-5-yl)benzonitrile (300 mg, 0.361 mmol, 1.00 equiv.), propargyl bromide (85.8 mg, 0.722 mmol, 2.00 equiv.) and sodium carbonate (115 mg, 0.722 mmol, 2.00 equiv.) were reacted in *N,N*-dimethylformamide (8 ml). The reaction mixture was extracted and purified by column chromatography over SiO$_2$ (dichloromethane/cyclohexane 3:1 to dichloromethane) to yield 219.1 mg of the title compound (70%) as a yellow solid.

R$_f$ (cyclohexane/ethyl acetate 1:1) = 0.65

¹H NMR (500 MHz, DMSO-d_6) δ 8.16 (d, *J* = 7.8 Hz, 2H), 7.88 (d, *J* = 8.2 Hz, 2H), 7.83 (t, *J* = 8.1 Hz, 4H), 7.65 (dd, *J* = 13.5, 7.8 Hz, 4H), 7.55 (d, *J* = 8.0 Hz, 2H), 7.50 – 7.41 (m, 4H), 7.29 (t, *J* = 7.5 Hz, 2H), 7.09 (dt, *J* = 24.3, 7.4 Hz, 4H), 6.98 – 6.87 (m, 4H), 6.77 (t, *J* = 7.4 Hz, 2H), 6.70 (t, *J* = 7.7 Hz, 2H), 4.83 (s, 2H), 3.22 (s, 1H).

¹³C NMR (126 MHz, CDCl$_3$) δ 157.8, 143,02, 143.0, 141.9, 140.8, 139.1, 139.0, 138.3, 137.5, 132.1, 126.1, 125.1, 124.7, 123.9, 123.0, 122.9, 122.8, 122.6, 120.7, 120.6, 120.4, 120.2, 120.0, 119.9, 119.6, 119.2, 117.4, 111.5, 111.5, 110.9, 110.7, 77.7, 73.6, 41.9.

IR (ATR) $\tilde{v}$ [cm^{-1}] = 3046 (vw), 1597 (vw), 1489 (w), 1478 (w), 1445 (m), 1333 (w), 1308 (w), 1221 (w), 1150 (w), 1119 (w), 1025 (vw), 924 (vw), 769 (vw), 742 (m), 720 (m), 617 (w), 577 (vw), 551 (vw), 528 (vw), 486 (vw), 440 (vw), 421 (w).

HRMS (FAB-MS, 3-NBA, C$_{59}$H$_{36}$N$_9$) calc. 870.3088, found 870.3086.

2-(5-(-2,3,4,6-Tetra(*9H*-carbazol-9-yl)-5-cyanophenyl)-*1H*-tetrazol-1-yl)acetamide[142]

According to the general procedure (GP 1), 2,3,4,6-tetra(carbazol-9-yl)-5-(*1H*-tetrazol-5-yl)benzonitrile (100 mg, 0.120 mmol, 1.00 equiv.), 2-bromoacetamide (24.9 mg, 0.180 mmol, 1.50 equiv.) and sodium carbonate (38.0 mg, 0.241 mmol, 2.00 equiv.) were reacted in *N,N*-dimethylformamide (5 mL). The reaction mixture was extracted and purified by column chromatography over SiO_2 (dichloromethane/methanol 95:5 to 90:10) to yield 68.8 mg of the title compound (64%) as a yellow solid and 5.0 mg of starting material.

R_f (dichloromethane/methanol 10:1) = 0.50

¹H NMR (400 MHz, DMSO-d_6) δ 8.16 (dt, J = 7.8, 1.0 Hz, 2H), 7.99 – 7.87 (m, 4H), 7.86 – 7.77 (m, 4H), 7.70 – 7.60 (m, 2H), 7.59 – 7.53 (m, 2H), 7.48 (ddd, J = 8.3, 7.2, 1.2 Hz, 2H), 7.45 – 7.33 (m, 4H), 7.29 (td, J = 7.6, 1.0 Hz, 2H), 7.14 – 7.02 (m, 4H), 6.97 – 6.88 (m, 4H), 6.77 (td, J = 7.4, 1.0 Hz, 2H), 6.69 (ddd, J = 8.4, 7.1, 1.3 Hz, 2H), 4.43 (s, 2H).

¹³C NMR (101 MHz, DMSO) δ 164.1, 162.3, 157.3, 143.9, 142.9, 141.9, 141.0, 139.2, 139.1, 138.3, 137.6, 132.4, 126.1, 125.1, 124.7, 123.9, 123.0, 122.9, 122.7, 122.5, 120.7, 120.5, 120.3, 120.1, 120.0, 119.9, 119.4, 119.2, 117.5, 111.5, 111.4, 111.1, 110.9, 53.5, 35.8, 30.8.

IR (ATR) $\tilde{v}$ [cm⁻¹] = 3421 (vw), 3048 (vw), 2922 (vw), 2851 (vw), 1708 (w), 1663 (w), 1624 (w), 1597 (w), 1490 (w), 1478 (w), 1450 (m), 1386 (w), 1332 (w), 1311 (w), 1223 (w), 1151 (w), 1119 (vw), 1092 (w), 1026 (vw), 928 (vw), 862 (vw), 824 (vw), 796 (vw), 743 (w), 718 (w), 657 (vw), 618 (vw), 582 (vw), 554 (vw), 528 (vw), 423 (w).

HRMS (FAB-MS, 3-NBA, $C_{58}H_{37}ON_{10}$) calc. 889.3146, found 889.3145.

2,3,4,6-Tetra(carbazol-9-yl)-5-(2-hexyl-2H-tetrazol-5-yl)benzonitrile[142]

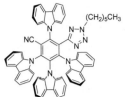

According to the general procedure (GP 1), 2,3,4,6-tetra(carbazol-9-yl)-5-(1H-tetrazol-5-yl)benzonitrile (500 mg, 0.602 mmol, 1.00 equiv.), 1-bromohexane (199 mg, 1.20 mmol, 2.00 equiv.) and sodium carbonate (199 mg, 1.20 mmol, 2.00 equiv.) were reacted in N,N-dimethylformamide (15 mL). The reaction mixture was extracted and purified by column chromatography over SiO₂ (dichloromethane/cyclohexane 1:1) to yield 394 mg of the title compound (72%) as a yellow solid.

R_f (cyclohexane/ethyl acetate 3:1) = 0.20

M.p. = 261 °C

¹H NMR (400 MHz, CDCl₃) δ 8.14 (dt, J = 7.8, 0.9 Hz, 2H), 7.81 – 7.75 (m, 2H), 7.68 – 7.62 (m, 2H), 7.51 (ddd, J = 8.3, 7.1, 1.3 Hz, 2H), 7.46 – 7.33 (m, 6H), 7.31 – 7.26 (m, 2H), 7.19 – 7.06 (m, 8H), 7.01 – 6.95 (m, 4H), 6.84 (td, J = 7.5, 0.9 Hz, 2H), 6.71 (ddd, J = 8.4, 7.2, 1.3 Hz, 2H), 3.77 (t, J = 7.1 Hz, 2H), 1.15 – 1.05 (m, 2H), 1.02 – 0.88 (m, 4H), 0.83 (t, J = 7.2 Hz, 3H), 0.59 – 0.50 (m, 2H).

¹³C NMR (101 MHz, CDCl₃) δ 156.7, 141.9, 141.7, 141.2, 140.0, 138.1, 137.9, 137.1, 136.0, 131.5, 125.7, 124.8, 124.5, 123.9, 123.2, 123.1, 122.8, 122.8, 120.6, 120.4, 120.0, 119.8, 119.8, 119.4, 119.0, 118.6, 116.8, 111.6, 109.5, 109.4, 109.2, 108.9, 53.2, 53.1, 52.9, 52.9, 52.7, 52.6, 52.4, 52.1, 51.9, 29.9, 29.4, 28.9, 27.6, 26.2, 24.4, 21.6, 12.9.

IR (ATR) $\tilde{v}$ [cm⁻¹] = 2922 (vw), 1597 (w), 1477 (w), 1446 (m), 1333 (w), 1308 (m), 1222 (m), 1150 (w), 1118 (w), 1026 (w), 923 (w), 743 (m), 720 (m), 618 (w), 598 (w), 576 (w), 550 (w), 537 (w), 486 (vw), 440 (w), 421 (w).

HRMS (FAB-MS, 3-NBA, C₆₂H₄₆N₉) calc. 916.3876, found 916.3878.

2,3,4,6-Tetra(9H-carbazol-9-yl)-5-(2-(4-nitrobenzyl)-2H-tetrazol-5-yl)benzonitrile[142]

According to the general procedure (GP 1), 2,3,4,6-tetra(carbazol-9-yl)-5-(1H-tetrazol-5-yl)benzonitrile (100 mg, 0.120 mmol, 1.00 equiv.), 4-nitrobenzyl bromide (39.0 mg, 0.180 mmol, 1.50 equiv.) and sodium carbonate (29.0 mg, 0.180 mmol, 1.50 equiv.) were reacted in N,N-dimethylformamide (5 mL). The reaction mixture was extracted and purified by column chromatography over SiO_2 (dichloromethane/cyclohexane 3:1 to dichloromethane) to yield 96.4 mg of the title compound (83%) as a yellow solid.

R_f (dichloromethane/cyclohexane 3:1) = 0.20

¹H NMR (400 MHz, CDCl₃) δ 8.05 (dd, J = 7.7, 1.1 Hz, 2H), 7.84 (d, J = 8.7 Hz, 2H), 7.71 – 7.65 (m, 2H), 7.55 – 7.49 (m, 2H), 7.48 – 7.44 (m, 2H), 7.40 (d, J = 8.0 Hz, 2H), 7.36 – 7.27 (m, 4H), 7.22 – 7.15 (m, 2H), 7.12 – 7.00 (m, 6H), 6.94 – 6.85 (m, 6H), 6.80 – 6.74 (m, 2H), 6.60 (ddd, J = 8.4, 7.2, 1.2 Hz, 2H), 6.40 – 6.36 (m, 2H), 1.56 (s, 2H).

¹³C NMR (101 MHz, CDCl₃) δ 158.5, 147.7, 143.1, 142.8, 142.0, 141.0, 138.9, 138.7, 137.7, 136.7, 131.9, 128.3, 126.7, 125.7, 125.4, 124.6, 124.3, 124.0, 123.9, 123.8, 123.8, 121.5, 121.3, 120.9, 120.8, 120.7, 120.4, 119.9, 119.6, 110.2, 110.0, 109.9, 109.7, 55.1, 29.8.

IR (ATR) $\tilde{v}$ [cm⁻¹] = 3045 (vw), 2919 (w), 2850 (vw), 1665 (w), 1598 (w), 1516 (w), 1490 (w), 1478 (w), 1444 (m), 1333 (m), 1309 (m), 1222 (m), 1149 (w), 1118 (w), 1015 (w), 926 (w), 848 (w), 814 (vw), 741 (m), 720 (m), 618 (w), 582 (w), 550 (w), 527 (w), 421 (w).

HRMS (FAB-MS, 3-NBA, $C_{63}H_{39}O_2N_{10}$) calc. 967.3252, found 967.3254.

2,3,4,6-Tetra(9*H*-carbazol-9-yl)-5-(2-(3,3,4,4,5,5,6,6,7,7,8,8,8-tridecafluorooctyl)-2*H*-tetrazol-5-yl)benzonitrile[142]

According to the general procedure (GP 1), 2,3,4,6-tetra(carbazol-9-yl)-5-(*1H*-tetrazol-5-yl)benzonitrile (100 mg, 0.120 mmol, 1.00 equiv.), 1*H*,1*H*,2*H*,2*H*-perfluorooctyl iodide (68.4 mg, 0.144 mmol, 1.20 equiv.) and sodium carbonate (38.0 mg, 0.241 mmol, 2.00 equiv.) were reacted in *N,N*-dimethylformamide (5 mL). The reaction mixture was extracted and purified by column chromatography over SiO$_2$ (ethyl acetate/cyclohexane 1:3) to yield 89.9 mg of the title compound (63%) as a yellow solid.

R$_f$ (ethyl acetate/cyclohexane 1:3) = 0.47

^{1}H NMR (400 MHz, CDCl$_3$) δ 7.96 (d, *J* = 7.7 Hz, 2H), 7.62 – 7.55 (m, 2H), 7.49 – 7.43 (m, 2H), 7.37 (ddd, *J* = 8.3, 7.2, 1.2 Hz, 2H), 7.29 (d, *J* = 8.1 Hz, 2H), 7.24 – 7.18 (m, 4H), 7.17 – 7.08 (m, 2H), 7.03 – 6.92 (m, 6H), 6.90 – 6.79 (m, 6H), 6.68 (td, *J* = 7.5, 0.9 Hz, 2H), 6.55 (ddd, *J* = 8.3, 7.2, 1.2 Hz, 2H), 4.01 – 3.95 (m, 2H), 1.48 (tt, *J* = 17.3, 8.0 Hz, 2H).

^{13}C NMR (101 MHz, CDCl$_3$) δ 158.4, 143.0, 142.8, 142.0, 141.0, 138.9, 138.7, 137.8, 136.8, 131.4, 126.7, 125.7, 125.5, 124.7, 124.4, 124.1, 123.9, 123.8, 121.5, 121.3, 121.0, 120.7, 120.4, 119.9, 119.7, 117.7, 112.3, 110.2, 110.0, 109.8, 44.7, 30.0.

^{19}F NMR (376 MHz, CDCl$_3$) δ -85.02, -118.61, -126.19, -127.14, -127.49, -130.42.

IR (ATR) $\tilde{v}$ [cm^{-1}] = 3048 (vw), 1772 (vw), 1598 (vw), 1490 (w), 1478 (w), 1445 (w), 1333 (w), 1310 (w), 1222 (w), 1142 (w), 1119 (w), 1026 (w), 925 (vw), 845 (vw), 808 (vw), 741 (m), 720 (w), 647 (vw), 617 (vw), 550 (w), 527 (w), 487 (vw), 421 (w).

HRMS (ESI, C$_{64}$H$_{36}$F$_{13}$N$_9$) calc. 1177.2886 [M]$^+$, found 1177.2873 [M]$^+$.

2,3,4,6-Tetra(9H-carbazol-9-yl)-5-(2-phenyl-2H-tetrazol-5-yl)benzonitrile[142]

According to the general procedure (GP 2), 2,3,4,6-tetra(carbazol-9-yl)-5-(1H-tetrazol-5-yl)benzonitrile (2.00 g, 2.41 mmol, 1.00 equiv.), [Cu(OH)(TMEDA)]₂Cl₂ (134 mg, 0.289 mmol, 0.12 equiv.), potassium carbonate (499 mg, 3.61 mmol, 1.5 equiv.) and phenylboronic acid (704 mg, 5.77 mmol, 2.4 equiv.) were reacted in dichloromethane (40 ml).

The reaction mixture was extracted and purified by column chromatography over SiO₂ (dichloromethane/cyclohexane 1:1 to dichloromethane) to yield 1.36 g (61%) of the title compound as a yellow solid and 364 mg (18%) of starting material.

R$_f$ (cyclohexane/ethyl acetate 3:1) = 0.39

^{1}H NMR (400 MHz, DMSO-d_6) δ 8.18 (d, J = 7.8 Hz, 2H), 7.97 (d, J = 8.1 Hz, 2H), 7.85 (dt, J = 7.4, 2.2 Hz, 4H), 7.71 (t, J = 7.0 Hz, 2H), 7.62 (d, J = 8.1 Hz, 2H), 7.54 – 7.44 (m, 5H), 7.42 – 7.35 (m, 4H), 7.30 (m, 4H), 7.17 – 7.05 (m, 4H), 7.03 – 6.89 (m, 4H), 6.86 – 6.71 (m, 4H).

^{13}C NMR (101 MHz, DMSO) δ 141.1, 139.7, 139.6, 139.1, 138.4, 137.9, 134.8, 130.1, 129.8, 126.2, 124.1, 123.4, 123.0, 123.0, 122.9, 122.6, 120.4, 120.1, 119.3, 118.9, 118.6, 111.5, 111.1, 110.9.

IR (ATR) $\tilde{v}$ [cm^{-1}] = 2921 (w), 1597 (w), 1491 (w), 1478 (w), 1446 (m), 1332 (w), 1309 (m), 1222 (m), 1150 (w), 1119 (w), 1000 (w), 915 (w), 742 (m), 720 (m), 679 (w), 617 (w), 597 (w), 577 (w), 557 (w), 527 (w), 487 (vw), 442 (w), 422 (w), 392 (vw).

HRMS (FAB-MS, 3-NBA, C₆₂H₃₈N₉) calc. 908.3245, found 908.3244.

2,3,4,6-Tetra(*9H*-carbazol-9-yl)-5-(2-(p-tolyl)-*2H*-tetrazol-5-yl)benzonitrile[142]

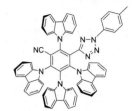

According to the general procedure (GP 2), 2,3,4,6-tetra(carbazol-9-yl)-5-(*1H*-tetrazol-5-yl)benzonitrile (150 mg, 0.180 mmol, 1.00 equiv.), [Cu(OH)(TMEDA)]$_2$Cl$_2$ (10.0 mg, 0.022 mmol, 0.12 equiv.), potassium carbonate (37.4 mg, 0.271 mmol, 1.50 equiv.) and p-tolylboronic acid (49.0 mg, 0.361 mmol, 2.00 equiv.) were reacted in dichloromethane (4 mL). The reaction mixture was extracted and purified by column chromatography over SiO$_2$ (dichloromethane/cyclohexane 1:1) to yield 88.1 mg (53%) of the title compound as a yellow solid.

R$_f$ (cyclohexane/ethyl acetate 3:1) = 0.40

¹H NMR (400 MHz, DMSO-d_6) δ 8.18 (d, *J* = 7.7 Hz, 2H), 7.97 (d, *J* = 8.3 Hz, 2H), 7.85 (d, *J* = 7.9 Hz, 4H), 7.70 (t, *J* = 6.8 Hz, 4H), 7.61 (d, *J* = 8.1 Hz, 2H), 7.54 – 7.42 (m, 4H), 7.29 (t, *J* = 7.5 Hz, 2H), 7.10 (td, *J* = 14.8, 13.2, 7.8 Hz, 6H), 6.94 (dt, *J* = 23.9, 7.2 Hz, 4H), 6.83 – 6.67 (m, 6H), 2.24 (s, 3H).

IR (ATR) $\tilde{v}$ [cm⁻¹] = 2922 (vw), 1597 (vw), 1490 (vw), 1477 (vw), 1446 (w), 1333 (w), 1308 (w), 1222 (w), 1150 (vw), 1119 (vw), 1002 (vw), 922 (vw), 821 (vw), 743 (w), 720 (w), 618 (vw), 595 (vw), 575 (vw), 552 (vw), 527 (vw), 488 (vw), 441 (vw), 424 (vw), 390 (vw).

HRMS (FAB-MS, 3-NBA, C$_{63}$H$_{40}$N$_9$) calc. 922.3401, found 922.3399.

2,3,4,6-Tetra(9H-carbazol-9-yl)-5-(2-(4-methoxyphenyl)-2H-tetrazol-5-yl)benzonitrile[142]

According to the general procedure (GP 2), 2,3,4,6-tetra(carbazol-9-yl)-5-(1H-tetrazol-5-yl)benzonitrile (100 mg, 0.120 mmol, 1.00 equiv.), [Cu(OH)(TMEDA)]$_2$Cl$_2$ (7.00 mg, 0.014 mmol, 0.12 equiv.), potassium carbonate (18.3 mg, 0.132 mmol, 1.10 equiv.) and 4-methoxyphenylboronic acid (37.0 mg, 0.241 mmol, 2.00 equiv.) were reacted in dichloromethane (2 mL). The reaction mixture was extracted and purified by column chromatography over SiO$_2$ (dichloromethane/cyclohexane 1:1 to dichloromethane) to yield 75.3 mg (67%) of the title compound as a yellow solid.

R_f (cyclohexane/ethyl acetate 3:1) = 0.35

^{1}H NMR (400 MHz, DMSO-d_6) δ 8.18 (d, J = 7.7 Hz, 2H), 7.96 (d, J = 8.2 Hz, 2H), 7.88 – 7.83 (m, 4H), 7.73 – 7.68 (m, 4H), 7.62 (d, J = 8.0 Hz, 2H), 7.47 (ddd, J = 9.2, 7.4, 1.4 Hz, 4H), 7.32 – 7.27 (m, 2H), 7.16 – 7.05 (m, 4H), 7.04 – 6.93 (m, 4H), 6.84 – 6.73 (m, 8H), 3.72 (s, 3H).

^{13}C NMR (101 MHz, DMSO) δ 150.9, 141.1, 139.6, 139.1, 138.4, 126.2, 125.2, 124.8, 124.0, 123.0, 122.9, 122.6, 120.7, 120.6, 120.3, 120.1, 119.4, 114.9, 114.7, 111.5, 110.9, 59.8, 55.4, 54.9.

IR (ATR) $\tilde{v}$ [cm^{-1}] = 2921 (w), 1720 (vw), 1596 (w), 1503 (w), 1477 (w), 1446 (m), 1370 (w), 1332 (w), 1309 (m), 1255 (w), 1222 (m), 1171 (w), 1150 (w), 1118 (w), 1026 (w), 1000 (w), 9223 (w), 832 (w), 742 (m), 720 (m), 638 (w), 617 (w), 593 (w), 578 (w), 553 (w), 527 (w), 485 (vw), 441 (w), 422 (w).

HRMS (FAB-MS, 3-NBA, C$_{63}$H$_{40}$ON$_9$) calc. 938.3350, found 938.3348.

9,9',9'',9'''-(-4,6-Bis(2-phenyl-*2H*-tetrazol-5-yl)benzene-1,2,3,5-tetrayl)tetrakis(carbazole)[142]

According to the general procedure (GP 2), 9,9',9'',9'''-(4,6-di(1*H*-tetrazol-5-yl)benzene-1,2,3,5-tetrayl)tetrakis(carbazole) (300 mg, 0.343 mmol, 1.00 equiv.), [Cu(OH)(TMEDA)]$_2$Cl$_2$ (38.0 mg, 0.082 mmol, 0.24 equiv.), potassium carbonate (142.3 mg, 1.03 mmol, 3.00 equiv.) and phenylboronic acid (209 mg, 1.716 mmol, 5.00 equiv.) were reacted in dichloromethane (25 mL). The reaction mixture was extracted and purified by column chromatography over SiO$_2$ (dichloromethane/cyclohexane 1:1 to dichloromethane) to yield 110 mg (31%) of the title compound as an off-white solid.

R$_f$ (cyclohexane/ethyl acetate 3:1) = 0.41

¹H NMR (400 MHz, DMSO-d_6) δ 8.02 (d, J = 7.7 Hz, 2H), 7.88 – 7.83 (m, 2H), 7.73 – 7.69 (m, 8H), 7.53 – 7.48 (m, 2H), 7.40 – 7.33 (m, 4H), 7.30 (td, J = 8.0, 7.3, 2.0 Hz, 6H), 7.16 – 7.11 (m, 2H), 7.01 (ddd, J = 8.4, 7.1, 1.3 Hz, 4H), 6.92 (t, J = 7.8 Hz, 4H), 6.86 – 6.78 (m, 8H).

¹³C NMR (101 MHz, DMSO) δ 161.0, 158.3, 141.2, 140.0, 139.1, 138.0, 134.9, 131.5, 131.3, 129.7, 129.5, 125.5, 124.9, 124.2, 122.7, 122.5, 120.1, 119.9, 119.8, 119.6, 119.4, 119.4, 119.3, 119.0, 118.4, 116.1, 111.2, 110.9.

IR (ATR) ṽ [cm⁻¹] = 3045 (vw), 2915 (vw), 1595 (w), 1491 (w), 1477 (w), 1446 (m), 1332 (w), 1309 (m), 1223 (m), 1148 (w), 1117 (w), 1070 (w), 1002 (w), 913 (w), 741 (m), 719 (m), 680 (m), 617 (w), 586 (w), 552 (w), 527 (w), 442 (vw), 422 (w).

HRMS (FAB-MS, 3-NBA, C$_{68}$H$_{43}$N$_{12}$) calc. 1027.3728, found 1027.3727.

9,9',9'',9'''-(-4,6-Bis(2-(prop-2-yn-1-yl)-2H-tetrazol-5-yl)benzene-1,2,3,5-tetrayl)tetrakis(9H-carbazole)

According to the general procedure (GP 1), 9,9',9'',9'''-(4,6-di(1H-tetrazol-5-yl)benzene-1,2,3,5-tetrayl)tetrakis(carbazole) (200 mg, 0.229 mmol, 1.00 equiv.), propargyl bromide (109 mg, 0.915 mmol, 4.00 equiv.) and sodium carbonate (97.0 mg, 0.915 mmol, 4.00 equiv.) were reacted in N,N-dimethylformamide (7 mL). The reaction mixture was extracted and purified by column chromatography over SiO_2 (dichloromethane) to yield 105 mg (48%) of the title compound as an off-white solid.

R_f (cyclohexane/ethyl acetate 3:1) = 0.41

¹H NMR (400 MHz, Methylene Chloride-d_2) δ 7.97 (dt, J = 7.7, 1.0 Hz, 2H), 7.69 – 7.63 (m, 4H), 7.45 – 7.41 (m, 2H), 7.33 – 7.16 (m, 12H), 7.02 – 6.96 (m, 8H), 6.84 (td, J = 7.4, 1.0 Hz, 2H), 6.75 (ddd, J = 8.4, 7.2, 1.3 Hz, 2H), 4.56 (d, J = 2.7 Hz, 4H), 2.17 (t, J = 2.6 Hz, 2H).

¹³C NMR (101 MHz, CD_2Cl_2) δ 159.6, 141.1, 140.9, 139.8, 139.6, 139.1, 137.5, 131.8, 126.6, 125.8, 125.1, 124.0, 123.9 123.9, 120.9, 120.9, 120.8, 120.6, 120.3, 119.9, 111.1, 110.6, 109.9, 75.8, 73.1, 42.7.

IR (ATR) $\tilde{v}$ [cm⁻¹] = 3274 (w), 1597 (w), 1491 (w), 1478 (w), 1447 (m), 1333 (w), 1311 (w), 1223 (m), 1151 (w), 1119 (w), 1022 (w), 928 (w), 827 (vw), 803 (vw), 770 (w), 744 (m), 721 (m), 682 (w), 617 (w), 573 (w), 550 (w), 527 (w), 478 (w), 445 (w), 422 (w).

HRMS (FAB-MS, 3-NBA, $C_{62}H_{39}N_{12}$) calc. 951.3421, found 951.3422.

9,9',9'',9'''-(-4,6-Bis(2-allyl-2*H*-tetrazol-5-yl)benzene-1,2,3,5-tetrayl)tetrakis(9*H*-carbazole)

According to the general procedure (GP 1), 9,9',9'',9'''-(4,6-di(1*H*-tetrazol-5-yl)benzene-1,2,3,5-tetrayl)tetrakis(carbazole) (300 mg, 0.343 mmol, 1.00 equiv.), allyl bromide (166 mg, 1.37 mmol, 4.00 equiv.) and sodium carbonate (146 mg, 1.37 mmol, 4.00 equiv.) were reacted in *N,N*-dimethylformamide (9 mL). The reaction mixture was extracted and purified by column chromatography over SiO_2 (dichloromethane) to yield 150.8 mg (46%) of the title compound as an off-white solid.

R_f (cyclohexane/ethyl acetate 3:1) = 0.43

^{1}H NMR (400 MHz, Methylene Chloride-d_2) δ 8.00 (d, *J* = 7.6 Hz, 2H), 7.72 – 7.64 (m, 4H), 7.47 – 7.41 (m, 2H), 7.35 – 7.16 (m, 12H), 7.04 – 6.95 (m, 8H), 6.84 (t, *J* = 7.4 Hz, 2H), 6.79 – 6.72 (m, 2H), 4.96 (ddt, *J* = 16.3, 10.9, 5.6 Hz, 2H), 4.80 (d, *J* = 10.3 Hz, 2H), 4.42 – 4.36 (m, 4H), 4.27 (dd, *J* = 17.0, 1.8 Hz, 2H).

^{13}C NMR (101 MHz, CD_2Cl_2) δ 159.2, 141.1, 140.6, 139.9, 139.4, 139.2, 137.5, 132.2, 129.4, 126.5, 125.8, 125.1, 123.9, 123.9, 123.9, 120.9, 120.8, 120.7, 120.6, 120.2, 119.9, 119.6, 111.2, 110.6, 110.0, 55.1, 54.5, 54.3, 54.0, 53.7, 53.5, 30.3, 1.35.

IR (ATR) $\tilde{v}$ [cm^{-1}] = 3046 (vw), 1597 (w), 1491 (w), 1478 (w), 1447 (m), 1333 (w), 1310 (m), 1223 (m), 1150 (w), 1025 (w), 979 (w), 922 (w), 827 (w), 775 (vw), 742 (m), 720 (m), 617 (w), 547 (w), 527 (w), 478 (vw), 421 (w).

HRMS (FAB-MS, 3-NBA, $C_{62}H_{43}N_{12}$) calc. 955.3734, found 955.3732.

9,9',9'',9'''-(4,6-Bis(2-hexyl-2*H*-tetrazol-5-yl)benzene-1,2,3,5-tetrayl)tetrakis(3,6-di-*tert*-butyl-9*H*-carbazole)

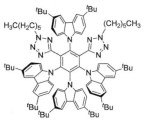

According to the general procedure (GP 1), 9,9',9'',9'''-(4,6-di(2*H*-tetrazol-5-yl)benzene-1,2,3,5-tetrayl)tetrakis(3,6-di-*tert*-butyl-9*H*-carbazole) (300 mg, 0.227 mmol, 1.00 equiv.), 1-bromohexane (150 mg, 0.906 mmol, 4.00 equiv.) and sodium carbonate (96.1 mg, 0.906 mmol, 4.00 equiv.) were reacted in *N,N*-dimethylformamide (10 mL). The reaction mixture was extracted and purified by column chromatography over SiO_2 (dichloromethane/cyclohexane 1:2 to 1:1) to yield 250 mg (0.168 mg, 74%) of the title compound as an off-white solid.

R_f (dichloromethane/cyclohexane 1:2) = 0.37

¹H NMR (400 MHz, Chloroform-*d*) δ 7.89 (s, 2H), 7.53 (s, 4H), 7.31 (d, *J* = 8.6 Hz, 2H), 7.24 (s, 2H), 7.14 (d, *J* = 8.6 Hz, 2H), 7.02 – 6.92 (m, 8H), 6.83 (d, *J* = 8.6 Hz, 2H), 6.61 (d, *J* = 8.7 Hz, 2H), 3.76 (t, *J* = 7.4 Hz, 4H), 1.40 – 1.36 (m, 18H), 1.29 – 1.26 (m, 36H), 1.24 (s, 18H), 1.09 (p, *J* = 7.3 Hz, 4H), 0.94 (dq, *J* = 29.4, 7.5 Hz, 8H), 0.86 – 0.78 (m, 6H), 0.65 (p, *J* = 7.4 Hz, 4H).

¹³C NMR (101 MHz, CDCl₃) δ 158.9, 142.6, 142.5, 142.3, 140.3, 139.4, 138.7, 138.2, 137.6, 137.3, 131.4, 124.2, 123.8, 123.6, 123.2, 122.8, 122.0, 115.8, 115.4, 114.8, 110.1, 109.7, 109.2, 52.2, 34.7, 34.5, 34.4, 32.2, 32.1, 32.0, 31.9, 30.7, 29.8, 28.4, 25.2, 22.2, 14.1.

IR (ATR) ṽ [cm⁻¹] = 3046 (vw), 2953 (w), 2566 (vw), 2470 (vw), 2315 (vw), 2167 (vw), 2105 (vw), 1980 (vw), 1859 (vw), 1739 (vw), 1585 (vw), 1472 (m), 1362 (w), 1295 (w), 1034 (w), 874 (w), 804 (m), 609 (w), 422 (w).

HRMS (ESI, C₁₀₀H₁₂₂N₁₂) calc. 1491.9988 [M+H]⁺, found 1491.9937 [M+H]⁺.

(5-(Tert-butyl)-1,3,4-oxadiazol-2-yl)-2,4,5,6-tetra(carbazol-9-yl)benzonitrile[180]

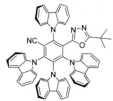

According to the general procedure (GP 3), 2,3,4,6-tetra(carbazol-9-yl)-5-(1H-tetrazol-5-yl)benzonitrile (250 mg, 0.300 mmol, 1.00 equiv.) and pivaloyl chloride (145 mg, 1.20 mmol, 4.00 equiv.) were reacted in trichloromethane (15 mL). The reaction mixture was extracted and purified by column chromatography on silica gel (dichloromethane) to yield 235 mg (0.265 mmol, 88%) of the title compound as a yellow solid.

R$_f$ (dichloromethane) = 0.42

M.p. = 330 °C

^{1}H NMR (400 MHz, Acetone-d_6) δ 8.19 (dt, J = 7.8, 1.0 Hz, 2H), 7.85 – 7.79 (m, 4H), 7.77 – 7.68 (m, 4H), 7.56 (dt, J = 8.2, 0.9 Hz, 2H), 7.53 – 7.45 (m, 4H), 7.42 (dt, J = 7.7, 0.9 Hz, 2H), 7.32 (ddd, J = 8.0, 7.3, 0.9 Hz, 2H), 7.08 (ddd, J = 6.5, 4.4, 1.6 Hz, 4H), 6.98 (ddd, J = 6.7, 4.8, 1.6 Hz, 4H), 6.79 (td, J = 7.5, 1.0 Hz, 2H), 6.71 (ddd, J = 8.4, 7.2, 1.3 Hz, 2H), 0.53 (s, 9H).

^{13}C NMR (101 MHz, acetone-d6) δ 173.7, 157.8, 145.2, 144.3, 143.4, 142.2, 140.5, 140.1, 139.3, 138.8, 130.8, 127.2, 126.2, 126.1, 125.2, 124.8, 124.7, 124.7, 124.4, 121.9, 121.9, 121.5, 121.3, 121.2, 120.8, 120.6, 120.1, 119.4, 113.3, 112.1, 112.0, 111.4, 111.3, 27.3.

IR (ATR) ṽ [cm^{-1}] = 3049 (vw), 2966 (vw), 1625 (vw), 1598 (w), 1552 (w), 1490 (w), 1478 (w), 1445 (m), 1334 (w), 1309 (w), 1222 (m), 1150 (w), 1120 (w), 1028 (w), 928 (w), 771 (vw), 742 (m), 721 (m), 616 (w), 549 (w), 528 (m), 505 (vw), 442 (w), 421 (w).

HRMS (FAB-MS, 3-NBA, C$_{61}$H$_{42}$N$_7$O) calc. 888.3451 [M+H]$^+$, found 888.3452 [M+H]$^+$.

Tetra(carbazol-9-yl)-5-(5-methyl-1,3,4-oxadiazol-2-yl)benzonitrile[180]

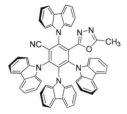

According to the general procedure (GP 3), 2,3,4,6-tetra(carbazol-9-yl)-5-(1*H*-tetrazol-5-yl)benzonitrile (1.00 g, 1.20 mmol, 1.00 equiv.) and acetyl chloride (189 mg, 2.40 mmol, 2.00 equiv.) were reacted in trichloromethane (45 mL). The reaction mixture was extracted and purified by column chromatography on silica gel (dichloromethane) to yield 987 mg (1.17 mmol, 93%) of the title compound as a yellow solid.

R_f (dichloromethane) = 0.45

^{1}H NMR (400 MHz, DMSO-d_6) δ 8.23 (d, *J* = 7.8 Hz, 2H), 7.95 (d, *J* = 8.2 Hz, 2H), 7.87 – 7.75 (m, 6H), 7.68 (d, *J* = 8.1 Hz, 2H), 7.61 (dd, *J* = 7.0, 1.9 Hz, 2H), 7.53 (ddd, *J* = 8.3, 7.2, 1.2 Hz, 2H), 7.46 (dd, *J* = 7.7, 1.3 Hz, 2H), 7.39 – 7.30 (m, 2H), 7.10 (dtd, *J* = 18.0, 7.3, 1.3 Hz, 4H), 7.00 (tt, *J* = 7.3, 5.7 Hz, 4H), 6.80 (td, *J* = 7.4, 1.1 Hz, 2H), 6.74 (td, *J* = 7.7, 1.4 Hz, 2H), 1.45 (s, 3H).

^{13}C NMR (101 MHz, DMSO) δ 163.2, 156.8, 143.8, 143.0, 142.0, 141.0, 139.4, 138.9, 138.2, 137.7, 128.6, 126.3, 125.2, 125.0, 124.1, 123.1, 123.0, 123.0, 122.7, 121.0, 120.9, 120.6, 120.5, 120.2, 120.0, 119.7, 119.3, 117.7, 112.6, 111.5, 111.4, 110.8, 110.8, 59.8, 54.9, 40.2, 40.1, 40.0, 39.9, 39.8, 39.7, 39.5, 39.3, 39.1, 38.9, 20.8, 14.1, 9.1.

IR (ATR) $\tilde{v}$ [cm^{-1}] = 3048 (vw), 2853 (vw), 2927 (vw), 1722 (w), 1574 (w), 1490 (w), 1477 (w), 1455 (m), 1371 (w), 1333 (w), 1309 (w), 1222 (m), 1151 (w), 1119 (w), 1043 (w), 1027 (w), 743 (m), 720 (m), 617 (w), 576 (w), 548 (w), 527 (w), 485 (vw), 560 (vw), 441 (w), 421 (w).

HRMS (FAB-MS, 3-NBA, $C_{58}H_{36}N_7O$) calc. 846.2983 [M+H]$^+$, found 846.2981 [M+H]$^+$.

2,3,4,6-Tetra(9*H*-carbazol-9-yl)-5-(5-undecyl-1,3,4-oxadiazol-2-yl)benzonitrile[180,202]

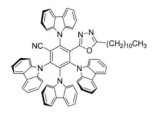

According to the general procedure (GP 3), 2,3,4,6-tetra(carbazol-9-yl)-5-(1*H*-tetrazol-5-yl)benzonitrile (250 mg, 0.300 mmol, 1.00 equiv.) and dodecanoyl chloride (263 mg, 1.20 mmol, 4.00 equiv.) were reacted in trichloromethane (15 mL). The reaction mixture was extracted and purified by column chromatography on silica gel (dichloromethane/cyclohexane 3:1) to yield 237 mg (0.241 mmol, 80%) of the title compound as a yellow solid.

R_f (dichloromethane/cyclohexane 2:1) = 0.32

¹H NMR (400 MHz, Chloroform-*d*) δ 8.00 (d, *J* = 7.8 Hz, 2H), 7.60 – 7.48 (m, 4H), 7.41 (t, *J* = 7.7 Hz, 2H), 7.33 (d, *J* = 8.1 Hz, 2H), 7.28 – 7.17 (m, 4H), 7.12 (dd, *J* = 10.5, 5.3 Hz, 2H), 7.03 (dd, *J* = 6.1, 3.1 Hz, 2H), 6.99 – 6.92 (m, 4H), 6.89 – 6.81 (m, 6H), 6.69 (t, *J* = 7.5 Hz, 2H), 6.56 (t, *J* = 7.7 Hz, 2H), 1.84 (d, *J* = 6.4 Hz, 2H), 1.30 – 1.06 (m, 12H), 0.98 (p, *J* = 6.8 Hz, 2H), 0.82 (t, *J* = 6.7 Hz, 3H), 0.58 (q, *J* = 3.6 Hz, 4H).

¹³C NMR (101 MHz, CDCl₃) δ 167.2, 157.0, 143.7, 142.9, 142.2, 141.1, 139.0, 138.6, 137.7, 136.7, 128.7, 126.8, 125.7, 125.7, 124.7, 124.4, 124.4, 124.1, 123.9, 121.6, 121.6, 121.3, 120.9, 120.8, 120.4, 120.2, 119.7, 117.7, 112.1, 110.2, 109.9, 109.6, 109.5, 32.1, 29.7, 29.7, 29.5, 29.3, 29.0, 28.4, 25.4, 24.4, 22.8, 14.3.

IR (ATR) $\tilde{v}$ [cm⁻¹] = 3049 (vw), 2921 (w), 2850 (w), 1598 (w), 1562 (w), 1490 (w), 1479 (w), 1445 (m), 1333 (w), 1309 (w), 1223 (m), 1151 (w), 1119 (w), 1028 (w), 925 (w), 741 (m), 720 (m), 618 (w), 577 (vw), 550 (w), 528 (w), 488 (vw), 422 (w), 397 (vw).

HRMS (FAB-MS, 3-NBA, C₆₈H₅₅N₇O) calc. 986.4546 [M+H]⁺; found 986.4547 [M+H]⁺.

(5-Benzhydryl-1,3,4-oxadiazol-2-yl)-2,4,5,6-tetra(carbazol-9-yl)benzonitrile[180,202]

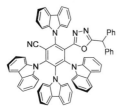

According to the general procedure (GP 3), 2,3,4,6-tetra(carbazol-9-yl)-5-(1H-tetrazol-5-yl)benzonitrile (250 mg, 0.300 mmol, 1.00 equiv.) and 2,2-diphenylacetyl chloride (277 mg, 1.20 mmol, 4.00 equiv.) were reacted in trichloromethane (15 mL). The reaction mixture was extracted and purified by column chromatography on silica gel (dichloromethane/cyclohexane 3:1) to yield 254 mg (0.255 mmol, 85%) of the title compound as a yellow solid.

R_f (dichloromethane/cyclohexane 3:1) = 0.75

^{1}H NMR (400 MHz, Methylene Chloride-d_2) δ 8.19 (dd, J = 7.6, 2.7 Hz, 2H), 7.84 – 7.73 (m, 2H), 7.68 (d, J = 7.7 Hz, 2H), 7.58 – 7.35 (m, 8H), 7.30 – 6.97 (m, 18H), 6.87 (td, J = 7.5, 4.4 Hz, 4H), 6.76 – 6.66 (m, 2H), 6.34 (dd, J = 7.8, 3.1 Hz, 4H), 4.88 (d, J = 3.6 Hz, 1H).

^{13}C NMR (101 MHz, CD$_2$Cl$_2$) δ 168.0, 158.1, 144.4, 143.5, 142.9, 141.4, 139.4, 139.1, 138.2, 138.1, 137.3, 129.2, 129.1, 128.5, 127.8, 127.4, 126.3, 126.2, 125.3, 124.7, 124.7, 124.4, 124.2, 122.1, 121.8, 121.6, 121.3, 120.8, 120.7, 120.1, 118.2, 112.6, 110.8, 110.5, 110.0, 110.0, 48.3, 27.5.

IR (ATR) $\tilde{\nu}$ [cm^{-1}] = 2921 (vw), 1598 (vw), 1551 (vw), 1490 (w), 1479 (w), 1444 (w), 1333 (w), 1309 (w), 1222 (w), 1151 (w), 1119 (w), 1028 (w), 925 (vw), 785 (vw), 741 (m), 720 (w), 696 (w), 617 (w), 550 (w), 527 (vw), 486 (vw), 421 (w).

HRMS (FAB-MS, 3-NBA, C$_{70}$H$_{43}$N$_7$O) calc. 998.3607 [M+H]$^+$; found 998.3605 [M+H]$^+$.

2,3,4,6-Tetra(carbazol-9-yl)-5-cyanophenyl)-1,3,4-oxadiazol-2-yl)propanoate[180,202]

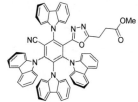

According to the general procedure (GP 3), 2,3,4,6-tetra(carbazol-9-yl)-5-(1*H*-tetrazol-5-yl)benzonitrile (250 mg, 0.300 mmol, 1.00 equiv.) and methyl 4-chloro-4-oxobutanoate (182 mg, 1.20 mmol, 4.00 equiv.) were reacted in trichloromethane (15 mL). The reaction mixture was extracted and purified by column chromatography on silica gel (dichloromethane to dichloromethane/ethyl acetate 97:3) to yield 208 mg (0.223 mmol, 74%) of the title compound as a yellow solid.

R_f (dichloromethane/cyclohexane 10:1) = 0.78

1H-NMR (300 MHz, CDCl$_3$) δ 8.12 (d, J = 7.7 Hz, 2H), 7.67 (ddd, J = 19.7, 5.5, 2.2 Hz, 4H), 7.52 (t, J = 7.6 Hz, 2H), 7.47 – 7.28 (m, 6H), 7.25 – 7.17 (m, 2H), 7.17 – 6.91 (m, 12H), 6.81 (t, J = 7.4 Hz, 2H), 6.67 (t, J = 7.7 Hz, 2H), 3.52 (s, 3H), 2.28 (dd, J = 9.5, 6.4 Hz, 2H), 1.77 (dd, J = 9.3, 6.5 Hz, 2H).

13C NMR (101 MHz, CDCl$_3$) δ 171.1, 165.5, 165.5, 157.3, 157.2, 143.8, 143.0, 142.3, 141.1, 141.0, 139.0, 139.0, 138.6, 138.6, 137.7, 136.8, 136.8, 126.9, 126.8, 125.8, 125.7, 125.7, 124.7, 124.7, 124.5, 124.4, 124.4, 124.4, 124.1, 124.1, 123.9, 123.9, 121.7, 121.7, 121.6, 121.3, 121.0, 120.8, 120.8, 120.4, 120.2, 119.7, 117.8, 112.1, 110.2, 110.2, 109.9, 109.9, 109.6, 109.6, 109.5, 51.9, 29.3, 29.3, 20.0, 20.0.

IR (ATR) $\tilde{v}$ [cm^{-1}] = 3046 (vw), 2947 (vw), 1373 (w), 1598 (vw), 1567 (w), 1490 (w), 1478 (w), 1447 (m), 1333 (w), 1309 (w), 1222 (m), 1151 (w), 1119 (w), 1027 (w), 926 (w), 743 (m), 720 (m), 617 (w), 577 (w), 549 (w), 527 (w), 486 (vw), 441 (w), 421 (w).

HRMS (FAB-MS, 3-NBA, C$_{61}$H$_{39}$N$_7$O$_3$) calc. 918.3193 [M+H]$^+$; found 918.3194 [M+H]$^+$.

2,3,4,6-Tetra(carbazol-9-yl)-5-(5-(chloromethyl)-1,3,4-oxadiazol-2-yl)benzonitrile[180,202]

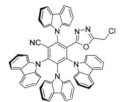

According to the general procedure (GP 3), 2,3,4,6-tetra(carbazol-9-yl)-5-(1H-tetrazol-5-yl)benzonitrile (300 mg, 0.361 mmol, 1.00 equiv.) and 2-chloroacetyl chloride (163 mg, 1.44 mmol, 4.00 equiv.) were reacted in trichloromethane (15 mL). The reaction mixture was extracted and purified by column chromatography on silica gel (dichloromethane/cyclohexane 3:2 to dichloromethane) to yield 246 mg (0.280 mmol, 78%) of the title compound as a yellow solid.

R_f (dichloromethane/cyclohexane 2:1) = 0.41

¹H NMR (400 MHz, Chloroform-d) δ 8.11 (dt, J = 7.8, 0.8 Hz, 2H), 7.73 – 7.66 (m, 2H), 7.66 – 7.60 (m, 2H), 7.52 (ddd, J = 8.3, 7.2, 1.2 Hz, 2H), 7.43 (d, J = 8.1 Hz, 2H), 7.40 – 7.32 (m, 4H), 7.23 – 7.17 (m, 2H), 7.16 – 7.09 (m, 2H), 7.09 – 7.02 (m, 4H), 7.02 – 6.90 (m, 6H), 6.82 (td, J = 7.5, 0.9 Hz, 2H), 6.66 (ddd, J = 8.4, 7.2, 1.2 Hz, 2H), 3.77 (s, 2H).

¹³C NMR (101 MHz, CDCl₃) δ 162.4, 158.4, 144.2, 143.2, 142.5, 141.0, 138.9, 138.5, 137.6, 136.7, 127.6, 126.9, 125.9, 125.7, 124.8, 124.5, 124.5, 124.2, 123.9, 121.8, 121.7, 121.4, 121.0, 120.9, 120.4, 120.3, 119.7, 117.6, 112.1, 110.2, 109.9, 109.5, 109.5, 31.7.

IR (ATR) ṽ [cm⁻¹] = 1597 (vw), 1544 (vw), 1490 (w), 1478 (w), 1443 (m), 1333 (w), 1309 (w), 1292 (w), 1221 (w), 1175 (w), 1149 (w), 1119 (w), 1027 (w), 984 (w), 929 (vw), 741 (m), 720 (m), 617 (w), 595 (w), 548 (w), 527 (w), 486 (vw), 426 (vw), 422 (w).

HRMS (FAB-MS, 3-NBA, C₅₈H₃₄³⁵ClN₇O) calc. 880.2592 [M+H]⁺; found 880.2590 [M+H]⁺.

2,3,4,6-Tetra(carbazol-9-yl)-5-(5-(perfluoroheptyl)-1,3,4-oxadiazol-2-yl)benzonitrile[180,202]

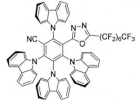

According to the general procedure (GP 3), 2,3,4,6-tetra(carbazol-9-yl)-5-(1H-tetrazol-5-yl)benzonitrile (83.0 mg, 0.100 mmol, 1.00 equiv.) and 2,2,3,3,4,4,5,5,6,6,7,7,8,8,8-pentadecafluorooctanoyl chloride (173 mg, 0.400 mmol, 4.00 equiv.) were reacted in trichloromethane (10 mL). The reaction mixture was extracted and purified by column chromatography on silica gel (dichloromethane/cyclohexane 3:1) to yield 129 mg (0.108 mmol, 90%) of the title compound as a yellow solid.

R$_f$ (dichloromethane/cyclohexane 1:1) = 0.65

M.p. = 348 °C

¹H NMR (400 MHz, Chloroform-d) δ 8.11 (d, J = 7.8 Hz, 2H), 7.71 (dd, J = 7.3, 1.5 Hz, 2H), 7.66 – 7.58 (m, 2H), 7.50 (ddd, J = 8.3, 7.1, 1.2 Hz, 2H), 7.43 – 7.32 (m, 6H), 7.22 – 7.17 (m, 2H), 7.13 – 7.02 (m, 6H), 7.02 – 6.91 (m, 6H), 6.82 (t, J = 7.4 Hz, 2H), 6.66 (ddd, J = 8.4, 7.2, 1.2 Hz, 2H).

¹³C NMR (101 MHz, CDCl₃) δ 160.6, 155.9, 145.6, 144.2, 143.6, 141.6, 139.5, 139.2, 138.3, 137.6, 127.7, 127.0, 126.7, 126.6, 125.6, 125.3, 125.1, 124.8, 122.7, 122.7, 122.4, 122.0, 121.8, 121.2, 121.2, 120.6, 118.6, 112.7, 111.0, 110.6, 110.0, 109.9, 32.8, 31.1, 30.6, 27.8, 23.6.

¹⁹F NMR (376 MHz, CDCl₃) δ -85.0, -117.9, -117.9, -126.3, -126.5, -127.0, -130.5.

IR (ATR) $\tilde{v}$ [cm⁻¹] = 3051 (vw), 2924 (w), 2853 (vw), 1599 (w), 1491 (w), 1479 (w), 1444 (m), 1333 (w), 1310 (w), 1204 (m), 1146 (m), 1028 (w), 963 (w), 926 (w), 741 (m), 720 (m), 669 (w), 643 (w), 618 (w), 558 (w), 528 (w), 487 (vw), 422 (w).

HRMS (APCI, C₆₄H₃₂F₁₅N₇O) calc. 1200.2502 [M+H]⁺, found 1200.2480 [M+H]⁺.

2,3,4,6-Tetra(carbazol-9-yl)-5-(5-mesityl-1,3,4-oxadiazol-2-yl)benzonitrile[180,202]

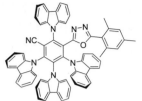

According to the general procedure (GP 3), 2,3,4,6-tetra(carbazol-9-yl)-5-(1H-tetrazol-5-yl)benzonitrile (250 mg, 0.300 mmol, 1.00 equiv.) and 2,4,6-trimethylbenzoyl chloride (220 mg, 1.20 mmol, 4.00 equiv.) were reacted in trichloromethane (15 mL). The reaction mixture was extracted and purified by column chromatography on silica gel (dichloromethane/cyclohexane 3:2) to yield the title compound as a yellow solid (230 mg, 0.242 mmol, 81%).

R_f (dichloromethane/cyclohexane 3:2) = 0.43

¹H NMR (400 MHz, Chloroform-d) δ 8.13 – 8.04 (m, 2H), 7.73 – 7.66 (m, 2H), 7.60 – 7.55 (m, 2H), 7.53 – 7.44 (m, 4H), 7.38 – 7.29 (m, 4H), 7.19 (ddd, J = 7.3, 5.0, 2.2 Hz, 4H), 7.06 (tt, J = 7.4, 5.9 Hz, 4H), 6.99 – 6.92 (m, 4H), 6.90 (d, J = 8.3 Hz, 2H), 6.79 (t, J = 7.5 Hz, 2H), 6.62 (ddd, J = 8.3, 7.2, 1.2 Hz, 2H), 6.58 (s, 2H), 2.14 (s, 3H), 1.04 (s, 6H).

¹³C NMR (101 MHz, CDCl₃) δ 164.6, 157.5, 144.0, 143.1, 142.6, 141.2, 140.9, 138.8, 138.8, 138.6, 137.6, 136.8, 129.5, 128.3, 127.0, 125.8, 125.7, 124.6, 124.5, 124.4, 124.2, 123.9, 121.7, 121.6, 121.3, 121.1, 120.8, 120.4, 120.1, 119.6, 119.5, 117.7, 112.0, 110.3, 110.0, 109.8, 109.4, 21.2, 19.2.

IR (ATR) $\tilde{v}$ [cm⁻¹] = 3044 (vw), 1596 (vw), 1490 (vw), 1478 (w), 1446 (w), 1333 (w), 1308 (w), 1221 (w), 1150 (w), 1016 (w), 855 (vw), 740 (m), 720 (m), 617 (vw), 577 (vw), 549 (vw), 528 (vw), 420 (w).

HRMS (FAB-MS, 3-NBA, C₆₆H₄₃N₇O) calc. 950.3607 [M+H]⁺; found 950.3606 [M+H]⁺.

2,3,4,6-Tetra(carbazol-9-yl)-5-(5-(2,6-dimethoxyphenyl)-1,3,4-oxadiazol-2-yl)benzonitrile[180]

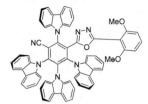

According to the general procedure (GP 3), 2,3,4,6-tetra(carbazol-9-yl)-5-(1H-tetrazol-5-yl)benzonitrile (200 mg, 0.240 mmol, 1.00 equiv.) and 2,6-dimethoxybenzoyl chloride (193 mg, 0.962 mmol, 4.00 equiv.) were reacted in trichloromethane (15 mL). The reaction mixture was extracted and purified by column chromatography on silica gel (dichloromethane to dichloromethane/ethyl acetate 95:5) to yield 74.0 mg (0.0765 mmol, 32%) of the title compound as a yellow solid.

R_f (dichloromethane) = 0.27

1H NMR (400 MHz, Methylene Chloride-d_2) δ 8.19 (d, J = 7.8 Hz, 2H), 7.80 – 7.72 (m, 2H), 7.71 – 7.64 (m, 2H), 7.65 – 7.53 (m, 4H), 7.42 (ddd, J = 8.0, 6.6, 1.5 Hz, 2H), 7.37 (d, J = 7.7 Hz, 2H), 7.31 – 7.16 (m, 5H), 7.15 – 7.05 (m, 4H), 7.02 (dtd, J = 10.1, 7.7, 5.1 Hz, 6H), 6.84 (t, J = 7.4 Hz, 2H), 6.73 – 6.64 (m, 2H), 6.28 (d, J = 8.5 Hz, 2H), 3.21 (s, 6H).

13C NMR (101 MHz, CD$_2$Cl$_2$) δ 161.2, 159.9, 158.4, 144.3, 143.6, 143.1, 141.8, 139.4, 139.1, 138.1, 137.4, 133.7, 130.6, 127.3, 126.2, 126.1, 125.2, 124.7, 124.6, 124.5, 124.1, 122.1, 121.9, 121.6, 121.3, 121.2, 120.8, 120.4, 120.0, 118.2, 112.5, 110.7, 110.6, 110.4, 110.1, 104.5, 103.7, 101.3, 56.1.

IR (ATR) $\tilde{v}$ [cm^{-1}] = 3046 (vw), 1721 (vw), 1594 (w), 1477 (w), 1443 (m), 1333 (w), 1308 (w), 1253 (w), 1221 (w), 1150 (w), 1111 (w), 1016 (w), 924 (w), 784 (vw), 740 (m), 719 (m), 617 (w), 550 (w), 527 (w), 421 (w).

HRMS (FAB-MS, 3-NBA, C$_{65}$H$_{42}$N$_7$O$_3$) calc. 968.3349 [M+H]$^+$, found 968.3353 [M+H]$^+$.

2,3,4,6-Tetra(carbazol-9-yl)-5-(5-(p-tolyl)-1,3,4-oxadiazol-2-yl)benzonitrile[180]

According to the general procedure (GP 3), 2,3,4,6-tetra(carbazol-9-yl)-5-(1H-tetrazol-5-yl)benzonitrile (200 mg, 0.240 mmol, 1.00 equiv.) and 4-methylbenzoyl chloride (112 mg, 0.721 mmol, 3.00 equiv.) were reacted in trichloromethane (15 mL). The reaction mixture was extracted and purified by column chromatography on silica gel (dichloromethane/cyclohexane 4:1) to yield 176 mg (0.192 mmol, 80%) of the title compound as a yellow solid.

R_f (dichloromethane/cyclohexane 9:1) = 0.45

1H NMR (400 MHz, Methylene Chloride-d_2) δ 8.16 (dt, J = 7.8, 1.0 Hz, 2H), 7.83 – 7.75 (m, 2H), 7.72 – 7.64 (m, 2H), 7.64 – 7.50 (m, 4H), 7.48 – 7.36 (m, 4H), 7.34 – 7.26 (m, 2H), 7.27 – 7.19 (m, 2H), 7.17 – 7.08 (m, 6H), 7.09 – 6.99 (m, 4H), 6.95 (d, J = 8.0 Hz, 2H), 6.88 (td, J = 7.5, 0.9 Hz, 2H), 6.75 (ddd, J = 8.4, 7.2, 1.3 Hz, 2H), 6.69 – 6.61 (m, 2H), 2.27 (s, 3H).

13C NMR (101 MHz, CD$_2$Cl$_2$) δ 165.3, 157.0, 144.1, 143.3, 143.3, 142.5, 141.7, 139.8, 139.1, 138.3, 137.7, 129.8, 128.9, 127.4, 126.9, 126.4, 126.2, 125.4, 124.7, 124.5, 124.2, 122.2, 122.1, 121.8, 121.5, 121.4, 120.8, 120.7, 120.2, 119.7, 118.7, 112.7, 110.8, 110.6, 110.0, 110.0, 21.8.

IR (ATR) $\tilde{v}$ [cm^{-1}] = 1598 (w), 1471 (w), 1477 (w), 1446 (m), 1332 (w), 1308 (w), 1222 (m), 1150 (w), 1119 (w), 1087 (w), 1015 (w), 925 (vw), 823 (w), 741 (m), 720 (m), 617 (w), 593 (vw), 575 (w), 557 (w), 527 (w), 506 (w), 490 (w), 442 (w), 424 (w), 386 (w).

HRMS (FAB-MS, 3-NBA, C$_{64}$H$_{40}$N$_7$O) calc. 922.3294 [M+H]$^+$, found 922.3293 [M+H]$^+$.

2,3,4,6-Tetra(carbazol-9-yl)-5-(5-(4-iodophenyl)-1,3,4-oxadiazol-2-yl)benzonitrile[180,216]

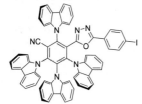

According to the general procedure (GP 3), 2,3,4,6-tetra(carbazol-9-yl)-5-(1H-tetrazol-5-yl)benzonitrile (250 mg, 0.300 mmol, 1.00 equiv.) and 4-iodobenzoyl chloride (240 mg, 0.901 mmol, 3.00 equiv.) were reacted in trichloromethane (15 mL). The reaction mixture was extracted and purified by column chromatography on silica gel (dichloromethane/cyclohexane 1:1 to dichloromethane) to yield 165 mg (0.160 mmol, 53%) of the title compound as a yellow solid.

R_f (dichloromethane/cyclohexane 1:1) = 0.26

¹H NMR (400 MHz, DMSO-d_6) δ 8.20 (d, J = 7.7 Hz, 2H), 8.04 (d, J = 8.2 Hz, 2H), 7.90 – 7.78 (m, 4H), 7.76 – 7.65 (m, 6H), 7.62 – 7.51 (m, 4H), 7.48 (d, J = 7.6 Hz, 2H), 7.34 (t, J = 7.5 Hz, 2H), 7.17 – 6.94 (m, 9H), 6.81 (t, J = 7.4 Hz, 2H), 6.75 (t, J = 7.6 Hz, 2H), 6.32 (d, J = 8.4 Hz, 2H).

¹³C NMR (101 MHz, DMSO) δ 156.9, 144.2, 143.2, 142.0, 141.1, 139.6, 138.9, 138.3, 138.1, 137.9, 128.1, 127.2, 126.5, 125.2, 124.2, 123.1, 123.1, 123.0, 122.7, 121.0, 120.7, 120.6, 120.3, 120.0, 119.4, 118.1, 112.6, 111.4, 110.7, 100.4.

IR (ATR) $\tilde{\nu}$ [cm⁻¹] = 2920 (m), 2850 (m), 1596 (w), 1446 (s), 1333 (m), 1308 (m), 1221 (m), 115 (m), 1118 (w), 1084 (w), 1004 (m), 921 (w), 826 (w), 741 (s), 720 (s), 617 (w), 594 (w), 577 (w), 551 (w), 527 (w), 503 (w), 419 (w).

HRMS (APCI, $C_{63}H_{37}IN_7O_2$) calc. 1034.2099 [M+H]⁺, found 1034.2108 [M+H]⁺.

2,3,4,6-Tetra(carbazol-9-yl)-5-(5-phenyl-1,3,4-oxadiazol-2-yl)benzonitrile[180]

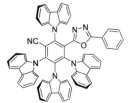

According to the general procedure (GP 3), 2,3,4,6-tetra(carbazol-9-yl)-5-(1H-tetrazol-5-yl)benzonitrile (200 mg, 0.240 mmol, 1.00 equiv.) and benzoyl chloride (67.6 mg, 0.481 mmol, 2.00 equiv.) were reacted in trichloromethane (10 mL). The reaction mixture was extracted and purified by column chromatography on silica gel (dichloromethane/cyclohexane 3:1 to 9:1) to yield 71.2 mg (0.785 mmol, 34%) of the title compound as a yellow solid.

R_f (dichloromethane/cyclohexane 3:1) = 0.33

^{1}H NMR (400 MHz, Methylene Chloride-d_2) δ 8.15 (d, J = 7.7 Hz, 2H), 7.79 (dd, J = 6.3, 2.5 Hz, 2H), 7.72 – 7.65 (m, 2H), 7.57 (dt, J = 14.3, 7.8 Hz, 4H), 7.49 – 7.37 (m, 4H), 7.37 – 7.27 (m, 3H), 7.26 – 7.21 (m, 2H), 7.19 – 7.08 (m, 9H), 7.04 (td, J = 5.8, 4.7, 3.3 Hz, 4H), 6.88 (t, J = 7.4 Hz, 2H), 6.75 (t, J = 7.6 Hz, 4H).

^{13}C NMR (101 MHz, CD$_2$Cl$_2$) δ 165.1, 157.2, 144.2, 143.4, 142.6, 141.7, 139.8, 139.0, 138.3, 137.7, 132.4, 129.1, 128.8, 127.4, 126.9, 126.4, 126.2, 125.4, 124.7, 124.5, 124.3, 122.5, 122.2, 122.1, 121.8, 121.5, 121.4, 120.8, 120.7, 120.2, 118.7, 112.7, 110.8, 110.6, 110.0, 110.0.

IR (ATR) $\tilde{v}$ [cm^{-1}] = 1598 (w), 1549 (w), 1489 (w), 1479 (w), 1447 (m), 1333 (w), 1309 (w), 1223 (m), 1151 (w), 1120 (w), 1089 (w), 1027 (w), 923 (w), 741 (m), 720 (m), 707 (m), 687 (w), 618 (w), 594 (w), 576 (w), 551 (w), 527 (w), 497 (vw), 442 (w), 422 (w).

HRMS (FAB-MS, 3-NBA, C$_{63}$H$_{38}$N$_7$O) calc. 908.3138 [M+H]$^+$, found 908.3140 [M+H]$^+$.

2,3,4,6-Tetra(carbazol-9-yl)-5-(5-vinyl-1,3,4-oxadiazol-2-yl)benzonitrile[180]

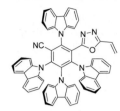

According to the general procedure (GP 3), 2,3,4,6-tetra(carbazol-9-yl)-5-(1H-tetrazol-5-yl)benzonitrile (200 mg, 0.240 mmol, 1.00 equiv.) and acryloyl chloride (87.0 mg, 0.962 mmol, 4.00 equiv.) were reacted in trichloromethane (15 mL). The reaction mixture was extracted and purified by column chromatography on silica gel (dichloromethane) to yield 179 mg (0.209 mmol, 87%) of the title compound as a yellow solid.

R_f (dichloromethane) = 0.68

^{1}H NMR (400 MHz, Methylene Chloride-d_2) δ 8.18 (d, J = 7.7 Hz, 2H), 7.83 – 7.76 (m, 2H), 7.75 – 7.68 (m, 2H), 7.58 (ddd, J = 8.3, 7.1, 1.2 Hz, 2H), 7.50 (d, J = 8.1 Hz, 2H), 7.47 – 7.36 (m, 4H), 7.31 – 7.25 (m, 2H), 7.21 – 7.15 (m, 2H), 7.15 – 7.07 (m, 6H), 7.07 – 7.00 (m, 4H), 6.91 – 6.84 (m, 2H), 6.75 (ddd, J = 8.4, 7.2, 1.3 Hz, 2H), 5.91 (dd, J = 17.7, 11.3 Hz, 1H), 5.24 (d, J = 11.3 Hz, 1H), 4.65 (d, J = 17.7 Hz, 1H).

^{13}C NMR (101 MHz, CD$_2$Cl$_2$) δ 164.0, 156.8, 144.2, 143.5, 142.7, 141.6, 139.7, 139.0, 138.3, 137.6, 128.6, 127.3, 126.3, 126.2, 126.2, 125.4, 125.4, 124.7, 124.5, 124.3, 122.2, 122.2, 122.1, 121.9, 121.5, 121.4, 121.4, 120.8, 120.7, 120.2, 118.6, 118.6, 112.7, 110.8, 110.6, 110.1, 110.0, 110.0.

IR (ATR) $\tilde{v}$ [cm^{-1}] = 3047 (vw), 3022 (vw), 1598 (vw), 1490 (w), 1478 (w), 1446 (m), 1333 (w), 1309 (w), 1222 (m), 1151 (w), 1119 (w), 1026 (w), 977 (w), 945 (w), 742 (m), 720 (m), 685 (w), 617 (w), 595 (w), 576 (w), 558 (w), 527 (w), 485 (vw), 441 (w), 421 (w).

HRMS (APCI, C$_{59}$H$_{35}$N$_7$O) calc. 858.2976 [M+H]$^+$, found 858.2943 [M+H]$^+$.

2,3,4,6-Tetra(carbazol-9-yl)-5-(5-(perfluorophenyl)-1,3,4-oxadiazol-2-yl)benzonitrile[180,202]

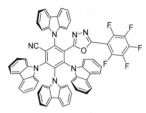

According to the general procedure (GP 3), 2,3,4,6-tetra(carbazol-9-yl)-5-(1*H*-tetrazol-5-yl)benzonitrile (250 mg, 0.300 mmol, 1.00 equiv.) and 2,3,4,5,6-pentafluorobenzoyl chloride (277 mg, 1.20 mmol, 4.00 equiv.) were reacted in trichloromethane (15 mL). The reaction mixture was extracted and purified by column chromatography on silica gel (dichloromethane/cyclohexane 3:1) to yield 247 mg (0.250 mmol, 87%) of the title compound as a yellow solid.

R_f (dichloromethane/cyclohexane 2:1) = 0.23

1H NMR (400 MHz, Chloroform-*d*) δ 8.05 (d, *J* = 7.7 Hz, 2H), 7.75 – 7.68 (m, 2H), 7.61 – 7.56 (m, 2H), 7.51 (ddd, *J* = 8.2, 7.0, 1.1 Hz, 2H), 7.46 (d, *J* = 8.0 Hz, 2H), 7.34 (td, *J* = 8.0, 1.2 Hz, 4H), 7.24 – 7.19 (m, 2H), 7.16 (dd, *J* = 6.3, 2.9 Hz, 2H), 7.07 (pd, *J* = 7.3, 1.3 Hz, 4H), 7.01 – 6.91 (m, 6H), 6.82 (t, *J* = 7.5 Hz, 2H), 6.66 (ddd, *J* = 8.3, 7.2, 1.2 Hz, 2H).

13C NMR (101 MHz, CDCl$_3$) δ 158.3, 144.4, 143.4, 142.8, 141.1, 139.0, 138.6, 137.7, 136.8, 127.3, 126.9, 125.8, 125.8, 124.8, 124.5, 124.5, 124.3, 124.0, 121.8, 121.8, 121.5, 121.0, 120.9, 120.4, 120.2, 119.8, 117.8, 112.0, 110.2, 109.9, 109.5.

19F NMR (376 MHz, CDCl$_3$) δ -139.9, -150.9, -164.1.

IR (ATR) $\tilde{v}$ [cm^{-1}] = 3043 (vw), 1597 (w), 1519 (w), 1496 (w), 1478 (w), 1445 (m), 1333 (w), 1308 (w), 1221 (m), 1150 (w), 1118 (w), 1096 (w), 996 (w), 831 (w), 742 (m), 719 (m), 618 (w), 595 (w), 576 (w), 550 (w), 527 (w), 485 (vw), 463 (vw), 422 (w).

HRMS (FAB-MS, 3-NBA, C$_{63}$H$_{32}$F$_5$N$_7$O) calc. 998.2667 [M+H]$^+$; found 998.2668 [M+H]$^+$.

2,3,4,6-Tetra(carbazol-9-yl)-5-(5-(thiophen-2-yl)-1,3,4-oxadiazol-2-yl)benzonitrile[180,202]

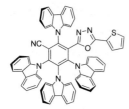

According to the general procedure (GP 3), 2,3,4,6-tetra(carbazol-9-yl)-5-(1*H*-tetrazol-5-yl)benzonitrile (250 mg, 0.300 mmol, 1.00 equiv.) and thiophene-2-carbonyl chloride (176 mg, 1.20 mmol, 4.00 equiv.) were reacted in trichloromethane (15 mL). The reaction mixture was extracted and purified by column chromatography on silica gel (dichloromethane/cyclohexane 3:1) to yield 217 mg (0.238 mmol, 79%) of the title compound as a yellow solid.

R_f (dichloromethane) = 0.62

¹H NMR (400 MHz, DMSO-d_6) δ 8.20 – 8.09 (m, 3H), 7.99 (d, J = 8.2 Hz, 2H), 7.77 (dd, J = 11.8, 7.7 Hz, 5H), 7.63 (ddt, J = 14.8, 11.0, 6.1 Hz, 7H), 7.52 (t, J = 7.7 Hz, 2H), 7.41 (d, J = 7.6 Hz, 2H), 7.32 (t, J = 7.5 Hz, 2H), 7.07 (p, J = 7.2 Hz, 4H), 6.97 (dq, J = 14.6, 7.2 Hz, 3H), 6.88 (t, J = 4.5 Hz, 1H), 6.78 (t, J = 7.4 Hz, 2H), 6.71 (t, J = 7.7 Hz, 2H), 6.25 (d, J = 3.7 Hz, 1H).

¹³C NMR (101 MHz, DMSO) δ 156.0, 142.9, 141.0, 139.4, 138.7, 138.1, 131.7, 129.5, 127.9, 126.2, 125.0, 125.0, 123.9, 123.1, 123.0, 123.0, 122.6, 122.2, 120.8, 120.5, 120.3, 120.1, 119.7, 119.6, 119.1, 111.2, 111.1, 110.5.

IR (ATR) $\tilde{v}$ [cm⁻¹] = 3048 (vw), 3022 (vw), 2920 (vw), 2853 (vw), 1582 (w), 1478 (w), 1448 (m), 1333 (w), 1309 (w), 1223 (m), 1151 (w), 1119 (w), 1027 (w), 922 (vw), 855 (vw), 741 (m), 720 (m), 618 (w), 576 (w), 550 (w), 527 (w), 486 (vw), 462 (vw), 442 (w), 422 (w).

HRMS (FAB-MS, 3-NBA, C₆₁H₃₅N₇OS) calc. 914.2702 [M+H]⁺; found 914.2704 [M+H]⁺.

2,3,4,6-Tetra(carbazol-9-yl)-5-(5-ethynyl-1,3,4-oxadiazol-2-yl)benzonitrile[180,202]

According to the general procedure (GP 4), 2,3,4,6-tetra(carbazol-9-yl)-5-(1*H*-tetrazol-5-yl)benzonitrile (250 mg, 0.300 mmol, 1.00 equiv.), propiolic acid (84.1 mg, 1.20 mmol, 4.00 equiv.) and *N,N'*-diisopropylcarbodiimide (152 mg, 1.20 mmol, 4.00 equiv.) were reacted in trichloromethane (15 mL). The reaction mixture was extracted and purified by column chromatography on silica gel (dichloromethane) to yield 178 mg (0.208 mmol, 70%) of the title compound as a yellow solid.

R_f (dichloromethane) = 0.63

^{1}H NMR (400 MHz, Chloroform-*d*) δ 8.13 (d, *J* = 7.8 Hz, 2H), 7.73 – 7.67 (m, 2H), 7.68 – 7.62 (m, 2H), 7.51 (ddd, *J* = 8.2, 7.1, 1.2 Hz, 2H), 7.39 (d, *J* = 8.5 Hz, 3H), 7.37 – 7.32 (m, 3H), 7.20 – 7.16 (m, 2H), 7.10 (dt, *J* = 7.6, 1.1 Hz, 3H), 7.07 – 7.03 (m, 2H), 7.03 – 6.90 (m, 6H), 6.85 – 6.77 (m, 2H), 6.65 (ddd, *J* = 8.3, 7.3, 1.2 Hz, 2H), 3.09 (s, 1H).

^{13}C NMR (101 MHz, CDCl$_3$) δ 157.5, 149.6, 144.3, 143.3, 142.7, 141.0, 138.9, 138.5, 137.6, 136.7, 127.1, 126.9, 125.9, 125.7, 124.8, 124.6, 124.5, 124.3, 124.0, 121.9, 121.8, 121.5, 121.1, 120.9, 120.4, 120.3, 119.7, 117.7, 112.0, 110.3, 109.9, 109.5, 109.4, 86.2, 65.9, 29.9.

IR (ATR) $\tilde{v}$ [cm^{-1}] = 3304 (vw), 3281 (vw), 3047 (vw), 2924 (vw), 2142 (vw), 1598 (w), 1490 (w), 1478 (w), 1447 (m), 1333 (w), 1308 (m), 1221 (m), 1172 (w), 1151 (w), 1119 (w), 1027 (w), 912 (w), 742 (m), 720 (m), 617 (w), 557 (w), 527 (w), 485 (vw), 439 (w), 420 (w).

HRMS (APCI, C$_{59}$H$_{33}$N$_7$O) calc. 856.2819 [M+H]$^+$; found 856.2803 [M+H]$^+$.

2,3,4,6-Tetra(carbazol-9-yl)-5-(5-(4-cyanophenyl)-1,3,4-oxadiazol-2-yl)benzonitrile[180,202]

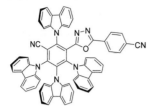

According to the general procedure (GP 3), 2,3,4,6-tetra(carbazol-9-yl)-5-(1H-tetrazol-5-yl)benzonitrile (250 mg, 0.300 mmol, 1.00 equiv.) and 4-cyanobenzoyl chloride (199 mg, 1.20 mmol, 4.00 equiv.) were reacted in trichloromethane (15 mL). The reaction mixture was extracted and purified by column chromatography on silica gel (dichloromethane/cyclohexane 1:1 to dichloromethane) to yield 257 mg (0.276 mmol, 92%) of the title compound as a yellow solid.

R_f (dichloromethane/cyclohexane 2:1) = 0.33

1H NMR (400 MHz, Chloroform-d) δ 8.07 (d, J = 7.7 Hz, 2H), 7.75 – 7.66 (m, 2H), 7.64 – 7.58 (m, 2H), 7.54 (ddd, J = 8.2, 7.1, 1.2 Hz, 2H), 7.48 (d, J = 8.1 Hz, 2H), 7.44 – 7.39 (m, 2H), 7.36 (ddt, J = 9.1, 7.1, 1.1 Hz, 4H), 7.22 (d, J = 8.0 Hz, 2H), 7.15 (dd, J = 6.2, 2.9 Hz, 2H), 7.13 – 7.02 (m, 4H), 7.01 – 6.94 (m, 6H), 6.85 – 6.79 (m, 4H), 6.67 (ddd, J = 8.4, 7.3, 1.2 Hz, 2H).

13C NMR (101 MHz, CDCl$_3$) δ 163.7, 144.9, 143.8, 142.9, 142.1, 140.0, 139.3, 138.5, 137.9, 133.2, 128.3, 127.8, 127.8, 126.8, 126.7, 126.5, 125.6, 125.3, 125.2, 125.0, 124.8, 122.6, 122.5, 122.2, 121.9, 121.7, 121.2, 121.2, 120.6, 118.6, 112.8, 111.0, 110.6, 110.2, 110.2.

IR (ATR) $\tilde{v}$ [cm^{-1}] = 2226 (vw), 1597 (vw), 1491 (w), 1447 (w), 1334 (w), 1310 (w), 1221 (w), 1015 (w), 925 (vw), 848 (vw), 742 (m), 721 (m), 609 (w), 583 (w), 555 (w), 528 (w), 422 (w). 48

HRMS (APCI, C$_{64}$H$_{36}$N$_8$O) calc. 933.3085 [M+H]$^+$, found 933.3072 [M+H]$^+$.

3-(5-(3,5-Bis(trifluoromethyl)phenyl)-1,3,4-oxadiazol-2-yl)-2,4,5,6-tetra(9H-carbazol-9-yl)benzonitrile

According to the general procedure (GP 3), 2,3,4,6-tetra(carbazol-9-yl)-5-(1H-tetrazol-5-yl)benzonitrile (250 mg, 0.300 mmol, 1.00 equiv.) and 3,5-bis(trifluoromethyl)benzoyl chloride (333 mg, 1.20 mmol, 4.00 equiv.) were reacted in trichloromethane (15 mL). The reaction mixture was extracted and purified by column chromatography on silica gel (dichloromethane/pentane 1:1 to dichloromethane) to yield 206 mg (0.198 mmol, 66%) of the title compound as a yellow solid.

R_f (dichloromethane) = 0.34

^{1}H NMR (400 MHz, Chloroform-d) δ 8.10 (d, J = 7.7 Hz, 2H), 7.84 (s, 1H), 7.75 – 7.69 (m, 2H), 7.65 – 7.59 (m, 2H), 7.55 (ddd, J = 8.2, 7.0, 1.2 Hz, 2H), 7.49 (d, J = 8.0 Hz, 2H), 7.40 – 7.35 (m, 4H), 7.34 (dd, J = 5.1, 1.3 Hz, 2H), 7.25 – 7.22 (m, 2H), 7.20 – 7.16 (m, 2H), 7.09 (pd, J = 7.3, 1.4 Hz, 4H), 7.03 – 6.96 (m, 6H), 6.84 (td, J = 7.5, 0.9 Hz, 2H), 6.69 (ddd, J = 8.3, 7.2, 1.2 Hz, 2H).

^{13}C NMR (101 MHz, CDCl$_3$) δ 162.2, 157.9, 144.2, 143.1, 142.3, 141.1, 139.1, 138.5, 137.7, 137.0, 133.0, 132.7, 132.4, 132.0, 127.2, 127.0, 126.6, 126.5, 126.0, 125.8, 125.3, 124.9, 124.5, 124.3, 124.1, 124.0, 123.9, 121.9, 121.8, 121.6, 121.2, 121.1, 121.0, 120.5, 120.4, 119.8, 118.1, 112.0, 110.2, 109.9, 109.3, 109.3.

^{19}F NMR (376 MHz, CDCl$_3$) δ -67.3.

IR (ATR) $\tilde{v}$ [cm^{-1}] = 3048 (vw), 2925 (vw), 2849 (vw), 1598 (w), 1449 (m), 1282 (m), 1135 (m), 1027 (w), 903 (w), 743 (s), 618 (w), 528 (w), 422 (w).

HRMS (APCI, C$_{65}$H$_{35}$F$_6$N$_7$O$_5$) calc. 1044.2880[M+H]$^+$, found 1044.2822 [M+H]$^+$.

2,3,4,6-Tetra(9H-carbazol-9-yl)-5-(5-(3,5-dinitrophenyl)-1,3,4-oxadiazol-2-yl)benzonitrile[202]

According to the general procedure (GP 3), 2,3,4,6-tetra(carbazol-9-yl)-5-(1H-tetrazol-5-yl)benzonitrile (250 mg, 0.300 mmol, 1.00 equiv.) and 3,5-dinitrobenzoyl chloride (277 mg, 1.20 mmol, 4.00 equiv.) were reacted in trichloromethane (15 mL). The reaction mixture was extracted and purified by column chromatography on silica gel (dichloromethane/cyclohexane 2:1 to dichloromethane) to yield 245 mg (0.246 mmol, 82%) of the title compound as an orange solid.

R_f (dichloromethane/cyclohexane 2:1) = 0.28

¹H NMR (400 MHz, DMSO-d_6) δ 8.80 (t, J = 2.1 Hz, 1H), 8.17 (dd, J = 7.8, 1.1 Hz, 2H), 8.08 (d, J = 8.2 Hz, 2H), 7.86 (dd, J = 7.2, 1.5 Hz, 2H), 7.85 – 7.81 (m, 2H), 7.76 – 7.65 (m, 8H), 7.56 (ddd, J = 8.3, 7.2, 1.2 Hz, 2H), 7.52 – 7.45 (m, 2H), 7.36 – 7.27 (m, 2H), 7.18 – 7.01 (m, 6H), 6.95 (td, J = 7.5, 1.0 Hz, 2H), 6.82 (td, J = 7.4, 1.0 Hz, 2H), 6.76 (ddd, J = 8.4, 7.2, 1.4 Hz, 2H).

¹³C NMR (101 MHz, DMSO) δ 160.7, 158.0, 148.6, 144.6, 143.4, 142.3, 141.1, 139.4, 138.8, 138.2, 138.0, 127.4, 126.5, 125.3, 125.2, 125.0, 124.2, 123.5, 123.1, 122.9, 122.7, 121.7, 121.2, 121.1, 120.9, 120.7, 120.4, 120.1, 119.9, 119.4, 118.0, 112.6, 111.4, 111.4, 110.8.

IR (ATR) $\tilde{v}$ [cm⁻¹] = 1599 (vw), 1537 (w), 1490 (w), 1478 (w), 1449 (m), 1342 (w), 1296 (w), 1220 (m), 1151 (w), 1027 (w), 982 (vw), 909 (w), 786 (vw), 763 (w), 741 (m), 719 (m), 651 (vw), 619 (w), 585 (vw), 553 (w), 528 (vw), 440 (w), 421 (w).

HRMS (FAB-MS, 3-NBA, C₆₃H₃₅N₉O₅) calc. 998.2839; found 998.2837.

3-(5-(Adamantan-2-yl)-1,3,4-oxadiazol-2-yl)-2,4,5,6-tetra(9H-carbazol-9-yl)benzonitrile[202]

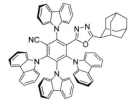

According to the general procedure (GP 3), 2,3,4,6-tetra(carbazol-9-yl)-5-(1H-tetrazol-5-yl)benzonitrile (250 mg, 0.300 mmol, 1.00 equiv.) and adamantane-2-carbonyl chloride (239 mg, 1.20 mmol, 4.00 equiv.) were reacted in trichloromethane (15 mL). The reaction mixture was extracted and purified by column chromatography on silica gel (dichloromethane/cyclohexane 1:1 to 2:1 to dichloromethane) to yield 156 mg (0.162 mmol, 54%) of the title compound as an orange solid.

R_f (dichloromethane/cyclohexane 2:1) = 0.30

¹H NMR (400 MHz, Chloroform-*d*) δ 8.11 (d, *J* = 7.7 Hz, 2H), 7.75 – 7.66 (m, 2H), 7.66 – 7.59 (m, 2H), 7.55 (t, *J* = 7.6 Hz, 2H), 7.48 (d, *J* = 8.1 Hz, 2H), 7.36 (dd, *J* = 12.4, 7.5 Hz, 4H), 7.25 – 7.19 (m, 2H), 7.18 – 7.11 (m, 2H), 7.07 (dd, *J* = 6.6, 3.0 Hz, 4H), 7.03 – 6.92 (m, 6H), 6.81 (t, *J* = 7.4 Hz, 2H), 6.68 (t, *J* = 7.6 Hz, 2H), 1.75 (t, *J* = 3.3 Hz, 3H), 1.56 (d, *J* = 12.0 Hz, 3H), 1.44 – 1.34 (m, 3H), 1.06 (d, *J* = 2.8 Hz, 6H).

¹³C NMR (101 MHz, CDCl₃) δ 172.7, 156.4, 143.6, 143.0, 142.0, 141.0, 139.1, 138.6, 137.7, 136.7, 129.1, 126.8, 125.7, 124.7, 124.4, 124.2, 123.9, 121.6, 121.5, 121.1, 120.9, 120.8, 120.4, 120.2, 119.7, 117.7, 112.1, 110.2, 109.9, 109.5, 109.5, 38.7, 36.0, 33.5, 27.4.

IR (ATR) ṽ [cm⁻¹] = 3047 (vw), 2904 (w), 2850 (w), 1598 (w), 1549 (w), 1490 (w), 1478 (w), 1444 (m), 1333 (w), 1308 (w), 1221 (m), 1150 (w), 1119 (w), 1057 (w), 1026 (w), 924 (w), 786 (vw), 741 (m), 720 (m), 617 (w), 578 (w), 551 (w), 528 (w), 495 (vw), 421 (w).

HRMS (FAB-MS, 3-NBA, C₆₇H₄₇N₇O) calc. 966.3920; found 966.3920.

2,3,4,6-Tetra(9H-carbazol-9-yl)-5-(5-(trichloromethyl)-1,3,4-oxadiazol-2-yl)benzonitrile[202]

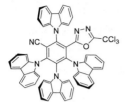

According to the general procedure (GP 3), 2,3,4,6-tetra(carbazol-9-yl)-5-(1H-tetrazol-5-yl)benzonitrile (250 mg, 0.300 mmol, 1.00 equiv.) and 2,2,2-trichloroacetyl chloride (219 mg, 1.20 mmol, 4.00 equiv.) were reacted in trichloromethane (15 mL). The reaction mixture was extracted and purified by column chromatography on silica gel (dichloromethane/cyclohexane 1:1 to 2:1) to yield 275 mg (0.290 mmol, 96%) of the title compound as an orange solid.

R_f (dichloromethane/cyclohexane 2:1) = 0.37

1H NMR (400 MHz, Chloroform-d) δ 8.09 (d, J = 7.8 Hz, 2H), 7.71 (dd, J = 7.1, 1.7 Hz, 2H), 7.65 – 7.60 (m, 2H), 7.53 (td, J = 7.6, 7.1, 1.2 Hz, 2H), 7.46 (d, J = 8.0 Hz, 2H), 7.36 (t, J = 7.6 Hz, 4H), 7.24 – 7.17 (m, 2H), 7.17 – 7.12 (m, 2H), 7.12 – 7.02 (m, 4H), 7.02 – 6.93 (m, 6H), 6.82 (t, J = 7.4 Hz, 2H), 6.69 – 6.63 (m, 2H).

13C NMR (101 MHz, CDCl3) δ 163.5, 159.3, 144.7, 143.4, 142.6, 140.9, 138.8, 138.5, 137.6, 136.8, 127.0, 126.0, 125.8, 124.8, 124.6, 124.5, 124.4, 124.0, 121.9, 121.9, 121.6, 121.2, 121.0, 120.5, 120.4, 119.8, 110.2, 109.9, 109.5, 109.4.

IR (ATR) $\tilde{v}$ [cm-1] = 3046 (vw), 1598 (w), 1544 (vw), 1490 (w), 1478 (w), 1444 (m), 1333 (w), 1309 (w), 1221 (w), 1151 (w), 1119 (w), 1028 (vw), 990 (vw), 910 (w), 846 (w), 822 (w), 790 (w), 741 (m), 721 (m), 617 (w), 550 (w), 528 (w), 489 (vw), 421 (w).

HRMS (FAB-MS, 3-NBA, $C_{58}H_{32}{}^{35}Cl_3N_7O$) calc. 947.1734; found. 947.1733.

Ethyl   5-(-2,3,4,6-tetra(9*H*-carbazol-9-yl)-5-cyanophenyl)-1,3,4-oxadiazole-2-carboxylate[202]

According to the general procedure (GP 3), 2,3,4,6-tetra(carbazol-9-yl)-5-(1*H*-tetrazol-5-yl)benzonitrile (250 mg, 0.300 mmol, 1.00 equiv.) and ethyl 2-chloro-2-oxoacetate (164 mg, 1.20 mmol, 4.00 equiv.) were reacted in trichloromethane (15 mL). The reaction mixture was extracted and purified by column chromatography on silica gel (dichloromethane/cyclohexane 2:1 to dichloromethane) to yield 233 mg (0.258 mmol, 86%) of the title compound as an orange solid.

R_f (dichloromethane/cyclohexane 2:1) = 0.47

1H NMR (400 MHz, Chloroform-*d*) δ 8.11 (d, *J* = 7.7 Hz, 2H), 7.71 (dd, *J* = 7.3, 1.5 Hz, 2H), 7.64 – 7.59 (m, 2H), 7.53 (ddd, *J* = 8.3, 7.1, 1.2 Hz, 2H), 7.44 (d, *J* = 8.1 Hz, 2H), 7.40 – 7.30 (m, 4H), 7.19 (dd, *J* = 7.2, 1.5 Hz, 2H), 7.16 – 7.11 (m, 2H), 7.11 – 7.01 (m, 4H), 7.01 – 6.96 (m, 4H), 6.94 (d, *J* = 8.2 Hz, 2H), 6.82 (t, *J* = 7.4 Hz, 2H), 6.66 (ddd, *J* = 8.3, 7.2, 1.2 Hz, 2H), 4.08 (q, *J* = 7.1 Hz, 2H), 1.10 (t, *J* = 7.1 Hz, 3H).

13C NMR (101 MHz, CDCl3) δ 159.0, 156.3, 152.6, 144.5, 143.4, 142.7, 141.1, 138.9, 138.5, 137.6, 136.8, 127.2, 126.9, 125.9, 125.8, 124.8, 124.6, 124.5, 124.3, 124.0, 121.9, 121.8, 121.5, 121.1, 120.9, 120.4, 120.3, 119.7, 117.8, 111.9, 110.2, 109.8, 109.5, 109.4, 29.9, 13.8.

IR (ATR) $\tilde{v}$ [cm-1] = 2921 (m), 2852 (w), 1753 (w), 1597 (w), 1490 (w), 1447 (m), 1333 (m), 1308 (m), 1267 (w), 1221 (m), 1193 (m), 1151 (m), 1119 (w), 1027 (w), 922 (w), 842 (w), 770 (vw), 741 (s), 719 (m), 646 (w), 617 (w), 594 (w), 576 (w), 551 (w), 527 (w), 485 (w), 440 (w), 422 (w).

HRMS (FAB-MS, 3-NBA, C60H37N7O3) calc. 904.3036; found 904.3035.

2,3,4,6-Tetra(9H-carbazol-9-yl)-5-(5-(4-(dimethylamino)phenyl)-1,3,4-oxadiazol-2-yl)benzonitrile

According to the general procedure (GP 4), 2,3,4,6-tetra(carbazol-9-yl)-5-(1H-tetrazol-5-yl)benzonitrile (250 mg, 0.300 mmol, 1.00 equiv.), 4-(dimethylamino)benzoyl chloride (199 mg, 1.20 mmol, 4.00 equiv.) and N,N'-diisopropylcarbodiimide (151 mg, 1.20 mmol, 4.00 equiv.) were reacted in trichloromethane (15 mL). The reaction mixture was extracted and purified by column chromatography on silica gel (dichloromethane to dichloromethane/ethyl acetate 99:1 to 98:1) to yield 147 mg (0.159 mmol, 53%) of the title compound as an orange solid.

R_f (dichloromethane) = 0.27

^{1}H NMR (400 MHz, Chloroform-d) δ 8.07 (td, J = 7.2, 7.1, 4.3 Hz, 2H), 7.70 (dq, J = 6.6, 4.1, 3.1, 3.1 Hz, 2H), 7.55 (ddt, J = 27.9, 8.9, 5.5, 5.5 Hz, 7H), 7.34 (td, J = 9.7, 8.6, 4.6 Hz, 4H), 7.25 – 7.20 (m, 2H), 7.20 – 7.13 (m, 2H), 7.07 (td, J = 5.2, 5.2, 2.5 Hz, 4H), 7.03 – 6.92 (m, 6H), 6.87 – 6.77 (m, 3H), 6.76 – 6.63 (m, 4H), 2.89 (s, 6H).

^{13}C NMR (101 MHz, CDCl$_3$) δ 164.4, 163.7, 161.8, 145.0, 143.8, 143.6, 143.0, 142.9, 142.1, 142.0, 141.3, 141.3, 139.2, 139.2, 138.6, 138.6, 137.8, 137.7, 137.1, 137.0, 137.0, 128.4, 128.3, 128.1, 127.9, 127.0, 127.0, 126.9, 125.9, 125.9, 125.9, 125.7, 124.8, 124.8, 124.5, 124.4, 124.4, 124.2, 124.0, 123.9, 121.8, 121.7, 121.6, 121.6, 121.4, 121.4, 121.3, 121.2, 121.1, 120.9, 120.9, 120.8, 120.4, 120.4, 120.3, 120.3, 119.8, 119.7, 118.1, 118.0, 112.1, 110.2, 110.0, 109.9, 109.5, 109.5, 109.4, 29.8.

IR (ATR) $\tilde{v}$ [cm^{-1}] = 3987 (w), 3802 (w), 3048 (w), 2921 (w), 2673 (w), 2313 (w), 2236 (w), 1891 (w), 1724 (w), 1607 (m), 1443 (m), 1309 (m), 1222 (m), 1027 (m), 925 (w), 821 (m), 740 (s), 528 (m), 420 (m).

HRMS (FAB-MS, 3-NBA, C$_{65}$H$_{43}$N$_8$O) calc. 951.3560, found 951.3560.

3-(5-(4-Benzoylphenyl)-1,3,4-oxadiazol-2-yl)-2,4,5,6-tetra(9H-carbazol-9-yl)benzonitrile[202]

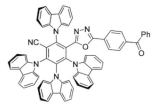

According to the general procedure (GP 4), 2,3,4,6-tetra(carbazol-9-yl)-5-(1H-tetrazol-5-yl)benzonitrile (250 mg, 0.300 mmol, 1.00 equiv.), 4-benzoylbenzoyl chloride (272 mg, 1.20 mmol, 4.00 equiv.) and N,N'-diisopropylcarbodiimide (151 mg, 1.20 mmol, 4.00 equiv.) were reacted in trichloromethane (15 mL). The reaction mixture was extracted and purified by column chromatography on silica gel (dichloromethane/cyclohexane 2:1 to dichloromethane) to yield 247 mg (0.244 mmol, 81%) of the title compound as a yellow solid.

R_f (dichloromethane) = 0.86

¹H NMR (400 MHz, DMSO-d_6) δ 8.21 (dt, J = 7.7, 1.0 Hz, 2H), 8.11 – 8.03 (m, 2H), 7.86 (ddd, J = 7.7, 3.1, 1.3 Hz, 4H), 7.76 – 7.70 (m, 6H), 7.67 (d, J = 7.4 Hz, 3H), 7.60 – 7.51 (m, 6H), 7.51 – 7.47 (m, 2H), 7.39 – 7.31 (m, 2H), 7.19 – 7.02 (m, 6H), 6.98 (td, J = 7.4, 1.0 Hz, 2H), 6.86 – 6.73 (m, 6H).

¹³C NMR (101 MHz, DMSO) δ 194.7, 162.8, 157.3, 144.3, 143.2, 142.1, 141.2, 139.8, 139.6, 138.9, 138.3, 138.1, 136.2, 133.2, 130.0, 129.7, 128.7, 128.0, 126.5, 125.8, 125.2, 124.3, 124.2, 123.1, 123.1, 123.1, 122.7, 121.0, 121.0, 120.7, 120.3, 120.0, 120.0, 119.4, 118.1, 112.6, 111.4, 111.4, 110.7.

IR (ATR) ṽ [cm⁻¹] = 2922 (vw), 1657 (w), 1597 (w), 1478 (w), 1446 (m), 1333 (w), 1308 (m), 1222 (m), 1149 (w), 1118 (w), 1086 (w), 1015 (w), 925 (w), 860 (vw), 741 (s), 720 (m), 700 (m), 662 (w), 618 (w), 596 (w), 577 (w), 551 (w), 528 (w), 440 (w), 423 (w).

HRMS (FAB-MS, 3-NBA, $C_{70}H_{41}N_7O_2$) calc. 1012.3400; found 1012.3401.

2,3,4,6-Tetrakis(3,6-di-*tert*-butyl-9*H*-carbazol-9-yl)-5-(5-(perfluoroheptyl)-1,3,4-oxadiazol-2-yl)benzonitrile

According to the general procedure (GP 3), 2,3,4,6-tetrakis(3,6-di-*tert*-butyl-9*H*-carbazol-9-yl)-5-(2*H*-tetrazol-5-yl)benzonitrile (250 mg, 0.195 mmol, 1.00 equiv.) and 2,2,3,3,4,4,5,5,6,6,7,7,8,8,8-pentadecafluorooctanoyl chloride (253 mg, 0.586 mmol, 3.00 equiv.) were reacted in trichloromethane (15 mL). The reaction mixture was extracted and purified by column chromatography on silica gel (dichloromethane/cyclohexane 2:1 to dichloromethane) to yield 254 mg (0.154 mmol, 79%) of the title compound as an orange solid.

R$_f$ (dichloromethane/cyclohexane 1:1) = 0.31

M.p. = 167 °C

^{1}H NMR (400 MHz, Chloroform-*d*) δ 8.04 (d, *J* = 1.8 Hz, 2H), 7.60 (d, *J* = 1.4 Hz, 2H), 7.58 (d, *J* = 1.9 Hz, 2H), 7.50 (d, *J* = 1.8 Hz, 1H), 7.48 (d, *J* = 1.9 Hz, 1H), 7.24 (q, *J* = 1.6 Hz, 4H), 7.01 (d, *J* = 1.3 Hz, 4H), 6.96 (dd, *J* = 8.6, 1.9 Hz, 2H), 6.89 (d, *J* = 8.6 Hz, 2H), 6.69 (d, *J* = 8.6 Hz, 2H), 6.59 (dd, *J* = 8.6, 1.9 Hz, 2H), 1.42 (s, 18H), 1.33 (s, 18H), 1.23 (s, 36H).

^{13}C NMR (101 MHz, CDCl$_3$) δ 160.2, 144.9, 144.5, 144.4, 144.3, 144.2, 143.5, 142.7, 139.3, 138.0, 137.1, 136.9, 136.5, 125.4, 124.8, 124.7, 124.5, 124.5, 124.4, 123.5, 123.3, 122.5, 117.3, 117.1, 116.4, 116.1, 115.3, 112.4, 109.8, 109.4, 109.1, 108.6, 34.9, 34.7, 34.6, 34.5, 32.0, 31.9, 31.9, 31.7.

^{19}F NMR (376 MHz, CDCl$_3$) δ -85.1, -117.9, -126.4, -126.5, -127.1, -130.5, -130.5.

IR (ATR) ṽ [cm^{-1}] = 2959 (w), 1734 (vw), 1587 (vw), 1471 (m), 1364 (w), 1203 (m), 1034 (w), 877 (w), 804 (w), 737 (w), 610 (w), 468 (w).

HRMS (APCI, C$_{96}$H$_{96}$F$_{15}$N$_7$O) calc. 1648.7510 [M+H]$^+$, found 1648.7421 [M+H]$^+$.

Tert-butyl   (S)-2-((((9H-fluoren-9-yl)methoxy)carbonyl)amino)-3-(5-(2,3,4,6-tetra(carbazol-9-yl)-5-cyanophenyl)-1,3,4-oxadiazol-2-yl)propanoate[180,202]

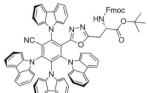

According to the general procedure (GP 4), 2,3,4,6-tetra(carbazol-9-yl)-5-(1H-tetrazol-5-yl)benzonitrile (100 mg, 0.120 mmol, 1.00 equiv.), (S)-3-((((9H-fluoren-9-yl)methoxy)carbonyl)amino)-4-(tert-butoxy)-4-oxobutanoic acid (98.9 mg, 0.240 mmol, 2.00 equiv.) and N,N'-diisopropylcarbodiimide (45.0 mg, 0.361 mmol, 3.00 equiv.) were reacted in trichloromethane (12 mL). The reaction mixture was extracted and purified by column chromatography on silica gel (dichloromethane/cyclohexane 3:2 to dichloromethane to dichloromethane/ethyl acetate 97:3 to 95:5) to yield 126 mg (0.106 mmol, 88%) of the title compound as a yellow solid.

R_f (dichloromethane/ethyl acetate 97:3) = 0.70

^{1}H NMR (400 MHz, Chloroform-d) δ 8.09 (t, J = 8.4 Hz, 2H), 7.81 (d, J = 7.6 Hz, 2H), 7.71 (dd, J = 17.9, 7.4 Hz, 2H), 7.60 (t, J = 7.6 Hz, 4H), 7.56 – 7.23 (m, 12H), 7.21 – 7.06 (m, 4H), 6.98 (tdt, J = 21.2, 12.1, 6.7 Hz, 7H), 6.82 (t, J = 7.4 Hz, 2H), 6.67 (q, J = 7.7 Hz, 2H), 4.94 (d, J = 8.0 Hz, 1H), 4.42 (dd, J = 10.5, 6.9 Hz, 1H), 4.32 (t, J = 8.9 Hz, 1H), 4.21 (t, J = 6.9 Hz, 1H), 3.87 (dt, J = 9.1, 5.5 Hz, 1H), 2.59 (dd, J = 16.3, 5.9 Hz, 1H), 2.35 (dd, J = 16.7, 4.8 Hz, 1H), 1.09 (d, J = 1.8 Hz, 9H).

^{13}C NMR (101 MHz, CDCl$_3$) δ 167.9, 163.7, 157.6, 155.6, 143.9, 143.9, 143.8, 143.1, 142.3, 141.5, 141.4, 141.3, 140.9, 139.1, 139.0, 138.6, 138.5, 137.7, 137.7, 136.7, 128.1, 127.9, 127.2, 126.9, 126.9, 125.9, 125.8, 125.7, 125.6, 125.3, 125.2, 124.8, 124.7, 124.5, 124.4, 124.4, 124.3, 124.2, 124.0, 123.9, 123.9, 121.8, 121.7, 121.6, 121.4, 121.4, 121.0, 120.9, 120.9, 120.5, 120.3, 120.2, 120.1, 119.7, 119.7, 117.7, 112.1, 110.2, 109.9, 109.7, 109.4, 109.4, 82.9, 67.1, 50.8, 47.2, 29.8, 27.6, 27.3.

IR (ATR) $\tilde{v}$ [cm^{-1}] = 3048 (vw), 2925 (vw), 1721 (w), 1598 (w), 1491 (w), 1479 (w), 1445 (m), 1369 (w), 1333 (w), 1309 (w), 1222 (m), 1150 (m), 1120 (w), 1028 (w), 926 (w), 843 (w), 740 (m), 721 (m), 618 (w), 592 (vw), 550 (w), 528 (w), 487 (vw), 422 (w).

HRMS (ESI, C$_{79}$H$_{56}$N$_8$O$_5$) calc. 1197.4446 [M+H]$^+$, found 1197.4419 [M+H]$^+$.

(2,4,5,6-Tetra(9*H*-carbazol-9-yl)-1,3-phenylene)bis(2-methyl-1,3,4-oxadiazole))[180]

According to the general procedure (GP 3), 9,9',9'',9'''-(-4,6-di(1*H*-tetrazol-5-yl)benzene-1,2,3,5-tetrayl)tetrakis(9*H*-carbazole) (200 mg, 0.229 mmol, 1.00 equiv.) and acetyl chloride (71.9 mg, 0.915 mmol, 4.00 equiv.) were reacted in trichloromethane (15 mL). The reaction mixture was extracted and purified by column chromatography on silica gel (dichloromethane to dichloromethane/ethyl acetate 9:1) to yield 151 mg (0.168 mmol, 73%) of the title compound as an off-white solid.

R_f (dichloromethane/ethyl acetate 9:1) = 0.57

1H NMR (400 MHz, DMSO-d_6) δ 8.10 (d, J = 7.8 Hz, 2H), 7.81 – 7.71 (m, 6H), 7.64 (dt, J = 8.1, 2.2 Hz, 5H), 7.52 – 7.45 (m, 2H), 7.35 (ddd, J = 8.3, 7.2, 1.2 Hz, 2H), 7.26 – 7.16 (m, 2H), 6.99 (pd, J = 7.2, 1.4 Hz, 8H), 6.79 (tt, J = 7.3, 5.6 Hz, 4H), 1.47 (s, 6H).

13C NMR (101 MHz, DMSO) δ 163.0, 157.4, 141.3, 140.9, 139.7, 138.8, 137.6, 128.4, 125.9, 125.0, 124.2, 122.8, 122.8, 122.6, 120.3, 120.1, 120.0, 119.6, 119.3, 111.6, 110.9, 110.1, 9.2.

IR (ATR) $\tilde{v}$ [cm⁻¹] = 2923 (vw), 1722 (vw), 1576 (w), 1491 (w), 1478 (w), 1446 (m), 1333 (w), 1308 (m), 1285 (w), 1221 (m), 1148 (w), 1120 (w), 1005 (w), 951 (w), 923 (w), 841 (vw), 742 (m), 720 (m), 618 (w), 577 (w), 544 (w), 527 (w), 473 (w), 422 (w), 402 (vw).

HRMS (FAB-MS, 3-NBA, $C_{60}H_{39}N_8O_2$) calc. 903.3197 [M+H]⁺, found 903.3196 [M+H]⁺.

5,5'-(-2,4,5,6-Tetra(9H-carbazol-9-yl)-1,3-phenylene)bis(2-(tert-butyl)-1,3,4-oxadiazole)[180]

According to the general procedure (GP 3), 9,9',9'',9'''-(-4,6-di(1H-tetrazol-5-yl)benzene-1,2,3,5-tetrayl)tetrakis(9H-carbazole) (263 mg, 0.300 mmol, 1.00 equiv.) and pivaloyl chloride (109 mg, 0.900 mmol, 3.00 equiv.) were reacted in trichloromethane (15 mL). The reaction mixture was extracted and purified by column chromatography on silica gel (dichloromethane to dichloromethane/ethyl acetate 98:2 to 97:3) to yield 110 mg (0.111 mmol, 37%) of the title compound as a yellow solid.

R_f (dichloromethane/ethyl acetate 97:3) = 0.51

M.p. = 380 °C

¹H NMR (400 MHz, Chloroform-d) δ 7.99 – 7.92 (m, 2H), 7.68 – 7.58 (m, 4H), 7.38 (d, J = 7.7 Hz, 2H), 7.35 – 7.31 (m, 2H), 7.29 (d, J = 8.0 Hz, 2H), 7.20 (ddd, J = 8.0, 7.0, 1.1 Hz, 2H), 7.15 (dd, J = 6.4, 2.8 Hz, 4H), 7.11 (d, J = 8.2 Hz, 2H), 7.00 – 6.91 (m, 8H), 6.83 – 6.77 (m, 2H), 6.69 (ddd, J = 8.4, 7.2, 1.3 Hz, 2H), 0.54 (s, 18H).

¹³C NMR (101 MHz, CDCl₃) δ 173.3, 157.4, 141.3, 140.6, 139.2, 138.4, 128.8, 126.6, 125.6, 124.8, 124.1, 123.9, 120.9, 120.8, 120.5, 120.5, 120.2, 119.7, 110.2, 109.6, 109.0, 31.7, 27.1.

IR (ATR) ṽ [cm⁻¹] = 2968 (vw), 2659 (vw), 2363 (vw), 2197 (vw), 1702 (vw), 1597 (vw), 1452 (w), 1333 (w), 1223 (w), 1027 (vw), 927 (vw), 743 (w), 548 (vw), 424 (vw).

HRMS (ESI, C₆₆H₅₀N₈O₂) calc. 987.4129 [M+H]⁺, found 987.4121 [M+H]⁺.

2,4,5,6-Tetra(carbazol-9-yl)-1,3-phenylene)bis(2-phenyl-1,3,4-oxadiazole)[180]

According to the general procedure (GP 3), 9,9',9'',9'''-(-4,6-di(1H-tetrazol-5-yl)benzene-1,2,3,5-tetrayl)tetrakis(9H-carbazole) (200 mg, 0.229 mmol, 1.00 equiv.) and benzoyl chloride (129 mg, 0.915 mmol, 4.00 equiv.) were reacted in trichloromethane (15 mL). The reaction mixture was extracted and purified by column chromatography on silica gel (dichloromethane to dichloromethane/ethyl acetate 9:1) to yield 133 mg (0.129 mmol, 67%) of the title compound as an off-white solid.

R_f (dichloromethane/ethyl acetate 9:1) = 0.72

1H NMR (400 MHz, Methylene Chloride-d_2) δ 8.03 (d, J = 7.8 Hz, 2H), 7.70 (dd, J = 6.8, 2.0 Hz, 4H), 7.54 – 7.48 (m, 2H), 7.45 (d, J = 3.7 Hz, 4H), 7.30 (dtd, J = 15.4, 8.1, 4.7 Hz, 10H), 7.14 (t, J = 7.8 Hz, 4H), 7.05 (pd, J = 7.3, 1.4 Hz, 8H), 6.90 (t, J = 7.4 Hz, 2H), 6.87 – 6.81 (m, 2H), 6.82 – 6.75 (m, 4H).

13C NMR (101 MHz, CD$_2$Cl$_2$) δ 165.0, 157.8, 142.0, 141.8, 140.0, 139.5, 139.0, 138.6, 132.3, 129.3, 129.1, 127.2, 126.9, 126.3, 125.5, 124.4, 124.2, 124.2, 122.6, 121.5, 121.4, 121.2, 121.2, 120.7, 120.2, 110.9, 110.1, 109.4.

IR (ATR) $\tilde{v}$ [cm^{-1}] = 3046 (vw), 1597 (w), 1548 (w), 1479 (w), 1448 (m), 1333 (w), 1309 (m), 1285 (w), 1220 (m), 1150 (w), 1120 (w), 1088 (w), 1017 (w), 923 (w), 791 (vw), 742 (s), 721 (m), 707 (m), 686 (m), 617 (w), 582 (w), 549 (w), 528 (w), 505 (w), 422 (w).

HRMS (FAB-MS, 3-NBA, C$_{70}$H$_{43}$N$_8$O$_2$) calc. 1027.3509 [M+H]$^+$, found 1027.3511 [M+H]$^+$.

2-(Perfluoroheptyl)-5-(2,3,4,6-tetra(carbazol-9-yl)-5-(5-(perfluoroheptyl)-1,3,4-oxadiazol-2-yl)phenyl)-1,3,4-oxadiazole[180]

According to the general procedure (GP 3), 9,9',9'',9'''-(-4,6-di(1H-tetrazol-5-yl)benzene-1,2,3,5-tetrayl)tetrakis(9H-carbazole) (200 mg, 0.229 mmol, 1.00 equiv.) and 2,2,3,3,4,4,5,5,6,6,7,7,8,8,8-pentadecafluorooctanoyl chloride (297 mg, 0.686 mmol, 3.00 equiv.) were reacted in trichloromethane (15 mL). The reaction mixture was extracted and purified by column chromatography on silica gel (dichloromethane) to yield 127 mg (0.0787 mmol, 34%) of the title compound as a yellow solid.

R_f (dichloromethane) = 0.41

M.p. = 222 °C

^{1}H NMR (400 MHz, Chloroform-d) δ 7.98 (d, J = 7.7 Hz, 1H), 7.69 – 7.60 (m, 4H), 7.40 (d, J = 7.7 Hz, 2H), 7.37 – 7.27 (m, 2H), 7.25 – 7.17 (m, 4H), 7.14 – 7.08 (m, 6H), 7.04 – 6.94 (m, 8H), 6.84 (t, J = 7.4 Hz, 2H), 6.71 (ddd, J = 8.4, 7.2, 1.2 Hz, 2H).

^{13}C NMR (101 MHz, CDCl$_3$) δ 160.3, 155.1, 143.1, 140.2, 138.8, 138.1, 137.6, 126.8, 126.3, 125.9, 125.0, 124.3, 124.2, 124.1, 121.6, 121.5, 121.0, 120.9, 120.4, 119.9, 110.0, 109.2, 108.1, 29.9.

^{19}F NMR (376 MHz, CDCl$_3$) δ -85.1, -118.0, -126.3, -126.5, -127.1, -130.5.

IR (ATR) $\tilde{v}$ [cm^{-1}] = 3054 (vw), 2930 (vw), 1599 (w), 1491 (w), 1480 (w), 1449 (m), 1334 (w), 1311 (w), 1199 (m), 1145 (m), 1083 (w), 1028 (w), 1004 (w), 957 (w), 791 (vw), 742 (s), 721 (m), 700 (w), 670 (w), 643 (w), 618 (w), 553 (w), 528 (w), 423 (w).

HRMS (APCI, C$_{72}$H$_{32}$F$_{30}$N$_8$O$_2$) calc. 1611.2242 [M+H]$^+$; found 1611.2226 [M+H]$^+$.

2-Methyl-5-(-2,3,4,6-tetra(9*H*-carbazol-9-yl)-5-(2*H*-tetrazol-5-yl)phenyl)-1,3,4-oxadiazole

In a vial, tetra(carbazol-9-yl)-5-(5-methyl-1,3,4-oxadiazol-2-yl)benzonitrile (500 mg, 0.592 mmol, 1.00 equiv.), sodium azide (154 mg, 2.37 mmol, 4.00 equiv.) and copper sulfate (19.0 mg, 0.118 mmol, 0.200 equiv.) were sealed and flushed with argon three times. Subsequently, dimethyl sulfoxide (15 mL) was added. The mixture was stirred at 140 °C for 16 h. Afterwards the solution was added to water, carefully acidified with HCl and extracted with dichloromethane (2 × 100 mL). The combined organic layers were washed with brine, dried over sodium sulfate and concentrated under reduced pressure. The crude product was purified by column chromatography over SiO_2 (dichloromethane to dichloromethane/ethyl acetate 98:2 to 95:5 to 1:1) to yield 513 mg of the title compound (0.580 mmol, 98%) as an off-white solid.

R_f (dichloromethane/ethyl acetate 1:1) = 0.68

¹H NMR (400 MHz, DMSO-d_6) δ 8.33 (s, 1H), 8.05 (d, J = 7.6 Hz, 2H), 7.76 (dd, J = 7.1, 1.6 Hz, 2H), 7.70 (d, J = 7.9 Hz, 2H), 7.66 (dd, J = 7.4, 1.5 Hz, 2H), 7.60 (dd, J = 12.0, 8.1 Hz, 6H), 7.45 (dd, J = 7.2, 1.6 Hz, 2H), 7.32 (ddd, J = 8.3, 7.1, 1.2 Hz, 2H), 7.17 (t, J = 7.5 Hz, 2H), 7.05 – 6.87 (m, 8H), 6.80 – 6.69 (m, 4H), 1.42 (s, 3H).

¹³C NMR (101 MHz, DMSO) δ 162.8, 140.9, 140.7, 139.8, 139.3, 138.7, 138.6, 137.4, 125.7, 125.0, 124.6, 124.0, 122.8, 122.6, 122.5, 122.5, 120.2, 120.1, 119.9, 119.8, 119.6, 119.4, 119.2, 111.6, 111.2, 110.8, 110.3, 79.2, 9.1.

IR (ATR) $\tilde{v}$ [cm⁻¹] = 3048 (vw), 2917 (vw), 2782 (vw), 1578 (w), 1449 (m), 1311 (w), 1225 (m), 1026 (w), 926 (vw), 844 (vw), 741 (m), 549 (w), 422 (w).

HRMS (FAB-MS, 3-NBA, $C_{58}H_{37}N_{10}O$) calc. 889.3152, found 889.3151.

2-Methyl-5-(-2,3,4,6-tetra(9H-carbazol-9-yl)-5-(2-hexyl-2H-tetrazol-5-yl)phenyl)-1,3,4-oxadiazole

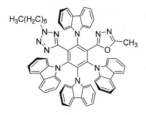

According to the general procedure (GP 1), 2-methyl-5-(-2,3,4,6-tetra(9H-carbazol-9-yl)-5-(2H-tetrazol-5-yl)phenyl)-1,3,4-oxadiazole (250 mg, 0.281 mmol, 1.00 equiv.), 1-bromohexane (93.0 mg, 0.562 mmol, 2.00 equiv.) and sodium carbonate (59.6 mg, 0.562 mmol, 2.00 equiv.) were reacted in N,N-dimethylformamide (7 mL). The reaction mixture was extracted and purified over SiO$_2$ (dichloromethane/cyclohexane 1:1 to 1:0 to dichloromethane/ethyl acetate 98:2) to yield 223 mg of the title compound (0.228 mmol, 81%) as an off-white solid.

R$_f$ (dichloromethane) = 0.29

^{1}H NMR (400 MHz, Chloroform-d) δ 7.98 (d, J = 7.7 Hz, 2H), 7.65 (dd, J = 6.3, 2.6 Hz, 2H), 7.61 – 7.52 (m, 2H), 7.38 – 7.23 (m, 8H), 7.23 – 7.09 (m, 8H), 7.02 – 6.96 (m, 3H), 6.96 – 6.92 (m, 3H), 6.79 (t, J = 7.4 Hz, 2H), 6.74 – 6.66 (m, 2H), 3.77 (t, J = 7.1 Hz, 2H), 1.58 (s, 3H), 1.07 (p, J = 7.4 Hz, 2H), 1.00 – 0.87 (m, 4H), 0.82 (d, J = 7.3 Hz, 3H), 0.52 (p, J = 7.7, 7.2 Hz, 2H).

^{13}C NMR (101 MHz, CDCl$_3$) δ 163.6, 158.3, 158.0, 141.1, 140.8, 140.3, 139.5, 139.1, 139.1, 138.5, 137.0, 132.1, 128.2, 126.4, 126.3, 125.5, 125.3, 124.6, 123.9, 123.7, 123.7, 123.6, 121.4, 121.0, 120.8, 120.6, 120.5, 120.4, 120.3, 120.2, 120.2, 120.0, 119.7, 119.5, 110.3, 110.3, 110.2, 110.1, 109.7, 109.7, 109.6, 109.3, 52.6, 30.8, 29.8, 28.5, 25.2, 22.2, 14.1, 9.9.

IR (ATR) $\tilde{v}$ [cm^{-1}] = 3965 (w), 3047 (w), 2923 (w), 2676 (w), 2321 (w), 2109 (w), 1888 (w), 1575 (w), 1446 (m), 1309 (m), 1223 (m), 1027 (w), 924 (w), 844 (w), 742 (m), 617 (w), 547 (m), 422 (m).

HRMS (FAB-MS, 3-NBA, C$_{64}$H$_{49}$N$_{10}$O) calc. 973.4091, found 973.4089.

5.1.3.2   Phthalimide-based TADF Emitter Project

General Procedure for the Synthesis of Phthalimides (GP 5)[238,239]

In a sealable vial or a round bottom flask equipped with a reflux condenser, 6-fluoro-phthalic anhydride (1.00 equiv.) was suspended in glacial acetic acid (1 mL/mmol anhydride). The corresponding amine (1.00 equiv.) was added slowly and the reaction mixture is stirred at 100 °C until completion (usually 3–4 h, monitored with GC/MS). After cooling to room temperature, the reaction mixture was neutralized by addition of saturated sodium carbonate solution. The reaction mixture was extracted with ethyl acetate or dichloromethane three times. The combined organic layers were washed two times with saturated sodium carbonate solution and brine. The organic layer was dried over sodium sulfate, reduced in vacuum and purified by silica column chromatography or used without purification.

General Procedure for the Nucleophilic Aromatic Substitution (GP 6)[238,239]

In a sealable vial, a 6-fluorophthalimide derivative (1.00 equiv.), a carbazole derivative (1.00 equiv.) and tripotassium phosphate (2.00 equiv.) were evacuated and backfilled with argon. Dry DMSO (4 ml/mmol carbazole) was added and the resulting mixture heated at 110 °C until completion (usually 4–16 h, monitored with TLC). After cooling to room temperature, the reaction mixture was poured into an excess of water and extracted with dichloromethane three times. The combined organic layers were washed with brine, dried over sodium sulfate, reduced in vacuum and purified by silica column chromatography or recrystallization.

2-Decyl-4-fluoroisoindoline-1,3-dione

According to the general procedure (GP 5), 6-fluorophthalic anhydride (4.98 g, 30.0 mmol, 1.00 equiv.) and decan-1-amine (4.72 g, 30.0 mmol, 1.00 equiv.) were reacted in glacial acetic acid (30 mL). The reaction mixture was extracted and purified by column chromatography over SiO_2 (dichloromethane) to yield 8.20 g of the title compound (26.7 mmol, 89%) as a colorless viscous oil.

^{1}H NMR (400 MHz, DMSO-d_6) δ 7.87 (ddd, J = 8.4, 7.3, 4.5 Hz, 1H), 7.71 (d, J = 7.3 Hz, 1H), 7.65 (t, J = 8.9 Hz, 1H), 3.52 (t, J = 7.1 Hz, 2H), 1.56 (p, J = 7.0 Hz, 2H), 1.23 (d, J = 15.1 Hz, 14H), 0.87 – 0.77 (m, 3H).

^{13}C NMR (101 MHz, DMSO-d_6) δ 166.9 (d, J = 2.9 Hz), 164.8 , 156.6 (d, J = 261.3 Hz), 137.4 (d, J = 7.8 Hz), 134.0 (d, J = 1.7 Hz), 122.4 (d, J = 19.9 Hz), 119.5 (d, J = 3.3 Hz), 117.4 (d, J = 12.7 Hz), 37.5 , 31.3 , 28.9 , 28.8 , 28.7 , 28.5 , 27.7 , 26.2 , 22.1 , 13.9 .

^{19}F NMR (376 MHz, DMSO) δ -119.8.

IR (ATR) $\tilde{v}$ [cm^{-1}] =2923 (m), 2854 (w), 1713 (s), 1611 (w), 1481 (m), 1395 (m), 1259 (m), 1174 (w), 1037 (w), 908 (w), 822 (w), 747 (m), 639 (w), 561 (w), 399 (vw).

HRMS (APCI, $C_{18}H_{24}FNO_2$) calc. 306.1864 [M+H]$^+$, found 306.1858 [M+H]$^+$.

4-(4-Fluoro-1,3-dioxoisoindolin-2-yl)benzonitrile

According to the general procedure (GP 5), 6-fluorophthalic anhydride (1.99 g, 12.0 mmol, 1.00 equiv.) and 4-aminobenzonitrile (1.42 g, 12.0 mmol, 1.00 equiv.) were reacted in glacial acetic acid (15 mL). The reaction mixture was extracted and purified by column chromatography over SiO_2 (dichloromethane) to yield 2.67 g of the title compound (10.0 mmol, 84%) as a colorless solid.

^{1}H NMR (400 MHz, Chloroform-d) δ 7.88 – 7.75 (m, 4H), 7.69 – 7.62 (m, 2H), 7.50 (ddd, J = 8.9, 7.9, 1.4 Hz, 1H).

^{13}C NMR (101 MHz, CDCl$_3$) δ 165.5, 165.4, 163.2, 159.6, 156.9, 150.4, 137.7, 137.6, 135.6, 134.0, 133.6, 133.1, 126.7, 123.5, 123.3, 120.5, 120.5, 118.3, 117.5, 117.3, 114.6, 111.8.

IR (ATR) $\tilde{v}$ [cm^{-1}] = 3989 (w), 3920 (w), 3691 (w), 3474 (w), 3367 (w), 3215 (w), 3109 (w), 2725 (w), 2445 (w), 2226 (w), 1730 (s), 1603 (m), 1479 (m), 1370 (m), 1259 (m), 1084 (m), 970 (m), 817 (m), 739 (m), 630 (m), 548 (m), 413 (m).

HRMS (APCI, $C_{15}H_8FN_2O_2$) calc. 267.0564 [M+H]$^+$, found 267.0560 [M+H]$^+$.

4-Fluoro-2-mesitylisoindoline-1,3-dione

According to the general procedure (GP 5), 6-fluorophthalic anhydride (9.97 g, 60.0 mmol, 1.00 equiv.) and 2,4,6-trimethylaniline (8.11 g, 60.0 mmol, 1.00 equiv.) were reacted in glacial acetic acid (60 mL). The reaction mixture was extracted and purified by column chromatography over SiO2 (dichloromethane) to yield 15.4 g of the title compound (54.3 mmol, 90%) as a colorless solid.

M.p. = 155 °C

^{1}H NMR (400 MHz, Chloroform-d) δ 7.83 – 7.75 (m, 2H), 7.46 (ddd, J = 9.0, 7.0, 2.1 Hz, 1H), 7.01 (s, 2H), 2.33 (s, 3H), 2.13 (s, 6H).

^{13}C NMR (101 MHz, Chloroform-d) δ 166.3 (d, J = 3.0 Hz), 164.1 (d, J = 1.6 Hz), 157.9 (d, J = 266.3 Hz), 139.7 , 137.0 (d, J = 7.6 Hz), 136.5 , 134.3 (d, J = 1.5 Hz), 129.5 , 126.8 , 122.8 (d, J = 19.8 Hz), 120.1 (d, J = 3.8 Hz), 117.9 (d, J = 12.4 Hz), 21.3 , 18.1 .

^{19}F NMR (376 MHz, CDCl3) δ -116.4.

IR (ATR) $\tilde{v}$ [cm^{-1}] = 2920 (vw), 1718 (m), 1609 (w), 1480 (m), 1375 (m), 1249 (w), 1111 (m), 974 (m), 749 (m), 636 (w), 573 (w), 411 (w).

HRMS (APCI, C$_{17}$H$_{14}$FN$_2$O$_2$) calc. 284.1081 [M+H]$^+$, found 284.1077 [M+H]$^+$.

2-Allyl-4-fluoroisoindoline-1,3-dione

According to the general procedure (GP 5), 6-fluorophthalic anhydride (1.66 g, 10.0 mmol, 1.00 equiv.) and prop-2-en-1-amine (571 mg, 10.0 mmol, 1.00 equiv.) were reacted in glacial acetic acid (10 mL). The reaction mixture was extracted and purified by column chromatography over SiO2 (dichloromethane) to yield 1.82 g of the title compound (8.93 mmol, 89%) as a colorless solid.

^{1}H NMR (400 MHz, Chloroform-d) δ 7.77 – 7.64 (m, 2H), 7.38 (td, J = 8.4, 1.0 Hz, 1H), 5.87 (ddt, J = 17.2, 10.2, 5.7 Hz, 1H), 5.32 – 5.12 (m, 2H), 4.28 (dt, J = 5.8, 1.5 Hz, 2H).

^{13}C NMR (101 MHz, Chloroform-d) δ 166.9 (d, J = 3.0 Hz), 164.7 , 157.7 (d, J = 265.9 Hz), 136.7 (d, J = 7.5 Hz), 134.4 (d, J = 1.6 Hz), 131.3 , 122.5 (d, J = 19.8 Hz), 119.7 (d, J = 3.7 Hz), 118.3 , 118.0 (d, J = 12.5 Hz), 40.3 .

^{19}F NMR (376 MHz, CDCl3) δ -117.1.

IR (ATR) $\tilde{v}$ [cm^{-1}] = 3469 (w), 3085 (w), 1702 (s), 1477 (s), 1390 (s), 1319 (m), 1242 (s), 1113 (s), 1008 (s), 948 (s), 820 (m), 748 (s), 641 (m), 568 (s), 380 (m).

HRMS (APCI, C$_{11}$H$_8$FNO$_2$) calc. 206.0612 [M+H]$^+$, found 206.0610 [M+H]$^+$.

4-Fluoro-2-(prop-2-yn-1-yl)isoindoline-1,3-dione

 According to the general procedure (GP 5), 6-fluorophthalic anhydride (1.66 g, 10.0 mmol, 1.00 equiv.) and prop-2-yn-1-amine (551 mg, 10.0 mmol, 1.00 equiv.) were reacted in glacial acetic acid (10 mL). The reaction mixture was extracted and purified by column chromatography over SiO_2 (dichloromethane) to yield 1.83 g of the title compound (9.01 mmol, 90%) as a colorless solid.

^{1}H NMR (400 MHz, Chloroform-d) δ 7.78 – 7.65 (m, 2H), 7.40 (ddd, J = 9.0, 8.1, 1.1 Hz, 1H), 4.44 (d, J = 2.5 Hz, 2H), 2.23 (t, J = 2.5 Hz, 1H).

^{13}C NMR (101 MHz, Chloroform-d) δ 165.9 (d, J = 3.0 Hz), 163.7 , 157.8 (d, J = 266.5 Hz), 137.0 (d, J = 7.6 Hz), 134.2 (d, J = 1.6 Hz), 122.8 (d, J = 19.7 Hz), 119.9 (d, J = 3.8 Hz), 117.9 (d, J = 12.5 Hz), 77.0 , 71.9 , 27.3.

^{19}F NMR (376 MHz, CDCl₃) δ -116.4.

IR (ATR) $\tilde{v}$ [cm^{-1}] = 3255 (m), 2969 (vw), 1708 (s), 1610 (m), 1482 (m), 1346 (m), 1254 (m), 1119 (m), 1024 (m), 952 (m), 822 (m), 746 (s), 628 (m), 563 (m), 420 (w).

HRMS (APCI, $C_{11}H_6FNO_2$) calc. 204.0455 [M+H]$^+$, found 204.0455 [M+H]$^+$.

2-(*Sec*-butyl)-4-fluoroisoindoline-1,3-dione

According to the general procedure (GP 5), 6-fluorophthalic anhydride (3.98 g, 24.0 mmol, 1.00 equiv.) and butan-2-amine (1.76 g, 24.0 mmol, 1.00 equiv.) were reacted in glacial acetic acid (24 mL). The reaction mixture was extracted and purified by column chromatography over SiO_2 (dichloromethane) to yield 3.60 g of the title compound (16.3 mmol, 68%) as a colorless oil.

^{1}H NMR (400 MHz, Chloroform-d) δ 7.69 (ddd, J = 8.2, 7.3, 4.3 Hz, 1H), 7.63 (dt, J = 7.3, 0.7 Hz, 1H), 7.35 (td, J = 8.5, 0.9 Hz, 1H), 4.32 – 4.15 (m, 1H), 2.02 (ddt, J = 13.8, 9.3, 7.4 Hz, 1H), 1.78 (dqd, J = 13.6, 7.4, 6.1 Hz, 1H), 1.46 (d, J = 6.9 Hz, 3H), 0.87 (t, J = 7.4 Hz, 3H).

^{13}C NMR (101 MHz, Chloroform-d) δ 167.5 (d, J = 2.9 Hz), 165.4 (d, J = 1.6 Hz), 157.6 (d, J = 265.5 Hz), 136.4 (d, J = 7.4 Hz), 134.4 (d, J = 1.7 Hz), 122.3 (d, J = 19.9 Hz), 119.4 (d, J = 3.7 Hz), 117.8 (d, J = 12.4 Hz), 49.5 , 26.8 , 18.4 , 11.4 .

^{19}F NMR (376 MHz, CDCl₃) δ -118.2.

IR (ATR) $\tilde{v}$ [cm^{-1}] = 2968 (w), 1707 (s), 1611 (m), 1480 (m), 1339 (s), 1249 (m), 1185 (m), 1056 (m), 986 (m), 885 (w), 822 (m), 748 (m), 637 (w), 574 (w), 402 (w).

HRMS (APCI, $C_{12}H_{12}FNO_2$) calc. 222.0925 [M+H]$^+$, found 222.0913 [M+H]$^+$.

2-(*Tert*-butyl)-4-fluoroisoindoline-1,3-dione

According to the general procedure (GP 5), 6-fluorophthalic anhydride (3.99 g, 24.0 mmol, 1.00 equiv.) and 2-methylpropan-2-amine (7.02 g, 96.0 mmol, 4.00 equiv.) were reacted in glacial acetic acid (24 mL). The reaction mixture was extracted and purified by column chromatography over SiO_2 (dichloromethane) to yield 3.21 g of the title compound (14.5 mmol, 60%) as a colorless oil.

M.p. = 87 °C

¹H NMR (400 MHz, Chloroform-*d*) δ 7.66 (ddd, *J* = 8.2, 7.3, 4.3 Hz, 1H), 7.58 (dt, *J* = 7.4, 0.7 Hz, 1H), 7.32 (td, *J* = 8.5, 0.8 Hz, 1H), 1.68 (s, 9H).

¹³C NMR (101 MHz, Chloroform-*d*) δ 168.5 (d, *J* = 2.9 Hz), 166.5 (d, *J* = 1.7 Hz), 157.5 (d, *J* = 265.4 Hz), 136.2 (d, *J* = 7.4 Hz), 134.7 , 122.0 (d, *J* = 19.9 Hz), 118.9 (d, *J* = 3.8 Hz), 117.9 (d, *J* = 11.8 Hz), 58.4 , 29.1.

¹⁹F NMR (376 MHz, CDCl₃) δ -119.5.

IR (ATR) ṽ [cm⁻¹] = 2988 (vw), 2677 (vw), 2333 (vw), 1696 (m), 1480 (w), 1311 (m), 1241 (m), 1103 (m), 988 (m), 823 (m), 744 (m), 666 (w), 569 (w), 403 (w).

HRMS (APCI, C₁₂H₁₂FNO₂) calc. 222.0925 [M+H]⁺, found 222.0913 [M+H]⁺.

2-Butyl-4-fluoroisoindoline-1,3-dione

According to the general procedure (GP 5), 6-fluorophthalic anhydride (4.93 g, 30.0 mmol, 1.00 equiv.) and butan-1-amine (2.19 g, 30.0 mmol, 1.00 equiv.) were reacted in glacial acetic acid (30 mL). The reaction mixture was extracted and purified by column chromatography over SiO_2 (dichloromethane) to yield 5.36 g of the title compound (24.2 mmol, 81%) as a colorless oil.

¹H NMR (400 MHz, Chloroform-*d*) δ 7.81 – 7.54 (m, 2H), 7.35 (td, *J* = 8.5, 1.1 Hz, 1H), 3.66 (td, *J* = 7.2, 1.3 Hz, 2H), 1.79 – 1.58 (m, 2H), 1.48 – 1.24 (m, 2H), 1.00 – 0.85 (m, 3H).

¹³C NMR (101 MHz, Chloroform-*d*) δ 167.4 (d, *J* = 3.0 Hz), 165.2 , 157.6 (d, *J* = 265.5 Hz), 136.5 (d, *J* = 7.6 Hz), 134.5 , 122.4 (d, *J* = 19.8 Hz), 119.5 (d, *J* = 3.8 Hz), 118.0 (d, *J* = 12.4 Hz), 38.1 , 30.6 , 20.1 , 13.7 .

¹⁹F NMR (376 MHz, CDCl₃) δ -117.6.

IR (ATR) ṽ [cm⁻¹] = 2959 (w), 1710 (s), 1610 (m), 1480 (m), 1394 (s), 1257 (m), 1173 (m), 1055 (m), 941 (m), 822 (m), 746 (m), 638 (w), 558 (w), 398 (w).

HRMS (APCI, C₁₂H₁₂FNO₂) calc. 222.0925 [M+H]⁺, found 222.0914 [M+H]⁺.

2-(4-(Dimethylamino)phenyl)-4-fluoroisoindoline-1,3-dione

According to the general procedure (GP 5), 6-fluorophthalic anhydride (1.66 g, 10.0 mmol, 1.00 equiv.) and *N,N*-dimethylbenzene-1,4-diamine (1.36 g, 10.0 mmol, 1.00 equiv.) were reacted in glacial acetic acid (10 mL). The reaction mixture was extracted and purified by column chromatography over SiO_2 (dichloromethane) to yield 2.32 g of the title compound (8.16 mmol, 82%) as an orange, crystalline solid.

^{1}H NMR (400 MHz, Chloroform-*d*) δ 7.80 – 7.67 (m, 2H), 7.41 (ddd, *J* = 9.0, 6.1, 2.9 Hz, 1H), 7.25 – 7.18 (m, 2H), 6.83 – 6.73 (m, 2H), 3.00 (s, 6H).

^{13}C NMR (101 MHz, Chloroform-*d*) δ 166.9 (d, *J* = 3.0 Hz), 164.6 (d, *J* = 1.7 Hz), 157.8 (d, *J* = 266.0 Hz), 150.4 , 136.8 (d, *J* = 7.5 Hz), 134.3 (d, *J* = 1.5 Hz), 127.6 , 122.6 (d, *J* = 19.8 Hz), 119.8 (d, *J* = 3.7 Hz), 119.7 , 117.8 (d, *J* = 12.4 Hz), 112.5 , 40.6 .

^{19}F NMR (376 MHz, CDCl₃) δ -117.1.

IR (ATR) ṽ [cm^{-1}] = 2895 (w), 2821 (vw), 2480 (vw), 2313 (vw), 2228 (vw), 2073 (vw), 1710 (m), 1608 (m), 1479 (m), 1352 (m), 1254 (m), 1098 (m), 968 (m), 806 (m), 739 (m), 553 (m), 403 (w).

HRMS (APCI, C₁₆H₁₃FN₂O₂) calc. 285.1034 [M+H]⁺, found 285.1019 [M+H]⁺.

2-(3,5-Bis(trifluoromethyl)phenyl)-4-fluoroisoindoline-1,3-dione

According to the general procedure (GP 5), 6-fluorophthalic anhydride (1.66 g, 10.0 mmol, 1.00 equiv.) and 3,5-bis(trifluoromethyl)aniline (2.29 g, 10.0 mmol, 1.00 equiv.) were reacted in glacial acetic acid (10 mL). The reaction mixture was extracted and purified by column chromatography over SiO_2 (dichloromethane) to yield 2.30 g of the title compound (6.10 mmol, 61%) as a colorless solid.

^{1}H NMR (400 MHz, Chloroform-*d*) δ 8.02 (d, *J* = 1.5 Hz, 2H), 7.91 (s, 1H), 7.90 – 7.79 (m, 2H), 7.51 (td, *J* = 8.3, 1.2 Hz, 1H).

^{13}C NMR (101 MHz, CDCl₃) δ 165.3, 165.3, 163.1, 163.1, 159.6, 157.0, 137.9, 137.8, 133.4, 133.4, 133.3, 133.1, 132.9, 132.6, 132.3, 126.4, 126.4, 126.4, 124.3, 123.6, 123.4, 121.9, 121.8, 121.8, 121.6, 120.6, 120.6, 117.3, 117.2.

^{19}F NMR (376 MHz, CDCl₃) δ -67.3, -115.5.

IR (ATR) ṽ [cm^{-1}] = 1713 (s), 1612 (m), 1469 (m), 1396 (s), 1277 (s), 1105 (vs), 986 (s), 860 (s), 740 (s), 560 (m), 395 (m).

HRMS (APCI, C₁₆H₆F₇NO₂) calc. 378.0360 [M+H]⁺, found 378.0338 [M+H]⁺.

2-(3,5-Dimethoxyphenyl)-4-fluoroisoindoline-1,3-dione

According to the general procedure (GP 5), 6-fluorophthalic anhydride (1.66 g, 10.0 mmol, 1.00 equiv.) and 3,5-dimethoxyaniline (1.53 g, 10.0 mmol, 1.00 equiv.) were reacted in glacial acetic acid (10 mL). The reaction mixture was extracted and purified by column chromatography over SiO_2 (dichloromethane) to yield 1.55 g of the title compound (5.14 mmol, 51%) as a colorless solid.

1H NMR (400 MHz, Chloroform-*d*) δ 7.79 (dd, *J* = 7.1, 3.4 Hz, 2H), 7.45 (ddd, *J* = 9.1, 6.9, 2.2 Hz, 1H), 6.57 (d, *J* = 2.2 Hz, 2H), 6.51 (t, *J* = 2.3 Hz, 1H), 3.81 (s, 6H).

13C NMR (101 MHz, Chloroform-*d*) δ 166.1 (d, *J* = 3.0 Hz), 163.9, 161.1, 158.0 (d, *J* = 266.6 Hz), 137.1 (d, *J* = 7.5 Hz), 134.0, 132.8, 122.9 (d, *J* = 19.7 Hz), 120.1 (d, *J* = 3.8 Hz), 117.7 (d, *J* = 12.2 Hz), 105.2, 101.0, 55.7.

19F NMR (376 MHz, CDCl₃) δ -116.7.

IR (ATR) $\tilde{v}$ [cm⁻¹] = 3081 (vw), 3011 (vw), 2942 (vw), 2845 (vw), 2724 (vw), 2613 (vw), 2440 (vw), 2325 (vw), 2108 (vw), 1717 (m), 1599 (m), 1473 (m), 1281 (m), 1156 (m), 911 (m), 820 (m), 742 (m), 561 (m), 409 (w).

HRMS (APCI, $C_{16}H_{12}FNO_4$) calc. 302.0823 [M+H]⁺, found 302.0807 [M+H]⁺.

2-(3,5-Di-*tert*-butylphenyl)-4-fluoroisoindoline-1,3-dione

According to the general procedure (GP 5), 6-fluorophthalic anhydride (831 mg, 5.00 mmol, 1.00 equiv.) and 3,5-di-*tert*-butylaniline (1.03 g, 5.00 mmol, 1.00 equiv.) were reacted in glacial acetic acid (5 mL). The reaction mixture was extracted and purified by column chromatography over SiO_2 (dichloromethane) to yield 1.74 g of the title compound (4.92 mmol, 98%) as a colorless solid.

1H NMR (400 MHz, Chloroform-*d*) δ 7.77 (qd, *J* = 4.1, 3.7, 2.3 Hz, 2H), 7.48 (t, *J* = 1.7 Hz, 1H), 7.44 (ddd, *J* = 9.0, 6.1, 3.1 Hz, 1H), 7.21 (d, *J* = 1.8 Hz, 2H), 1.35 (s, 18H).

13C NMR (101 MHz, Chloroform-*d*) δ 166.6 (d, *J* = 2.9 Hz), 164.4 (d, *J* = 1.7 Hz), 158.0 (d, *J* = 266.2 Hz), 151.9, 136.9 (d, *J* = 7.5 Hz), 134.2 (d, *J* = 1.4 Hz), 130.7, 122.7 (d, *J* = 19.7 Hz), 122.7, 121.2, 119.9 (d, *J* = 3.8 Hz), 117.8 (d, *J* = 12.3 Hz), 35.1, 31.5.

19F NMR (376 MHz, CDCl₃) δ -117.1.

IR (ATR) $\tilde{v}$ [cm⁻¹] = 2963 (w), 1715 (s), 1595 (w), 1373 (m), 1251 (m), 1097 (m), 992 (w), 903 (w), 744 (m), 669 (w), 564 (w), 407 (w).

HRMS (APCI, $C_{22}H_{24}FNO_2$) calc. 354.1864 [M+H]⁺, found 354.1846 [M+H]⁺.

4-Fluoro-2-(3,4,5-trimethoxyphenyl)isoindoline-1,3-dione

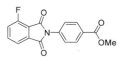 According to the general procedure (GP 5), 6-fluorophthalic anhydride (2.49 g, 15.0 mmol, 1.00 equiv.) and 3,4,5-trimethoxyaniline (2.75 g, 15.0 mmol, 1.00 equiv.) were reacted in glacial acetic acid (15 mL). The reaction mixture was extracted and purified by column chromatography over SiO_2 (dichloromethane) to yield 2.75 g of the title compound (8.30 mmol, 55%) as a colorless solid.

^{1}H NMR (400 MHz, Chloroform-d) δ 7.84 – 7.72 (m, 2H), 7.45 (ddd, J = 8.9, 7.6, 1.5 Hz, 1H), 6.63 (s, 2H), 3.88 (s, 3H), 3.86 (s, 6H).

^{13}C NMR (101 MHz, Chloroform-d) δ 166.3 (d, J = 3.0 Hz), 164.1 (d, J = 1.5 Hz), 158.0 (d, J = 266.7 Hz), 153.6 , 138.1 , 137.2 (d, J = 7.4 Hz), 133.9 , 126.8 , 122.9 (d, J = 19.6 Hz), 120.1 (d, J = 3.8 Hz), 117.6 (d, J = 12.2 Hz), 104.4 , 61.0 , 56.3.

^{19}F NMR (376 MHz, CDCl$_3$) δ -116.6.

IR (ATR) $\tilde{v}$ [cm^{-1}] = 3080 (vw), 2947 (vw), 2831 (vw), 1711 (m), 1598 (w), 1469 (w), 1404 (w), 1306 (w), 1238 (w), 1114 (m), 998 (w), 927 (w), 822 (m), 746 (w), 682 (w), 616 (w), 407 (w).

HRMS (APCI, $C_{17}H_{14}FNO_5$) calc. 332.0929 [M+H]$^+$, found 332.0914 [M+H]$^+$.

4-(4-Fluoro-1,3-dioxoisoindolin-2-yl)benzoate

According to the general procedure (GP 5), 6-fluorophthalic anhydride (1.66 g, 10.0 mmol, 1.00 equiv.) and methyl 4-aminobenzoate (1.51 g, 10.0 mmol, 1.00 equiv.) were reacted in glacial acetic acid (10 mL). The reaction mixture was extracted and purified by column chromatography over SiO_2 (dichloromethane) to yield 2.25 g of the title compound (7.52 mmol, 75%) as a colorless solid.

^{1}H NMR (400 MHz, Chloroform-d) δ 8.22 – 8.14 (m, 2H), 7.86 – 7.77 (m, 2H), 7.61 – 7.54 (m, 2H), 7.47 (ddd, J = 8.9, 7.2, 1.9 Hz, 1H), 3.95 (s, 3H).

^{13}C NMR (101 MHz, CDCl$_3$) δ 166.4, 165.8, 165.8, 163.6, 159.5, 156.8, 137.4, 137.3, 135.5, 133.8, 130.6, 129.7, 126.1, 123.2, 123.0, 120.3, 120.3, 117.6, 117.5, 52.5.

IR (ATR) $\tilde{v}$ [cm^{-1}] = 3081 (vw), 2957 (vw), 2848 (vw), 2725 (vw), 2594 (vw), 2458 (vw), 2324 (vw), 1715 (m), 1602 (w), 1477 (w), 1374 (m), 1279 (m), 1087 (m), 966 (m), 854 (m), 764 (m), 632 (w), 562 (m), 490 (w), 391 (w).

HRMS (ESI, $C_{16}H_{10}FNO_4$) calc. 300.0667 [M+H]$^+$, found 300.0663 [M+H]$^+$.

2-(3-Bromophenyl)-4-fluoroisoindoline-1,3-dione

 According to the general procedure (GP 5), 6-fluorophthalic anhydride (8.61 g, 51.8 mmol, 1.00 equiv.) and 3-bromoaniline (8.91 g, 51.8 mmol, 1.00 equiv.) were reacted in glacial acetic acid (52 mL). The reaction mixture was extracted and purified by column chromatography over SiO_2 (dichloromethane) to yield 15.2 g of the title compound (47.5 mmol, 92%) as a colorless solid.

^{1}H NMR (400 MHz, Chloroform-*d*) δ 7.84 – 7.75 (m, 2H), 7.63 (t, *J* = 1.9 Hz, 1H), 7.55 (dt, *J* = 6.9, 2.0 Hz, 1H), 7.47 (ddd, *J* = 8.9, 7.5, 1.7 Hz, 1H), 7.43 – 7.35 (m, 2H).

^{13}C NMR (101 MHz, CDCl$_3$) δ 165.8, 165.8, 163.6, 163.6, 159.5, 156.8, 137.4, 137.3, 133.8, 132.6, 131.5, 130.5, 129.7, 125.2, 123.2, 123.0, 122.6, 120.3, 120.3, 117.6, 117.5.

IR (ATR) $\tilde{v}$ [cm^{-1}] = 3983 (w), 3900 (w), 3814 (w), 3732 (w), 3627 (w), 3470 (w), 3076 (w), 2727 (w), 2453 (w), 2351 (w), 2105 (w), 1970 (w), 1716 (s), 1574 (m), 1476 (s), 1376 (m), 1258 (m), 1106 (m), 973 (m), 863 (m), 740 (s), 674 (m), 561 (m), 433 (m).

4-(9*H*-Carbazol-9-yl)-2-decylisoindoline-1,3-dione

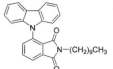 According to the general (GP 6), 2-decyl-4-fluoroisoindoline-1,3-dione (4.89 g, 16.0 mmol, 1.00 equiv.), 9*H*-carbazole (2.68 g, 16.0 mmol, 1.00 equiv.) and tripotassium phosphate (6.78 g, 32.0 mmol, 2.00 equiv.) were reacted in dry DMSO (64 mL). The reaction mixture was extracted and purified by column chromatography over SiO$_2$ (dichloromethane/cyclohexane 1:1) to yield 4.04 g of the title compound (8.93 mmol, 56%) as an off-white solid.

R$_f$ (dichloromethane/cyclohexane 1:1) = 0.48

M.p. = 110 °C

^{1}H NMR (400 MHz, DMSO-d_6) δ 8.27 – 8.22 (m, 2H), 8.07 – 8.01 (m, 2H), 7.96 (dd, *J* = 7.0, 1.9 Hz, 1H), 7.38 (ddd, *J* = 8.3, 7.1, 1.3 Hz, 2H), 7.30 (td, *J* = 7.4, 1.0 Hz, 2H), 7.24 (d, *J* = 8.1 Hz, 2H), 3.46 (t, *J* = 7.1 Hz, 2H), 1.50 (q, *J* = 7.0 Hz, 2H), 1.19 (s, 14H), 0.86 – 0.76 (m, 3H).

^{13}C NMR (101 MHz, DMSO) δ 167.3, 165.5, 140.3, 136.2, 134.3, 134.3, 132.9, 126.8, 126.0, 123.0, 122.7, 120.4, 120.4, 110.3, 37.5, 31.3, 28.9, 28.8, 28.6, 28.5, 27.7, 26.4, 26.2, 22.1, 14.0.

IR (ATR) $\tilde{v}$ [cm^{-1}] = 3065 (w), 2923 (m), 2852 (m), 1703 (s), 1604 (m), 1450 (m), 1393 (m), 1314 (m), 1228 (m), 1158 (m), 1031 (m), 927 (w), 826 (w), 749 (s), 551 (m), 423 (m).

EA (C$_{30}$H$_{32}$N$_2$O$_2$) calc. C: 79.61, H: 7.13, N: 6.19; found C: 79.60, H: 7.00, N: 6.18.

4-(3-Bromo-9*H*-carbazol-9-yl)-2-mesitylisoindoline-1,3-dione

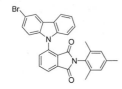

According to the general procedure (GP 6), 4-fluoro-2-mesitylisoindoline-1,3-dione (561 mg, 2.00 mmol, 1.00 equiv.), 3-bromo-9*H*-carbazole (493 mg, 2.00 mmol, 1.00 equiv.) and tripotassium phosphate (849 mg, 4.00 mmol, 2.00 equiv.) were reacted in dry DMSO (8 mL). The reaction mixture was extracted and purified by column chromatography over SiO$_2$ (dichloromethane/cyclohexane 1:2 to 1:1 to dichloromethane) to yield 979 mg of the title compound (1.92 mmol, 96%) as an off-white solid.

R_f (dichloromethane/cyclohexane 1:2) = 0.27

¹H NMR (400 MHz, Chloroform-*d*) δ 8.24 (d, *J* = 1.9 Hz, 1H), 8.13 – 8.05 (m, 2H), 8.02 (t, *J* = 7.6 Hz, 1H), 7.95 (dd, *J* = 8.1, 1.1 Hz, 1H), 7.47 (dd, *J* = 8.7, 1.9 Hz, 1H), 7.42 (ddd, *J* = 8.3, 7.2, 1.3 Hz, 1H), 7.32 (ddd, *J* = 8.0, 7.3, 1.0 Hz, 1H), 7.23 (dd, *J* = 8.2, 0.9 Hz, 1H), 7.09 (d, *J* = 8.7 Hz, 1H), 6.94 (d, *J* = 3.0 Hz, 2H), 2.28 (s, 3H), 2.11 (s, 3H), 2.09 (s, 3H).

¹³C NMR (101 MHz, CDCl$_3$) δ 166.5, 164.5, 140.9, 139.5, 139.3, 136.3, 136.2, 136.1, 134.6, 134.4, 134.3, 129.4, 129.4, 128.8, 126.9, 126.8, 126.6, 125.8, 123.4, 123.4, 123.1, 121.3, 120.9, 113.8, 111.5, 110.2, 21.2, 18.2, 18.2.

IR (ATR) $\tilde{v}$ [cm^{-1}] = 2922 (w), 2849 (w), 1713 (s), 1605 (w), 1442 (m), 1373 (s), 1229 (m), 1113 (m), 881 (m), 748 (m), 547 (m), 421 (w).

HRMS (APCI, C$_{29}$H$_{21}$^{79}BrN$_2$O$_2$) calc. 509.0859 [M+H]$^+$, found 509.0859 [M+H]$^+$.

4-(9*H*-Carbazol-9-yl)-2-mesitylisoindoline-1,3-dione

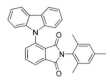

According to the general procedure (GP 6), 4-fluoro-2-mesitylisoindoline-1,3-dione (2.27 g, 8.00 mmol, 1.00 equiv.), 9*H*-carbazole (1.34 g, 8.00 mmol, 1.00 equiv.) and tripotassium phosphate (3.39 g, 16.0 mmol, 2.00 equiv.) were reacted in dry DMSO (32 mL). The reaction mixture was extracted and purified by column chromatography over SiO_2 (dichloromethane/cyclohexane 1:1 to dichloromethane) to yield 3.20 g of the title compound (7.44 mmol, 93%) as a yellow solid.

R_f (dichloromethane/cyclohexane 1:1) = 0.35

M.p. = 258 °C

¹H NMR (400 MHz, Chloroform-*d*) δ 8.14 (d, *J* = 7.8 Hz, 2H), 8.09 (dd, *J* = 6.6, 1.7 Hz, 1H), 8.02 – 7.94 (m, 2H), 7.40 (ddd, *J* = 8.3, 7.2, 1.3 Hz, 2H), 7.32 (td, *J* = 7.5, 1.0 Hz, 2H), 7.26 – 7.23 (m, 2H), 6.94 (s, 2H), 2.28 (s, 3H), 2.13 (s, 6H).

¹³C NMR (101 MHz, CDCl₃) δ 166.7, 164.5, 140.5, 139.4, 136.3, 135.9, 134.8, 134.6, 134.5, 129.3, 126.9, 126.5, 126.1, 124.1, 123.0, 120.9, 120.7, 110.1, 21.2, 18.2.

IR (ATR) $\tilde{v}$ [cm⁻¹] = 3990 (w), 3902 (w), 3481 (w), 3370 (w), 3022 (w), 2919 (w), 2483 (w), 1714 (s), 1605 (m), 1477 (m), 1378 (m), 1303 (m), 1226 (m), 1117 (m), 1004 (m), 854 (m), 717 (m), 549 (m), 417 (m).

HRMS (APCI, $C_{29}H_{22}N_2O_2$) calc. 431.1754 [M+H]⁺, found 431.1755 [M+H]⁺.

EA ($C_{29}H_{22}N_2O_2$) calc. C: 80.91, H: 5.15, N: 6.51; found C: 80.26, H: 5.11, N: 6.36.

9-(2-Mesityl-1,3-dioxoisoindolin-4-yl)-9*H*-carbazole-3-carbonitrile

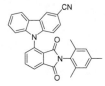 According to the general procedure (GP 6), 4-fluoro-2-mesitylisoindoline-1,3-dione (1.13 g, 4.00 mmol, 1.00 equiv.), 9*H*-carbazole-3-carbonitrile (769 mg, 4.00 mmol, 1.00 equiv.) and tripotassium phosphate (1.70 g, 8.00 mmol, 2.00 equiv.) were reacted in dry DMSO (16 mL). The reaction mixture was extracted and purified by column chromatography over SiO_2 (dichloromethane/cyclohexane 1:1 to dichloromethane) to yield 1.53 g of the title compound (3.36 mmol, 84%) as an off-white solid.

R_f (dichloromethane/cyclohexane 1:1) = 0.17

M.p. = 259 °C

¹H NMR (400 MHz, Chloroform-*d*) δ 8.42 (d, *J* = 1.5 Hz, 1H), 8.19 – 8.11 (m, 2H), 8.06 (t, *J* = 7.7 Hz, 1H), 7.95 (d, *J* = 8.0 Hz, 1H), 7.63 (dd, *J* = 8.5, 1.6 Hz, 1H), 7.49 (ddd, *J* = 8.3, 7.2, 1.2 Hz, 1H), 7.42 – 7.34 (m, 1H), 7.30 – 7.21 (m, 2H), 6.95 – 6.92 (m, 2H), 2.27 (s, 3H), 2.12 (s, 3H), 2.09 (s, 3H).

¹³C NMR (101 MHz, CDCl₃) δ 166.2, 164.4, 142.3, 141.3, 139.6, 136.4, 136.2, 136.1, 134.6, 134.5, 133.3, 129.4, 129.3, 129.3, 127.6, 126.9, 126.6, 125.5, 124.2, 124.0, 122.8, 122.1, 121.1, 120.2, 110.8, 110.4, 103.8, 29.8, 21.1, 18.1, 18.1.

IR (ATR) $\tilde{v}$ [cm⁻¹] = 3950 (w), 3822 (w), 3762 (vw), 3692 (vw), 3482 (w), 3065 (w), 2920 (w), 2484 (vw), 2220 (w), 1718 (m), 1601 (w), 1477 (m), 1372 (m), 1231 (m), 1116 (m), 1037 (w), 882 (m), 813 (m), 748 (m), 614 (m), 550 (m), 418 (w).

HRMS (APCI, C₃₀H₂₂N₃O₂) calc. 456.1707 [M+H]⁺, found 456.1706 [M+H]⁺.

4-(3-Bromo-9H-carbazol-9-yl)-2-decylisoindoline-1,3-dione

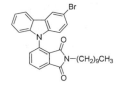 According to the general (GP 6), 2-decyl-4-fluoroisoindoline-1,3-dione (458 mg, 1.50 mmol, 1.00 equiv.), 3-bromo-9H-carbazole (369 mg, 1.50 mmol, 1.00 equiv.) and tripotassium phosphate (636 mg, 3.00 mmol, 2.00 equiv.) were reacted in dry DMSO (6 mL). The reaction mixture was extracted and purified by column chromatography over SiO_2 (dichloromethane/cyclohexane 1:1) to yield 585 mg of the title compound (1.10 mmol, 73%) as a white solid.

R_f (dichloromethane/cyclohexane 1:1) = 0.42

¹H NMR (400 MHz, Chloroform-*d*) δ 8.26 (d, *J* = 1.9 Hz, 1H), 8.12 – 8.06 (m, 1H), 8.00 (dd, *J* = 7.3, 1.0 Hz, 1H), 7.91 (t, *J* = 7.6 Hz, 1H), 7.82 (dd, *J* = 8.0, 1.0 Hz, 1H), 7.47 (dd, *J* = 8.7, 2.0 Hz, 1H), 7.42 (ddd, *J* = 8.4, 7.2, 1.2 Hz, 1H), 7.35 – 7.30 (m, 1H), 7.16 (d, *J* = 8.1 Hz, 1H), 7.02 (d, *J* = 8.7 Hz, 1H), 3.63 – 3.55 (m, 2H), 1.59 (q, *J* = 6.9 Hz, 2H), 1.34 – 1.17 (m, 14H), 0.87 (t, *J* = 6.8 Hz, 3H).

¹³C NMR (101 MHz, CDCl₃) δ 167.5, 165.6, 141.0, 139.4, 135.7, 134.8, 134.3, 133.8, 128.8, 127.1, 126.8, 125.7, 123.4, 123.0, 122.9, 121.2, 120.8, 113.6, 111.5, 110.2, 38.4, 32.0, 29.6, 29.6, 29.4, 29.2, 28.5, 27.0, 22.8, 14.2.

IR (ATR) $\tilde{v}$ [cm⁻¹] = 3062 (w), 2921 (m), 2851 (m), 1704 (s), 1607 (m), 1394 (m), 1325 (m), 1229 (m), 1025 (m), 744 (s), 548 (m), 421 (m).

HRMS (APCI, $C_{30}H_{31}{}^{79}BrN_2O_2$) calc. 531.1642 [M+H]⁺, found 531.1641 [M+H]⁺.

9-(2-Decyl-1,3-dioxoisoindolin-4-yl)-9*H*-carbazole-3-carbonitrile

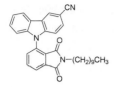

According to the general (GP 6), 2-decyl-4-fluoroisoindoline-1,3-dione (611 mg, 2.00 mmol, 1.00 equiv.), 9*H*-carbazole-3-carbonitrile (385 mg, 2.00 mmol, 1.00 equiv.) and tripotassium phosphate (848 mg, 4.00 mmol, 2.00 equiv.) were reacted in dry DMSO (8 mL). The reaction mixture was extracted and purified by column chromatography over SiO_2 (dichloromethane/cyclohexane 1:2 to 1:1 to 3:1) to yield 677 mg of the title compound (1.42 mmol, 71%) as a white solid.

R_f (dichloromethane/cyclohexane 1:2) = 0.29

^{1}H NMR (400 MHz, Chloroform-*d*) δ 8.46 (d, *J* = 1.5 Hz, 1H), 8.17 (d, *J* = 7.8 Hz, 1H), 8.06 (d, *J* = 7.4 Hz, 1H), 7.98 (t, *J* = 7.6 Hz, 1H), 7.84 (d, *J* = 7.9 Hz, 1H), 7.63 (dd, *J* = 8.5, 1.5 Hz, 1H), 7.48 (t, *J* = 7.7 Hz, 1H), 7.40 (t, *J* = 7.5 Hz, 1H), 7.22 – 7.13 (m, 2H), 3.59 (t, *J* = 7.4 Hz, 2H), 1.69 – 1.49 (m, 2H), 1.35 – 1.04 (m, 14H), 0.86 (t, *J* = 6.9 Hz, 3H).

^{13}C NMR (101 MHz, CDCl$_3$) δ 167.3, 165.5, 142.5, 141.5, 135.9, 134.9, 134.3, 132.9, 129.4, 127.6, 127.4, 125.5, 124.1, 123.6, 122.7, 122.0, 121.1, 120.2, 110.8, 110.4, 103.7, 38.5, 32.0, 29.6, 29.6, 29.4, 29.2, 28.5, 26.9, 22.8, 14.2.

IR (ATR) $\tilde{v}$ [cm^{-1}] = 2922 (m), 2852 (m), 2220 (m), 1711 (s), 1597 (m), 1476 (m), 1365 (s), 1232 (s), 1161 (m), 1026 (m), 891 (m), 747 (s), 612 (m), 551 (m), 417 (m).

HRMS (APCI, C$_{31}$H$_{31}$N$_3$O$_2$) calc. 478.2489 [M+H]$^+$, found 478.2488 [M+H]$^+$.

EA (C$_{31}$H$_{31}$N$_3$O$_2$) calc. C: 77.96, H: 6.54, N: 8.80; found C: 77.88, H: 6.48, N: 8.76.

2-(3-Bromophenyl)-4-(9H-carbazol-9-yl)isoindoline-1,3-dione

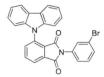

According to the general procedure (GP 6), 2-(3-bromophenyl)-4-fluoroisoindoline-1,3-dione (12.8 g, 40.0 mmol, 1.00 equiv.), 9H-carbazole (7.68 g, 40.0 mmol, 1.00 equiv.) and tripotassium phosphate (17.0 g, 80.0 mmol, 2.00 equiv.) were reacted in dry DMSO (160 mL). The reaction mixture was extracted and purified by recrystallization in toluene to yield 10.8 g of the title compound (23.2 mmol, 58%) as a yellow solid.

^{1}H NMR (400 MHz, Chloroform-d) δ 8.17 – 8.12 (m, 2H), 8.08 (dd, J = 7.2, 1.2 Hz, 1H), 7.98 (t, J = 7.6 Hz, 1H), 7.92 (dd, J = 8.0, 1.2 Hz, 1H), 7.58 (t, J = 1.9 Hz, 1H), 7.44 (dt, J = 8.0, 1.5 Hz, 1H), 7.41 – 7.22 (m, 7H), 7.17 (d, J = 8.1 Hz, 2H).

^{13}C NMR (101 MHz, CDCl$_3$) δ 166.1, 164.0, 140.7, 136.3, 135.3, 135.1, 134.1, 132.7, 131.2, 130.2, 129.3, 126.4, 126.2, 124.9, 124.1, 123.4, 122.4, 121.0, 120.7, 109.9.

IR (ATR) $\tilde{v}$ [cm^{-1}] = 3941 (w), 3473 (w), 3068 (w), 2919 (w), 2845 (w), 1716 (s), 1588 (w), 1477 (m), 1365 (m), 1229 (m), 1078 (m), 1007 (w), 880 (m), 749 (s), 546 (m), 425 (w).

HRMS (APCI, C$_{26}$H$_{15}$^{79}BrN$_2$O$_2$) calc. 467.0390 [M+H]$^+$, found 467.0391 [M+H]$^+$.

2-Allyl-4-(9*H*-carbazol-9-yl)isoindoline-1,3-dione

According to the general procedure (GP 6), 2-allyl-4-fluoroisoindoline-1,3-dione (821 mg, 4.00 mmol, 1.00 equiv.), 9*H*-carbazole (669 mg, 4.00 mmol, 1.00 equiv.) and tripotassium phosphate (1.70 g, 8.00 mmol, 2.00 equiv.) were reacted in dry DMSO (16 mL). The reaction mixture was extracted and purified by column chromatography over SiO_2 (dichloromethane/cyclohexane 1:1 to 2:1) to yield 1.02 g of the title compound (2.92 mmol, 73%) as a yellow solid.

R_f (dichloromethane/cyclohexane 1:1) = 0.33

¹H NMR (400 MHz, Chloroform-*d*) δ 8.16 (dt, *J* = 7.6, 0.9 Hz, 2H), 8.00 (dd, *J* = 7.2, 1.2 Hz, 1H), 7.91 (t, *J* = 7.6 Hz, 1H), 7.86 (dd, *J* = 8.0, 1.2 Hz, 1H), 7.40 (ddd, *J* = 8.3, 7.2, 1.3 Hz, 2H), 7.33 (td, *J* = 7.5, 1.1 Hz, 2H), 7.20 – 7.09 (m, 2H), 5.83 (ddt, *J* = 17.1, 10.2, 5.9 Hz, 1H), 5.23 (dq, *J* = 17.2, 1.4 Hz, 1H), 5.16 (dq, *J* = 10.1, 1.2 Hz, 1H), 4.23 (dt, *J* = 5.9, 1.4 Hz, 2H).

¹³C NMR (101 MHz, CDCl₃) δ 167.2, 165.1, 140.7, 135.7, 134.7, 134.6, 134.5, 131.4, 127.0, 126.1, 124.0, 122.8, 120.9, 120.7, 118.4, 110.0, 40.4.

IR (ATR) $\tilde{v}$ [cm⁻¹] = 3062 (vw), 2936 (vw), 1704 (m), 1609 (w), 1479 (m), 1388 (m), 1314 (m), 1233 (m), 1074 (w), 938 (m), 747 (m), 552 (w), 421 (w).

HRMS (APCI, $C_{23}H_{16}N_2O_2$) calc. 353.1285 [M+H]⁺, found 353.1265 [M+H]⁺.

EA ($C_{23}H_{16}N_2O_2$) calc. C: 78.39, H: 4.58, N: 7.95; found C: 78.52, H: 4.52, N: 7.93.

4-(9*H*-Carbazol-9-yl)-2-(propa-1,2-dien-1-yl)isoindoline-1,3-dione

 According to the general procedure (GP 6), 4-fluoro-2-(prop-2-yn-1-yl)isoindoline-1,3-dione (813 mg, 4.00 mmol, 1.00 equiv.), 9*H*-carbazole (669 mg, 4.00 mmol, 1.00 equiv.) and tripotassium phosphate (1.70 g, 8.00 mmol, 2.00 equiv.) were reacted in dry DMSO (16 mL). The reaction mixture was extracted and purified by column chromatography over SiO$_2$ (dichloromethane/cyclohexane 1:1 to 2:1) to yield 230 mg of the title compound (0.641 mmol, 16%) as a yellow solid.

R$_f$ (dichloromethane/cyclohexane 1:1) = 0.30

^{1}H NMR (400 MHz, Chloroform-*d*) δ 8.15 (dt, *J* = 7.4, 0.9 Hz, 2H), 8.00 (dd, *J* = 7.3, 1.1 Hz, 1H), 7.92 (t, *J* = 7.7 Hz, 1H), 7.85 (dd, *J* = 7.9, 1.1 Hz, 1H), 7.39 (ddd, *J* = 8.2, 7.2, 1.3 Hz, 2H), 7.32 (td, *J* = 7.5, 1.0 Hz, 2H), 7.14 (d, *J* = 8.2 Hz, 2H), 6.73 (t, *J* = 6.7 Hz, 1H), 5.39 (d, *J* = 6.7 Hz, 2H).

^{13}C NMR (101 MHz, CDCl$_3$) δ 204.3, 165.0, 162.8, 140.7, 136.0, 135.0, 134.8, 134.5, 126.2, 124.1, 123.0, 120.9, 120.7, 109.9, 88.3, 85.8.

IR (ATR) $\tilde{v}$ [cm^{-1}] = 3470 (vw), 3232 (vw), 3033 (vw), 2927 (vw), 2807 (vw), 2678 (vw), 2560 (vw), 2472 (vw), 2393 (vw), 2323 (vw), 2164 (vw), 2100 (vw), 1950 (vw), 1712 (m), 1602 (w), 1448 (m), 1375 (m), 1226 (w), 1103 (w), 1019 (w), 941 (w), 869 (w), 745 (m), 622 (w), 547 (w), 422 (w).

HRMS (APCI, C$_{23}$H$_{14}$N$_2$O$_2$) calc. 351.1128 [M+H]$^+$, found 351.1111 [M+H]$^+$.

4-(3,6-Dibromo-9*H*-carbazol-9-yl)-2-mesitylisoindoline-1,3-dione

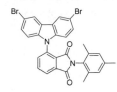

According to the general procedure (GP 6), 4-fluoro-2-mesitylisoindoline-1,3-dione (1.13 g, 4.00 mmol, 1.00 equiv.), 3,6-dibromo-9*H*-carbazole (1.30 g, 4.00 mmol, 1.00 equiv.) and tripotassium phosphate (1.70 g, 8.00 mmol, 2.00 equiv.) were reacted in dry DMSO (16 mL). The reaction mixture was extracted and purified recrystallization in toluene to yield 1.42 g of the title compound (2.41 mmol, 60%) as an off-white solid.

¹H NMR (400 MHz, Chloroform-*d*) δ 8.19 (d, *J* = 1.9 Hz, 2H), 8.12 (dd, *J* = 7.4, 1.0 Hz, 1H), 8.03 (t, *J* = 7.7 Hz, 1H), 7.92 (dd, *J* = 8.0, 1.0 Hz, 1H), 7.49 (dd, *J* = 8.7, 1.9 Hz, 2H), 7.08 (d, *J* = 8.7 Hz, 2H), 6.94 (s, 2H), 2.27 (s, 3H), 2.09 (s, 6H).

¹³C NMR (101 MHz, CDCl₃) δ 166.3, 164.4, 139.7, 139.6, 136.3, 136.2, 134.7, 134.3, 133.7, 129.6, 129.4, 126.7, 124.7, 123.7, 123.6, 114.1, 111.7, 21.2, 18.2.

IR (ATR) $\tilde{v}$ [cm⁻¹] = 3291 (vw), 2917 (vw), 2731 (vw), 2656 (vw), 2594 (vw), 2479 (vw), 2415 (vw), 2346 (vw), 2160 (vw), 1861 (vw), 1715 (m), 1605 (w), 1374 (m), 1282 (w), 1116 (w), 1019 (w), 860 (w), 800 (m), 655 (w), 550 (w), 419 (w).

HRMS (APCI, C₂₉H₂₀⁷⁹Br₂N₂O₂) calc. 586.9964 [M+H]⁺, found 586.9937 [M+H]⁺.

4-(3-Iodo-9*H*-carbazol-9-yl)-2-mesitylisoindoline-1,3-dione

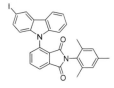

According to the general procedure (GP 6), 4-fluoro-2-mesitylisoindoline-1,3-dione (1.13 g, 4.00 mmol, 1.00 equiv.), 3-iodo-9*H*-carbazole (1.17 g, 4.00 mmol, 1.00 equiv.) and tripotassium phosphate (1.70 g, 8.00 mmol, 2.00 equiv.) were reacted in dry DMSO (16 mL). The reaction mixture was extracted and purified by column chromatography over SiO$_2$ (dichloromethane/cyclohexane 1:1 to 3:1) to yield 1.92 g of the title compound (3.44 mmol, 86%) as a yellow solid.

R_f (dichloromethane/cyclohexane 1:1) = 0.39

^{1}H NMR (400 MHz, DMSO-d_6) δ 8.65 (d, J = 1.7 Hz, 1H), 8.29 (d, J = 7.8 Hz, 1H), 8.19 – 8.15 (m, 2H), 8.14 – 8.09 (m, 1H), 7.66 (dd, J = 8.6, 1.8 Hz, 1H), 7.43 (ddd, J = 8.3, 7.0, 1.2 Hz, 1H), 7.36 – 7.27 (m, 2H), 7.18 (d, J = 8.6 Hz, 1H), 6.97 (d, J = 2.4 Hz, 2H), 2.24 (s, 3H), 2.04 (s, 3H), 2.01 (s, 3H).

^{13}C NMR (101 MHz, DMSO) δ 166.1, 164.4, 140.2, 139.4, 138.6, 137.0, 136.2, 136.2, 134.7, 134.0, 133.9, 132.8, 129.0, 128.8, 128.8, 126.9, 126.8, 126.2, 125.6, 123.6, 121.8, 121.0, 120.9, 112.7, 110.3, 83.7, 20.6, 17.6, 17.5.

IR (ATR) $\tilde{v}$ [cm^{-1}] = 2915 (vw), 1713 (m), 1605 (w), 1489 (m), 1372 (m), 1229 (w), 1120 (w), 1009 (w), 880 (m), 742 (m), 547 (w), 416 (w).

HRMS (APCI, C$_{29}$H$_{21}$IN$_2$O$_2$) calc. 557.0720 [M+H]$^+$, found 557.0691 [M+H]$^+$.

4-(3,6-Di-*tert*-butyl-9*H*-carbazol-9-yl)-2-mesitylisoindoline-1,3-dione

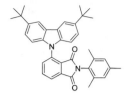 According to the general procedure (GP 6), 4-fluoro-2-mesitylisoindoline-1,3-dione (708 mg, 2.50 mmol, 1.00 equiv.), 3,6-di-*tert*-butyl-9*H*-carbazole (698 mg, 2.50 mmol, 1.00 equiv.) and tripotassium phosphate (1.06 g, 5.00 mmol, 2.00 equiv.) were reacted in dry DMSO (10 mL). The reaction mixture was extracted and purified by column chromatography over SiO2 (dichloromethane/cyclohexane 1:1) to yield 1.01 g of the title compound (1.85 mmol, 74%) as a yellow solid.

R$_f$ (dichloromethane/cyclohexane 1:1) = 0.41

M.p. = 160 °C

^{1}H NMR (400 MHz, Chloroform-*d*) δ 8.14 (d, *J* = 1.9 Hz, 2H), 8.06 (dd, *J* = 5.2, 3.1 Hz, 1H), 7.96 – 7.92 (m, 2H), 7.44 (dd, *J* = 8.6, 1.9 Hz, 2H), 7.17 (d, *J* = 8.6 Hz, 2H), 6.97 (s, 2H), 2.30 (s, 3H), 2.16 (s, 6H), 1.47 (s, 18H).

^{13}C NMR (101 MHz, CDCl3) δ 166.8, 164.7, 143.9, 139.4, 138.9, 136.4, 135.8, 135.4, 134.5, 134.2, 129.4, 127.0, 125.9, 124.3, 123.7, 122.5, 116.7, 109.7, 34.9, 32.1, 21.2, 18.2.

IR (ATR) $\tilde{v}$ [cm^{-1}] = 2954 (w), 1725 (m), 1610 (w), 1486 (m), 1373 (m), 1296 (w), 1115 (w), 1034 (w), 879 (w), 751 (w), 678 (w), 613 (w), 414 (w).

HRMS (APCI, C37H38N2O2) calc. 543.3006 [M+H]$^+$, found 543.2977 [M+H]$^+$.

EA (C37H38N2O2) calc. C: 81.88, H: 7.06, N: 5.16; found C: 80.64, H: 7.06, N: 4.99.

2-(*Tert*-butyl)-4-(9*H*-carbazol-9-yl)isoindoline-1,3-dione

 According to the general procedure (GP 6), 2-(*tert*-butyl)-4-fluoroisoindoline-1,3-dione (1.77 g, 8.00 mmol, 1.00 equiv.), 9*H*-carbazole (1.34 g, 8.00 mmol, 1.00 equiv.) and tripotassium phosphate (3.39 g, 16.0 mmol, 2.00 equiv.) were reacted in dry DMSO (32 mL). The reaction mixture was extracted and purified by column chromatography over SiO_2 (dichloromethane/pentane 1:1) to yield 1.78 g of the title compound (4.83 mmol, 60%) as an off-white solid.

R_f (dichloromethane/cyclohexane 1:1) = 0.44

M.p. = 200 °C

¹H NMR (400 MHz, Chloroform-*d*) δ 8.16 (d, *J* = 7.7 Hz, 2H), 7.93 (dd, *J* = 7.3, 1.2 Hz, 1H), 7.86 (t, *J* = 7.6 Hz, 1H), 7.79 (dd, *J* = 7.8, 1.2 Hz, 1H), 7.40 (ddd, *J* = 8.3, 7.2, 1.3 Hz, 2H), 7.32 (td, *J* = 7.4, 1.0 Hz, 2H), 7.16 (d, *J* = 8.2 Hz, 2H), 1.63 (s, 9H).

¹³C NMR (101 MHz, CDCl₃) δ 168.8, 166.7, 140.8, 135.3, 134.9, 134.3, 133.9, 126.9, 126.0, 123.9, 122.2, 120.7, 120.7, 110.1, 58.3, 29.1.

IR (ATR) ṽ [cm⁻¹] = 2926 (vw), 1708 (m), 1606 (w), 1478 (w), 1310 (m), 1227 (w), 1097 (w), 1009 (w), 868 (w), 750 (m), 554 (w), 422 (vw).

HRMS (APCI, C₂₄H₂₀N₂O₂) calc. 369.1598 [M+H]⁺, found 369.1578 [M+H]⁺.

EA (C₂₄H₂₀N₂O₂) calc. C: 78.24, H: 5.47, N: 7.60; found C: 77.93, H: 5.41, N: 7.46.

2-(*Sec*-butyl)-4-(9*H*-carbazol-9-yl)isoindoline-1,3-dione

According to the general procedure (GP 6), 2-(*sec*-butyl)-4-fluoroisoindoline-1,3-dione (885 mg, 4.00 mmol, 1.00 equiv.), 9*H*-carbazole (669 mg, 4.00 mmol, 1.00 equiv.) and tripotassium phosphate (1.70 g, 8.00 mmol, 2.00 equiv.) were reacted in dry DMSO (16 mL). The reaction mixture was extracted and purified by column chromatography over SiO$_2$ (dichloromethane/cyclohexane 1:1) to yield 1.17 g of the title compound (3.18 mmol, 79%) as an off-white solid.

R$_f$ (dichloromethane/cyclohexane 1:1) = 0.40

M.p. = 150 °C

^{1}H NMR (400 MHz, Chloroform-*d*) δ 8.20 – 8.13 (m, 2H), 7.97 (dd, *J* = 7.2, 1.2 Hz, 1H), 7.89 (t, *J* = 7.5 Hz, 1H), 7.84 (dd, *J* = 8.0, 1.2 Hz, 1H), 7.45 – 7.37 (m, 2H), 7.36 – 7.27 (m, 2H), 7.18 (d, *J* = 8.1 Hz, 2H), 4.21 (dp, *J* = 9.2, 6.8 Hz, 1H), 1.99 (ddq, *J* = 14.8, 9.1, 7.4 Hz, 1H), 1.81 – 1.68 (m, 1H), 1.43 (d, *J* = 6.9 Hz, 3H), 0.87 (t, *J* = 7.4 Hz, 3H).

^{13}C NMR (101 MHz, CDCl$_3$) δ 167.8, 165.7, 140.7, 135.5, 134.6, 134.3, 134.3, 126.7, 126.0, 124.0, 124.0, 122.5, 120.8, 120.6, 110.0, 110.0, 49.4, 26.8, 18.4, 11.3.

IR (ATR) ṽ [cm^{-1}] = 2964 (w), 1701 (m), 1606 (w), 1477 (m), 1328 (m), 1227 (m), 1049 (m), 876 (w), 748 (s), 551 (m), 409 (w).

HRMS (APCI, C$_{24}$H$_{20}$N$_2$O$_2$) calc. 369.1598 [M+H]$^+$, found 369.1592 [M+H]$^+$.

EA (C$_{24}$H$_{20}$N$_2$O$_2$) calc. C: 78.24, H: 5.47, N: 7.60; found C: 78.31, H: 5.48, N: 7.46.

2-Butyl-4-(9*H*-carbazol-9-yl)isoindoline-1,3-dione

According to the general procedure (GP 6), 2-butyl-4-fluoroisoindoline-1,3-dione (885 mg, 4.00 mmol, 1.00 equiv.), 9*H*-carbazole (669 mg, 4.00 mmol, 1.00 equiv.) and tripotassium phosphate (1.70 g, 8.00 mmol, 2.00 equiv.) were reacted in dry DMSO (16 mL). The reaction mixture was extracted and purified by column chromatography over SiO$_2$ (dichloromethane/cyclohexane 1:1) to yield 1.26 g of the title compound (3.42 mmol, 85%) as an off-white solid.

R$_f$ (dichloromethane/cyclohexane 1:1) = 0.56

M.p. = 166 °C

¹H NMR (300 MHz, Chloroform-*d*) δ 8.21 – 8.12 (m, 2H), 7.99 (dd, *J* = 7.1, 1.4 Hz, 1H), 7.91 (t, *J* = 7.5 Hz, 1H), 7.85 (dd, *J* = 7.9, 1.4 Hz, 1H), 7.40 (ddd, *J* = 8.2, 7.2, 1.4 Hz, 2H), 7.32 (td, *J* = 7.4, 1.2 Hz, 2H), 7.17 (d, *J* = 8.1 Hz, 2H), 3.62 (t, *J* = 7.3 Hz, 2H), 1.66 – 1.51 (m, 2H), 1.31 (dq, *J* = 14.7, 7.3 Hz, 2H), 0.89 (t, *J* = 7.3 Hz, 3H).

¹³C NMR (101 MHz, CDCl$_3$) δ 167.8, 165.6, 140.8, 135.6, 134.9, 134.5, 127.1, 126.1, 124.1, 122.7, 120.9, 120.6, 110.0, 38.1, 30.6, 20.2, 13.7.

IR (ATR) $\tilde{v}$ [cm⁻¹] = 2955 (w), 1703 (s), 1603 (w), 1477 (m), 1395 (m), 1332 (m), 1226 (m), 1053 (m), 932 (w), 864 (w), 749 (m), 555 (w), 424 (w).

HRMS (APCI, C$_{24}$H$_{20}$N$_2$O$_2$) calc. 369.1598 [M+H]⁺, found 369.1577 [M+H]⁺.

EA (C$_{24}$H$_{20}$N$_2$O$_2$) calc. C: 78.24, H: 5.47, N: 7.60; found C: 77.94, H: 5.41, N: 7.52.

4-(4-(9H-Carbazol-9-yl)-1,3-dioxoisoindolin-2-yl)benzonitrile

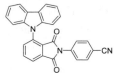 According to the general procedure (GP 6), 4-(4-fluoro-1,3-dioxoisoindolin-2-yl)benzonitrile (1.07 g, 4.00 mmol, 1.00 equiv.), 9H-carbazole (0.669 g, 4.00 mmol, 1.00 equiv.) and tripotassium phosphate (1.70 g, 8.00 mmol, 2.00 equiv.) were reacted in dry DMSO (16 mL). The reaction mixture was extracted and purified by column chromatography over SiO$_2$ (dichloromethane/cyclohexane 1:1 to dichloromethane) to yield 0.948 g of the title compound (2.29 mmol, 57%) as a yellow solid.

R$_f$ (dichloromethane/cyclohexane 1:1) = 0.21

M.p. = 261 °C

^{1}H NMR (400 MHz, Chloroform-d) δ 8.16 (d, J = 7.7 Hz, 2H), 8.13 (dd, J = 7.2, 1.2 Hz, 1H), 8.04 (t, J = 7.6 Hz, 1H), 7.99 (dd, J = 8.0, 1.2 Hz, 1H), 7.72 – 7.67 (m, 2H), 7.63 – 7.59 (m, 2H), 7.40 (ddd, J = 8.3, 7.2, 1.3 Hz, 2H), 7.33 (td, J = 7.5, 1.1 Hz, 2H), 7.18 (d, J = 8.0 Hz, 2H).

^{13}C NMR (101 MHz, CDCl$_3$) δ 165.8, 163.6, 140.6, 136.6, 135.7, 135.6, 135.4, 133.9, 132.9, 126.4, 126.2, 124.1, 123.6, 121.1, 120.8, 118.3, 111.4, 109.9.

IR (ATR) $\tilde{v}$ [cm^{-1}] = 3065 (vw), 2918 (vw), 2849 (vw), 2704 (vw), 2585 (vw), 2457 (vw), 2321 (vw), 2231 (vw), 1905 (vw), 1712 (vw), 1604 (m), 1476 (w), 1365 (m), 1221 (w), 1091 (w), 1003 (w), 825 (w), 753 (m), 548 (w), 423 (w).

HRMS (APCI, C$_{27}$H$_{15}$N$_3$O$_2$) calc. 414.1237 [M+H]$^+$, found 414.1213 [M+H]$^+$.

4-(9*H*-Carbazol-9-yl)-2-(4-(dimethylamino)phenyl)isoindoline-1,3-dione

According to the general procedure (GP 6), 2-(4-(dimethylamino)phenyl)-4-fluoroisoindoline-1,3-dione (569 mg, 2.00 mmol, 1.00 equiv.), 9*H*-carbazole (334 mg, 2.00 mmol, 1.00 equiv.) and tripotassium phosphate (848 mg, 4.00 mmol, 2.00 equiv.) were reacted in dry DMSO (8 mL). The reaction mixture was extracted and purified by column chromatography over SiO₂ (dichloromethane/cyclohexane 2:1 to dichloromethane to dichloromethane/ethyl acetate 50:1 to 10:1) to yield 615 mg of the title compound (1.43 mmol, 71%) as an orange solid.

R$_f$ (dichloromethane/ethyl acetate 10:1) = 0.78

M.p. = 262 °C

¹H NMR (400 MHz, Chloroform-*d*) δ 8.14 (d, *J* = 7.7 Hz, 2H), 8.07 (dd, *J* = 7.3, 1.1 Hz, 1H), 7.99 – 7.92 (m, 1H), 7.90 (dd, *J* = 8.0, 1.1 Hz, 1H), 7.39 (ddd, *J* = 8.3, 7.2, 1.3 Hz, 2H), 7.30 (td, *J* = 7.5, 1.0 Hz, 2H), 7.24 – 7.17 (m, 4H), 6.72 (d, *J* = 9.0 Hz, 2H), 2.94 (s, 6H).

¹³C NMR (101 MHz, CDCl₃) δ 167.1, 164.9, 140.8, 135.8, 134.8, 134.7, 134.6, 127.3, 126.9, 126.1, 124.0, 123.0, 120.8, 120.6, 112.5, 110.1, 40.7.

IR (ATR) $\tilde{v}$ [cm⁻¹] = 3060 (vw), 2800 (vw), 2575 (vw), 2477 (vw), 2322 (vw), 2217 (vw), 1711 (m), 1608 (w), 1478 (w), 1355 (w), 1231 (w), 1093 (m), 880 (w), 817 (w), 748 (m), 554 (w), 421 (w).

HRMS (APCI, C₂₈H₂₁N₃O₂) calc. 432.1707 [M+H]⁺, found 432.1682 [M+H]⁺.

EA (C₂₈H₂₁N₃O₂) calc. C: 77.94, H: 4.91, N: 9.74; found C: 77.71, H: 4.81, N: 9.65.

2-(3,5-Bis(trifluoromethyl)phenyl)-4-(9*H*-carbazol-9-yl)isoindoline-1,3-dione

According to the general procedure (GP 6), 2-(3,5-bis(trifluoromethyl)phenyl)-4-fluoroisoindoline-1,3-dione (943 mg, 2.50 mmol, 1.00 equiv.), 9*H*-carbazole (418 mg, 2.50 mmol, 1.00 equiv.) and tripotassium phosphate (1.06 g, 5.00 mmol, 2.00 equiv.) were reacted in dry DMSO (10 mL). The reaction mixture was extracted and purified by column chromatography over SiO_2 (dichloromethane/cyclohexane 1:1) to yield 436 mg of the title compound (0.831 mmol, 33%) as a yellow solid.

R_f (dichloromethane/cyclohexane 1:1) = 0.56

^{1}H NMR (400 MHz, Chloroform-*d*) δ 8.19 – 8.12 (m, 3H), 8.07 (t, *J* = 7.6 Hz, 1H), 8.01 (dd, *J* = 8.0, 1.1 Hz, 1H), 7.98 (s, 2H), 7.85 (s, 1H), 7.41 (ddd, *J* = 8.3, 7.3, 1.3 Hz, 2H), 7.34 (td, *J* = 7.5, 1.1 Hz, 2H), 7.20 (d, *J* = 8.2 Hz, 2H).

^{13}C NMR (101 MHz, CDCl$_3$) δ 165.7, 163.6, 140.6, 136.8, 135.7, 135.6, 133.7, 133.1, 133.1, 132.8, 132.4, 132.1, 126.3, 126.3, 126.1, 124.3, 124.1, 123.7, 121.6, 121.6, 121.5, 121.5, 121.2, 120.9, 109.8.

IR (ATR) $\tilde{v}$ [cm^{-1}] = 1726 (m), 1603 (w), 1473 (w), 1390 (m), 1276 (m), 1125 (m), 875 (m), 748 (m), 636 (m), 549 (w), 404 (w).

HRMS (APCI, C$_{28}$H$_{14}$F$_6$N$_2$O$_2$) calc. 525.1032 [M+H]$^+$, found 525.1004 [M+H]$^+$.

EA (C$_{28}$H$_{14}$F$_6$N$_2$O$_2$) calc. C: 64.13, H: 2.69, N: 5.34; found C: 64.33, H: 2.51, N: 5.34.

4-(9*H*-Carbazol-9-yl)-2-(3,5-dimethoxyphenyl)isoindoline-1,3-dione

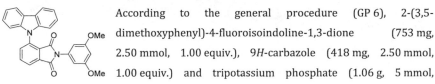 According to the general procedure (GP 6), 2-(3,5-dimethoxyphenyl)-4-fluoroisoindoline-1,3-dione (753 mg, 2.50 mmol, 1.00 equiv.), 9*H*-carbazole (418 mg, 2.50 mmol, 1.00 equiv.) and tripotassium phosphate (1.06 g, 5 mmol, 2.00 equiv.) were reacted in dry DMSO (10 mL). The reaction mixture was extracted and purified by column chromatography over SiO$_2$ (dichloromethane/cyclohexane 2:1 to dichloromethane) to yield 676 mg of the title compound (1.51 mmol, 60%) as a yellow solid.

R$_f$ (dichloromethane) = 0.38

^{1}H NMR (400 MHz, Chloroform-*d*) δ 8.16 (d, *J* = 7.7 Hz, 2H), 8.10 (dd, *J* = 7.2, 1.1 Hz, 1H), 7.99 (dd, *J* = 8.0, 7.2 Hz, 1H), 7.93 (dd, *J* = 8.0, 1.1 Hz, 1H), 7.40 (ddd, *J* = 8.3, 7.2, 1.3 Hz, 2H), 7.32 (td, *J* = 7.5, 1.0 Hz, 2H), 7.21 (dt, *J* = 8.1, 0.9 Hz, 2H), 6.55 (d, *J* = 2.3 Hz, 2H), 6.44 (t, *J* = 2.3 Hz, 1H), 3.74 (s, 6H).

^{13}C NMR (101 MHz, CDCl$_3$) δ 166.4, 164.2, 160.9, 140.7, 136.1, 135.1, 135.0, 134.3, 132.9, 126.6, 126.2, 124.1, 123.2, 120.9, 120.7, 110.0, 105.0, 100.8, 55.6.

IR (ATR) ṽ [cm^{-1}] = 3052 (vw), 2933 (vw), 2839 (vw), 1719 (s), 1594 (m), 1448 (m), 1282 (m), 1153 (m), 881 (m), 746 (m), 656 (m), 548 (w), 422 (w).

HRMS (APCI, C$_{28}$H$_{20}$N$_2$O$_4$) calc. 449.1496 [M+H]$^+$, found 449.1471 [M+H]$^+$.

EA (C$_{28}$H$_{20}$N$_2$O$_4$) calc. C: 74.99, H: 4.50, N: 6.25; found C: 74.82, H: 4.44, N: 6.17.

4-(9H-Carbazol-9-yl)-2-(3,5-di-*tert*-butylphenyl)isoindoline-1,3-dione

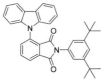

According to the general procedure (GP 6), 2-(3,5-di-*tert*-butylphenyl)-4-fluoroisoindoline-1,3-dione (530 mg, 1.50 mmol, 1.00 equiv.), 9H-carbazole (251 mg, 1.50 mmol, 1.00 equiv.) and tripotassium phosphate (636 g, 3.00 mmol, 2.00 equiv.) were reacted in dry DMSO (6 mL). The reaction mixture was extracted and purified by column chromatography over SiO_2 (dichloromethane/pentane 1:1 to 2:1) to yield 541 mg of the title compound (1.08 mmol, 72%) as a yellow solid.

R_f (dichloromethane/cyclohexane 1:1) = 0.41

1H NMR (400 MHz, Chloroform-*d*) δ 8.16 (d, *J* = 7.6 Hz, 2H), 8.13 – 8.07 (m, 2H), 7.98 (dd, *J* = 8.0, 7.2 Hz, 1H), 7.92 (dd, *J* = 8.0, 1.1 Hz, 1H), 7.46 – 7.29 (m, 6H), 7.28 – 7.21 (m, 3H), 7.17 (d, *J* = 1.7 Hz, 2H), 1.31 (s, 18H).

13C NMR (101 MHz, CDCl3) δ 167.0, 164.7, 151.7, 140.8, 139.6, 135.9, 134.9, 134.9, 134.5, 130.7, 126.8, 126.2, 125.9, 124.1, 123.4, 123.1, 122.6, 121.2, 120.9, 120.7, 120.4, 119.5, 110.7, 110.1, 35.1, 31.5.

IR (ATR) $\tilde{v}$ [cm⁻¹] = 3051 (vw), 2959 (w), 2865 (w), 1719 (m), 1595 (w), 1449 (m), 1362 (m), 1229 (w), 1099 (w), 1014 (w), 862 (w), 746 (m), 648 (m), 550 (w), 422 (w).

HRMS (APCI, $C_{34}H_{32}N_2O_2$) calc. 501.2537 [M+H]⁺, found 501.2513 [M+H]⁺.

4-(3,6-Diacetyl-9*H*-carbazol-9-yl)-2-mesitylisoindoline-1,3-dione

According to the general procedure (GP 6), 4-fluoro-2-mesitylisoindoline-1,3-dione (567 mg, 2.00 mmol, 1.00 equiv.), 1,1'-(9*H*-carbazole-3,6-diyl)bis(ethan-1-one) (503 mg, 2.00 mmol, 1.00 equiv.) and tripotassium phosphate (848 mg, 4.00 mmol, 2.00 equiv.) were reacted in dry DMSO (8 mL). The reaction mixture was extracted and purified by column chromatography over SiO$_2$ (dichloromethane/pentane 1:1 to dichloromethane to dichloromethane/ethyl acetate 50:1) to yield 695 mg of the title compound (1.36 mmol, 68%) as a white solid.

R$_f$ (dichloromethane/ethyl acetate 50:1) = 0.67

M.p. = 320 °C

¹H NMR (400 MHz, Chloroform-*d*) δ 8.82 – 8.76 (m, 2H), 8.22 – 8.16 (m, 1H), 8.15 – 8.05 (m, 3H), 7.98 (dd, *J* = 8.0, 1.0 Hz, 1H), 7.26 – 7.23 (m, 2H), 6.92 (s, 2H), 2.72 (s, 6H), 2.26 (s, 3H), 2.08 (s, 6H).

¹³C NMR (101 MHz, CDCl$_3$) δ 197.4, 166.2, 164.4, 144.1, 139.7, 136.5, 136.2, 134.7, 134.4, 133.1, 131.3, 129.4, 127.4, 127.0, 126.6, 124.3, 123.8, 122.2, 110.2, 26.9, 21.2, 18.1.

IR (ATR) ṽ [cm⁻¹] = 3390 (vw), 3308 (vw), 3071 (vw), 2917 (vw), 2731 (vw), 2488 (vw), 2411 (vw), 2323 (vw), 1661 (m), 1596 (m), 1486 (m), 1376 (m), 1256 (m), 1120 (m), 957 (w), 817 (m), 752 (m), 573 (m), 423 (w).

HRMS (APCI, C$_{33}$H$_{26}$N$_2$O$_4$) calc. 515.1965 [M+H]⁺, found 515.1950 [M+H]⁺.

2-(*Tert*-butyl)-4-(3,6-diacetyl-9*H*-carbazol-9-yl)isoindoline-1,3-dione

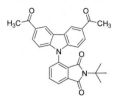

According to the general procedure (GP 6), 2-(*tert*-butyl)-4-fluoroisoindoline-1,3-dione (442 mg, 2.00 mmol, 1.00 equiv.), 1,1'-(9*H*-carbazole-3,6-diyl)bis(ethan-1-one) (503 g, 2.00 mmol, 1.00 equiv.) and tripotassium phosphate (848 g, 4.00 mmol, 2.00 equiv.) were reacted in dry DMSO (8 mL). The reaction mixture was extracted and purified by column chromatography over SiO₂ (dichloromethane/ethyl acetate 20:1) to yield 521 mg of the title compound (1.16 mmol, 58%) as a white solid.

**R*f* (dichloromethane/ethyl acetate 20:1) = 0.55

¹H NMR (400 MHz, Chloroform-*d*) δ 8.83 (d, *J* = 1.7 Hz, 2H), 8.09 (dd, *J* = 8.7, 1.7 Hz, 2H), 8.02 (dd, *J* = 7.4, 1.0 Hz, 1H), 7.96 (t, *J* = 7.6 Hz, 1H), 7.80 (dd, *J* = 7.8, 1.1 Hz, 1H), 7.17 (d, *J* = 8.6 Hz, 2H), 2.74 (s, 6H), 1.59 (s, 9H).

¹³C NMR (101 MHz, CDCl₃) δ 197.5, 168.4, 166.5, 144.3, 135.7, 135.0, 134.2, 132.1, 131.1, 127.4, 127.3, 123.6, 123.5, 122.3, 110.2, 58.6, 29.1, 26.9.

IR (ATR) $\tilde{v}$ [cm⁻¹] = 3909 (vw), 3831 (vw), 3689 (vw), 3608 (vw), 3462 (vw), 3308 (vw), 3074 (vw), 2983 (vw), 2924 (vw), 2680 (vw), 2306 (vw), 2099 (vw), 1995 (vw), 1912 (vw), 1662 (m), 1594 (m), 1487 (m), 1362 (m), 1252 (m), 1106 (m), 1014 (w), 811 (m), 745 (m), 573 (m), 503 (w), 423 (w).

HRMS (APCI, C₂₈H₂₄N₂O₄) calc. 453.1809 [M+H]⁺, found 453.1790 [M+H]⁺.

EA (C₂₈H₂₄N₂O₄) calc. C: 74.32, H: 5.35, N: 6.19; found C: 74.45, H: 5.33, N: 6.09.

9-(2-(*Tert*-butyl)-1,3-dioxoisoindolin-4-yl)-9*H*-carbazole-3-carbonitrile

According to the general procedure (GP 6), 2-(*tert*-butyl)-4-fluoroisoindoline-1,3-dione (664 mg, 3.00 mmol, 1.00 equiv.), 9*H*-carbazole-3-carbonitrile (577 mg, 3.00 mmol, 1.00 equiv.) and tripotassium phosphate (1.27 g, 6.00 mmol, 2.00 equiv.) were reacted in dry DMSO (12 mL). The reaction mixture was extracted and purified by column chromatography over SiO_2 (dichloromethane) to yield 874 mg of the title compound (2.22 mmol, 74%) as a white solid.

R_f (dichloromethane) = 0.38

¹H NMR (300 MHz, Chloroform-*d*) δ 8.48 – 8.43 (m, 1H), 8.20 – 8.12 (m, 1H), 8.00 (dd, *J* = 7.4, 1.2 Hz, 1H), 7.93 (t, *J* = 7.6 Hz, 1H), 7.78 (dd, *J* = 7.7, 1.2 Hz, 1H), 7.64 (dd, *J* = 8.6, 1.6 Hz, 1H), 7.48 (ddd, *J* = 8.3, 7.2, 1.3 Hz, 1H), 7.39 (td, *J* = 7.5, 1.1 Hz, 1H), 7.21 – 7.10 (m, 2H), 1.60 (s, 9H).

¹³C NMR (101 MHz, CDCl₃) δ 168.5, 166.6, 142.6, 141.5, 135.6, 134.9, 134.3, 132.4, 129.3, 127.6, 127.2, 125.6, 124.0, 123.2, 122.6, 121.9, 121.1, 120.3, 110.8, 110.4, 103.5, 58.5, 29.1.

IR (ATR) $\tilde{v}$ [cm⁻¹] = 3983 (vw), 3822 (vw), 3753 (vw), 3633 (vw), 3467 (vw), 3367 (vw), 3064 (vw), 2976 (w), 2664 (vw), 2387 (vw), 2323 (vw), 2220 (w), 1704 (m), 1597 (w), 1477 (w), 1310 (w), 1232 (w), 1102 (w), 1012 (w), 870 (w), 746 (m), 613 (w), 551 (w), 421 (w).

HRMS (APCI, $C_{25}H_{19}N_3O_2$) calc. 394.1550 [M+H]⁺, found 394.1543 [M+H]⁺.

2-(*Tert*-butyl)-4-(3,6-di-*tert*-butyl-9*H*-carbazol-9-yl)isoindoline-1,3-dione

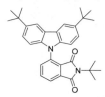 According to the general procedure (GP 6), 2-(*tert*-butyl)-4-fluoroisoindoline-1,3-dione (442 mg, 2.00 mmol, 1.00 equiv.), 3,6-di-*tert*-butyl-9*H*-carbazole (558 mg, 2.00 mmol, 1.00 equiv.) and tripotassium phosphate (848 mg, 4.00 mmol, 2.00 equiv.) were reacted in dry DMSO (8 mL). The reaction mixture was extracted and purified by column chromatography over SiO$_2$ (dichloromethane/pentane 1:4 to 1:3) to yield 761 mg of the title compound (1.58 mmol, 79%) as a yellow solid.

R$_f$ (dichloromethane/pentane 1:4) = 0.28

M.p. = 241 °C

¹H NMR (400 MHz, Chloroform-*d*) δ 8.14 (dd, *J* = 2.0, 0.6 Hz, 2H), 7.88 (dd, *J* = 7.2, 1.1 Hz, 1H), 7.82 (t, *J* = 7.6 Hz, 1H), 7.75 (dd, *J* = 7.9, 1.1 Hz, 1H), 7.43 (dd, *J* = 8.6, 1.9 Hz, 2H), 7.07 (dd, *J* = 8.6, 0.6 Hz, 2H), 1.64 (s, 9H), 1.47 (s, 18H).

¹³C NMR (101 MHz, CDCl$_3$) δ 169.0, 166.9, 143.5, 139.3, 135.1, 134.8, 134.6, 134.1, 126.5, 124.0, 123.7, 121.7, 116.7, 109.6, 58.2, 34.9, 32.1, 29.2.

IR (ATR) ṽ [cm⁻¹] = 3994 (vw), 3866 (vw), 3669 (vw), 3460 (vw), 3311 (vw), 3178 (vw), 2962 (w), 2660 (vw), 2321 (vw), 1706 (m), 1612 (w), 1487 (m), 1360 (m), 1294 (m), 1233 (m), 1100 (m), 1008 (m), 867 (m), 803 (m), 679 (w), 556 (m), 410 (w).

HRMS (APCI, C$_{32}$H$_{36}$N$_2$O$_2$) calc. 481.2850 [M+H]⁺, found 481.2839 [M+H]⁺.

4-(4-(3,6-Di-*tert*-butyl-9*H*-carbazol-9-yl)-1,3-dioxoisoindolin-2-yl)benzonitrile

According to the general procedure (GP 6), 4-(4-fluoro-1,3-dioxoisoindolin-2-yl)benzonitrile (1.60 g, 6.00 mmol, 1.00 equiv.), 3,6-di-*tert*-butyl-9*H*-carbazole (1.68 g, 6.00 mmol, 1.00 equiv.) and tripotassium phosphate (2.55 g, 12.0 mmol, 2.00 equiv.) were reacted in dry DMSO (24 mL). The reaction mixture was extracted and purified by column chromatography over SiO2 (dichloromethane/pentane 1:1 to dichloromethane) to yield 1.39 g of the title compound (2.64 mmol, 44%) as a yellow solid.

R_f (dichloromethane/pentane 1:1) = 0.17

1H NMR (400 MHz, Chloroform-*d*) δ 8.18 – 8.13 (m, 2H), 8.09 (dd, *J* = 7.1, 1.3 Hz, 1H), 8.01 – 7.89 (m, 2H), 7.75 – 7.68 (m, 2H), 7.68 – 7.60 (m, 2H), 7.44 (dd, *J* = 8.6, 2.0 Hz, 2H), 7.15 – 7.07 (m, 2H), 1.46 (s, 18H).

13C NMR (101 MHz, CDCl3) δ 165.9, 163.8, 144.0, 139.0, 136.4, 136.1, 135.8, 135.3, 133.8, 132.9, 126.5, 125.7, 124.2, 123.8, 123.1, 118.3, 116.9, 111.4, 109.4, 34.9, 32.1.

IR (ATR) $\tilde{v}$ [cm^{-1}] = 3979 (w), 3873 (w), 3735 (w), 3495 (w), 3399 (w), 3248 (vw), 3049 (w), 2953 (w), 2659 (w), 2322 (vw), 2228 (w), 1719 (m), 1603 (w), 1486 (m), 1362 (m), 1087 (w), 1004 (w), 808 (m), 744 (m), 676 (w), 555 (m), 411 (w).

HRMS (APCI, C35H31N3O2) calc. 526.2489 [M+H]+, found 526.2480 [M+H]+.

4-(9*H*-Carbazol-9-yl)-2-(3,4,5-trimethoxyphenyl)isoindoline-1,3-dione

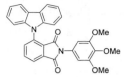

According to the general procedure (GP 6), 4-fluoro-2-(3,4,5-trimethoxyphenyl)isoindoline-1,3-dione (663 mg, 2.00 mmol, 1.00 equiv.), 9*H*-carbazole (335 mg, 2.00 mmol, 1.00 equiv.) and tripotassium phosphate (850 mg, 4.00 mmol, 2.00 equiv.) were reacted in dry DMSO (8.00 mL). The reaction mixture was extracted and purified by column chromatography over SiO$_2$ (dichloromethane to dichloromethane/ethyl acetate 99:1) to yield 694 mg of the title compound (1.45 mmol, 72%) as a yellow solid.

R$_f$ (dichloromethane/ethyl acetate 99:1) = 0.35

1H NMR (400 MHz, Chloroform-*d*) δ 8.16 (d, *J* = 7.7 Hz, 2H), 8.11 (dd, *J* = 7.2, 1.1 Hz, 1H), 8.01 (dd, *J* = 8.0, 7.2 Hz, 1H), 7.95 (dd, *J* = 8.0, 1.1 Hz, 1H), 7.40 (ddd, *J* = 8.3, 7.2, 1.3 Hz, 2H), 7.37 – 7.28 (m, 2H), 7.21 (dt, *J* = 8.2, 0.9 Hz, 2H), 6.60 (s, 2H), 3.84 (s, 3H), 3.81 (s, 6H).

13C NMR (101 MHz, CDCl$_3$) δ 166.6, 164.4, 153.4, 140.7, 137.9, 136.1, 135.2, 135.0, 134.2, 126.9, 126.5, 126.2, 124.0, 123.2, 120.9, 120.7, 110.0, 104.3, 61.0, 56.3.

IR (ATR) $\tilde{v}$ [cm^{-1}] = 3819 (vw), 3652 (vw), 3555 (vw), 3479 (vw), 3394 (vw), 3065 (w), 2924 (w), 2654 (vw), 2314 (vw), 1720 (m), 1595 (w), 1449 (m), 1296 (m), 1229 (m), 1125 (m), 1003 (w), 882 (w), 747 (m), 654 (m), 549 (w), 422 (w).

HRMS (APCI, C$_{29}$H$_{22}$N$_2$O$_5$) calc. 479.1601 [M+H]$^+$, found 479.1591 [M+H]$^+$.

Methyl 4-(4-(9*H*-carbazol-9-yl)-1,3-dioxoisoindolin-2-yl)benzoate

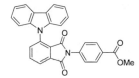

According to the general procedure (GP 6), methyl 4-(4-fluoro-1,3-dioxoisoindolin-2-yl)benzoate (2.10 g, 7.00 mmol, 1.00 equiv.), 9*H*-carbazole (1.17 g, 7.00 mmol, 1.00 equiv.) and tripotassium phosphate (2.97 g, 14.0 mmol, 2.00 equiv.) were reacted in dry DMSO (28 mL). The reaction mixture was extracted and purified by column chromatography over SiO_2 (dichloromethane/pentane to dichloromethane) to yield 2.11 g of the title compound (4.73 mmol, 68%) as a yellow solid.

R$_f$ (dichloromethane/pentane 1:1) = 0.21

^{1}H NMR (400 MHz, Chloroform-*d*) δ 8.16 (dt, *J* = 7.6, 0.9 Hz, 2H), 8.14 – 8.06 (m, 3H), 8.02 (dd, *J* = 8.0, 7.2 Hz, 1H), 7.96 (dd, *J* = 8.0, 1.1 Hz, 1H), 7.56 – 7.50 (m, 2H), 7.40 (ddd, *J* = 8.3, 7.2, 1.3 Hz, 2H), 7.32 (td, *J* = 7.5, 1.1 Hz, 2H), 7.20 (dt, *J* = 8.1, 0.9 Hz, 2H), 3.91 (s, 3H).

^{13}C NMR (101 MHz, CDCl$_3$) δ 166.4, 166.1, 163.9, 140.7, 136.4, 135.6, 135.4, 135.2, 134.1, 130.4, 129.4, 126.5, 126.3, 126.2, 125.9, 124.1, 123.4, 121.0, 120.8, 109.9, 52.4.

IR (ATR) $\tilde{v}$ [cm^{-1}] = 3063 (vw), 2944 (vw), 2838 (vw), 2679 (vw), 2584 (vw), 2480 (vw), 2394 (vw), 2322 (vw), 2099 (vw), 1725 (m), 1603 (w), 1479 (w), 1364 (m), 1280 (m), 1111 (w), 883 (w), 748 (m), 550 (w), 423 (w).

HRMS (ESI, C$_{28}$H$_{18}$N$_2$O$_4$) calc. 447.1339 [M+H]$^+$, found 447.1334 [M+H]$^+$.

4-(9,9-Dimethylacridin-10(9H)-yl)-2-mesitylisoindoline-1,3-dione

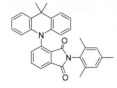

In a sealable vial, 4-fluoro-2-mesitylisoindoline-1,3-dione (567 mg, 2.00 mmol, 1.00 equiv.), 9,9-dimethyl-9,10-dihydroacridine (419 mg, 2.00 mmol, 1.00 equiv.) and tripotassium phosphate (849 mg, 4.00 mmol, 2.00 equiv.) were evacuated and backfilled with argon. Dry DMSO (8 ml) was added and the resulting mixture heated at 110 °C for 16 h. After cooling to room temperature, the reaction mixture was poured into an excess of water and extracted with dichloromethane three times. The combined organic layers were washed with brine, dried over sodium sulfate, reduced in vacuum. The crude product was purified by column chromatography over SiO_2 (dichloromethane/pentane 1:1 to dichloromethane) to yield 577 mg of the title compound (1.24 mmol, 62%) as an orange solid.

R_f (dichloromethane) = 0.33

¹H NMR (400 MHz, Chloroform-*d*) δ 8.12 (dd, *J* = 7.4, 1.0 Hz, 1H), 8.05 (t, *J* = 7.6 Hz, 1H), 7.78 (dd, *J* = 7.8, 1.0 Hz, 1H), 7.52 – 7.43 (m, 2H), 6.97 – 6.92 (m, 4H), 6.90 (s, 2H), 6.22 – 6.15 (m, 2H), 2.26 (s, 3H), 2.02 (s, 6H), 1.73 (s, 6H).

¹³C NMR (101 MHz, CDCl₃) δ 166.7, 164.2, 140.0, 139.6, 139.4, 138.7, 136.7, 136.4, 135.6, 130.6, 129.3, 127.0, 126.4, 125.9, 123.9, 121.3, 113.0, 36.2, 21.2, 18.1.

IR (ATR) $\tilde{v}$ [cm⁻¹] = 3480 (vw), 2916 (vw), 2333 (vw), 1725 (w), 1589 (w), 1475 (w), 1325 (w), 1114 (w), 880 (w), 747 (w), 564 (w), 425 (vw).

HRMS (ESI, C₃₂H₂₈N₂O₂) calc. 472.2145 [M]⁺, found 472.2140 [M]⁺.

2-Mesityl-4-(3-(*p*-tolyl)-9*H*-carbazol-9-yl)isoindoline-1,3-dione

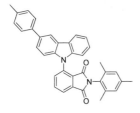

In a sealable vial, 4-(3-bromo-9*H*-carbazol-9-yl)-2-mesitylisoindoline-1,3-dione (153 mg, 0.300 mmol, 1.00 equiv.), *p*-tolylboronic acid (61.3 mg, 0.450 mmol, 1.50 equiv.), tripotassium phosphate (191 mg, 0.900 mmol, 3.00 equiv.), Pd₂(dba)₃ (6.00 mg, 6.55 µmol, 0.022 equiv.) and SPhos (4.90 mg, 11.9 µmol, 0.04 equiv.) were evacuated and backfilled with argon. Toluene (2.3 mL) and water (0.2 mL) were added and the resulting mixture was degassed by bubbling argon through the liquid phase for 10–15 min. Subsequently, the mixture was heated at 110 °C for 16 hours. After cooling to room temperature, an excess of water was added, and the mixture was extracted with dichloromethane. The combined organic layers were washed with brine, dried over sodium sulfate and reduced in vacuum. The crude product was purified by column chromatography over SiO_2 (dichloromethane/cyclohexane 1:1 to dichloromethane) to yield 110 mg of the title compound (0.211 mmol, 70%) as a yellow solid.

R_f (dichloromethane/cyclohexane 1:1) = 0.31

¹H NMR (400 MHz, Chloroform-*d*) δ 8.35 (d, *J* = 1.7 Hz, 1H), 8.20 (d, *J* = 7.7 Hz, 1H), 8.10 (dd, *J* = 5.9, 2.4 Hz, 1H), 8.02 – 7.95 (m, 2H), 7.68 – 7.60 (m, 3H), 7.44 (ddd, *J* = 8.3, 7.2, 1.3 Hz, 1H), 7.37 – 7.26 (m, 5H), 6.97 (s, 2H), 2.45 (s, 3H), 2.30 (s, 3H), 2.17 (s, 3H), 2.16 (s, 3H).

¹³C NMR (101 MHz, CDCl₃) δ 166.6, 164.5, 140.9, 139.8, 139.4, 139.0, 136.5, 136.3, 136.0, 134.7, 134.5, 134.5, 134.4, 129.6, 129.3, 129.3, 127.3, 126.9, 126.4, 126.2, 125.5, 124.6, 124.3, 123.0, 121.0, 120.7, 118.9, 110.3, 110.2, 21.2, 21.2, 18.2, 18.1.

IR (ATR) ṽ [cm⁻¹] = 3974 (w), 3745 (w), 3647 (w), 3581 (w), 3477 (w), 3022 (w), 2921 (w), 2726 (w), 1718 (s), 1601 (w), 1490 (m), 1372 (m), 1227 (m), 1119 (m), 1028 (w), 883 (m), 748 (m), 552 (m).

HRMS (APCI, C₃₆H₂₈N₂O₂) calc. 521.2224 [M+H]⁺, found 521.2218 [M+H]⁺.

4-(3-Butyl-9*H*-carbazol-9-yl)-2-mesitylisoindoline-1,3-dione

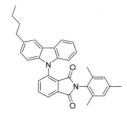

In a sealable vial, 4-(3-bromo-9*H*-carbazol-9-yl)-2-mesitylisoindoline-1,3-dione (510 mg, 1.00 mmol, 1.00 equiv.), butylboronic acid (153 mg, 1.50 mmol, 1.50 equiv.), tripotassium phosphate (638 mg, 3.00 mmol, 3.00 equiv.), Pd$_2$(dba)$_3$ (18.3 mg, 20.0 µmol, 0.02 equiv.) and SPhos (16.4 mg, 40.0 µmol, 0.04 equiv.) were evacuated and backfilled with argon. Toluene (5.3 mL) and water (0.7 mL) were added and the resulting mixture was degassed by bubbling argon through the liquid phase for 10–15 min. Subsequently, the mixture was heated at 110 °C for 3 hours. After cooling to room temperature, an excess of water was added, and the mixture was extracted with dichloromethane. The combined organic layers were washed with brine, dried over sodium sulfate and reduced in vacuum. The crude product was purified by column chromatography over SiO$_2$ (dichloromethane/cyclohexane 1:1 to 2:1) to yield 437 mg of the title compound (0.898 mmol, 90%) as a yellow solid.

R$_f$ (dichloromethane/cyclohexane 1:1) = 0.35

¹H NMR (400 MHz, Chloroform-*d*) δ 8.13 (d, *J* = 7.6 Hz, 1H), 8.09 – 8.04 (m, 1H), 7.98 – 7.92 (m, 3H), 7.43 – 7.35 (m, 1H), 7.31 (td, *J* = 7.4, 1.0 Hz, 1H), 7.27 – 7.21 (m, 2H), 7.17 (d, *J* = 8.3 Hz, 1H), 6.96 (s, 2H), 2.86 – 2.73 (m, 2H), 2.30 (s, 3H), 2.15 (s, 6H), 1.78 – 1.68 (m, 2H), 1.45 (h, *J* = 7.3 Hz, 2H), 0.99 (t, *J* = 7.3 Hz, 3H).

¹³C NMR (101 MHz, CDCl$_3$) δ 166.7, 164.6, 140.7, 139.4, 138.9, 136.3, 135.9, 135.6, 135.1, 134.5, 134.4, 129.3, 127.0, 126.8, 126.2, 125.8, 124.2, 124.2, 122.7, 120.8, 120.5, 120.0, 110.1, 109.8, 35.8, 34.5, 22.6, 21.2, 18.2, 14.2.

IR (ATR) ṽ [cm^{-1}] = 3952 (w), 3823 (w), 3753 (w), 3652 (w), 3477 (w), 2923 (w), 2487 (w), 2350 (w), 1721 (m), 1607 (w), 1486 (m), 1371 (m), 1228 (m), 1114 (m), 1005 (w), 883 (m), 746 (m), 551 (m), 424 (w).

HRMS (APCI, C$_{33}$H$_{30}$N$_2$O$_2$) calc. 487.2380 [M+H]$^+$, found 487.2377 [M+H]$^+$.

EA (C$_{33}$H$_{30}$N$_2$O$_2$) calc. C: 81.45, H: 6.21, N: 5.76; found C: 81.44, H: 6.17, N: 5.68.

4-(9*H*-Carbazol-9-yl)-2-(4'-methyl-[1,1'-biphenyl]-3-yl)isoindoline-1,3-dione

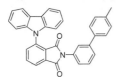 In a sealable vial, 2-(3-bromophenyl)-4-(9*H*-carbazol-9-yl)isoindoline-1,3-dione (140 mg, 0.300 mmol, 1.00 equiv.), *p*-tolylboronic acid (61.3 mg, 0.450 mmol, 1.50 equiv.), tripotassium phosphate (191 mg, 0.900 mmol, 3.00 equiv.), $Pd_2(dba)_3$ (5.00 mg, 5.46 µmol, 0.018 equiv.) and SPhos (5.00 mg, 12.2 µmol, 0.04 equiv.) were evacuated and backfilled with argon. Toluene (1.8 mL) and water (0.2 mL) were added and the resulting mixture was degassed by bubbling argon through the liquid phase for 10–15 min. Subsequently, the mixture was heated at 110 °C for 16 hours. After cooling to room temperature, an excess of water was added, and the mixture was extracted with dichloromethane. The combined organic layers were washed with brine, dried over sodium sulfate and reduced in vacuum. The crude product was purified by column chromatography over SiO_2 (dichloromethane/cyclohexane 1:1 to 2:1) to yield 45.0 mg of the title compound (0.0940 mmol, 31%) as a yellow solid.

R_f (dichloromethane/cyclohexane 1:1) = 0.41

1H NMR (400 MHz, Chloroform-*d*) δ 8.20 – 8.08 (m, 3H), 8.01 (t, *J* = 7.6 Hz, 1H), 7.95 (dd, *J* = 8.0, 1.1 Hz, 1H), 7.59 (t, *J* = 1.9 Hz, 1H), 7.54 (dt, *J* = 7.8, 1.5 Hz, 1H), 7.50 – 7.37 (m, 5H), 7.33 (qd, *J* = 7.2, 1.4 Hz, 3H), 7.21 (dd, *J* = 8.0, 5.4 Hz, 4H), 2.37 (s, 3H).

13C NMR (101 MHz, CDCl₃) δ 166.6, 164.4, 142.4, 140.8, 137.6, 137.4, 136.1, 135.1, 135.1, 134.4, 131.9, 129.6, 129.4, 127.2, 126.8, 126.7, 126.2, 125.1, 124.9, 124.1, 123.3, 120.9, 120.7, 110.0, 21.2.

IR (ATR) $\tilde{v}$ [cm⁻¹] = 3984 (w), 3856 (w), 3633 (w), 3479 (w), 3024 (w), 2921 (w), 2332 (w), 1714 (s), 1601 (w), 1479 (m), 1368 (s), 1230 (m), 1106 (m), 1007 (w), 754 (m), 626 (m), 529 (m), 425 (m).

HRMS (APCI, $C_{33}H_{22}N_2O_2$) calc. 479.1754 [M+H]⁺, found 479.1750 [M+H]⁺.

4-(3-Acetyl-6-(p-tolyl)-9H-carbazol-9-yl)-2-mesitylisoindoline-1,3-dione

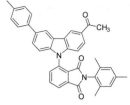

In a sealable vial, 4-(3-acetyl-6-bromo-9H-carbazol-9-yl)-2-mesitylisoindoline-1,3-dione (275 mg, 0.499 mmol, 1.00 equiv.), p-tolylboronic acid (102 mg, 0.748 mmol, 1.50 equiv.), tripotassium phosphate (317 mg, 1.49 mmol, 3.00 equiv.), Pd₂(dba)₃ (9.00 mg, 9.83 µmol, 0.02 equiv.) and SPhos (8.20 mg, 20.0 µmol, 0.04 equiv.) were evacuated and backfilled with argon. Toluene (4 mL) and water (0.4 mL) were added and the resulting mixture was degassed by bubbling argon through the liquid phase for 10–15 min. Subsequently, the mixture was heated at 110 °C for 3 hours. After cooling to room temperature, an excess of water was added, and the mixture was extracted with dichloromethane. The combined organic layers were washed with brine, dried over sodium sulfate and reduced in vacuum. The crude product was purified by column chromatography over SiO₂ (dichloromethane/cyclohexane 1:1 to dichloromethane to dichloromethane/ethyl acetate 50:1 to 25:1) to yield 262 mg of the title compound (0.466 mmol, 93%) as a yellow solid.

R_f (dichloromethane/ethyl acetate 50:1) = 0.61

¹H NMR (400 MHz, Chloroform-d) δ 8.82 (d, J = 1.6 Hz, 1H), 8.39 – 8.35 (m, 1H), 8.15 (dd, J = 7.3, 1.0 Hz, 1H), 8.08 – 8.03 (m, 2H), 7.99 (dd, J = 8.0, 1.1 Hz, 1H), 7.67 (dd, J = 8.5, 1.8 Hz, 1H), 7.63 – 7.57 (m, 2H), 7.30 (d, J = 8.2 Hz, 3H), 7.22 (d, J = 8.7 Hz, 1H), 6.95 – 6.88 (m, 2H), 2.73 (s, 3H), 2.43 (s, 3H), 2.27 (s, 3H), 2.13 (s, 3H), 2.10 (s, 3H).

¹³C NMR (101 MHz, CDCl₃) δ 197.6, 166.4, 164.5, 143.8, 140.7, 139.6, 138.6, 136.8, 136.3, 136.3, 136.2, 135.4, 134.7, 134.4, 133.9, 130.7, 129.7, 129.4, 129.4, 127.3, 126.8, 126.8, 126.7, 126.3, 124.6, 124.0, 123.7, 122.1, 119.2, 110.5, 109.9, 26.8, 21.2, 21.2, 18.2, 18.1.

IR (ATR) ṽ [cm⁻¹] = 3478 (vw), 3297 (vw), 3017 (vw), 2917 (vw), 2727 (vw), 2487 (vw), 2320 (vw), 2099 (vw), 1722 (m), 1597 (w), 1483 (m), 1370 (m), 1261 (m), 1111 (w), 1030 (w), 884 (w), 801 (w), 679 (w), 551 (w), 423 (vw).

HRMS (APCI, C₃₈H₃₀N₂O₃) calc. 563.2329 [M+H]⁺, found 563.2305 [M+H]⁺.

EA (C₃₈H₃₀N₂O₃) calc. C: 81.12, H: 5.37, N: 4.98; found C: 80.26, H: 5.29, N: 4.84.

4-(3-Decyl-6-dodecanoyl-9*H*-carbazol-9-yl)-2-mesitylisoindoline-1,3-dione

In a sealable vial, 4-(3-bromo-6-undecanoyl-9*H*-carbazol-9-yl)-2-mesitylisoindoline-1,3-dione (346 mg, 0.500 mmol, 1.00 equiv.), decylboronic acid (140 mg, 0.752 mmol, 1.50 equiv.), tripotassium phosphate (318 mg, 1.49 mmol, 3.00 equiv.), Pd$_2$(dba)$_3$ (9.00 mg, 9.83 μmol, 0.02 equiv.) and SPhos (8.20 mg, 20.0 μmol, 0.04 equiv.) were evacuated and backfilled with argon. Toluene (3 mL) and water (0.3 mL) were added and the resulting mixture was degassed by bubbling argon through the liquid phase for 10-15 min. Subsequently, the mixture was heated at 110 °C for 3 hours. After cooling to room temperature, an excess of water was added, and the mixture was extracted with dichloromethane. The combined organic layers were washed with brine, dried over sodium sulfate and reduced in vacuum. The crude product was purified by column chromatography over SiO$_2$ (dichloromethane/cyclohexane 2:1 to dichloromethane to dichloromethane/ethyl acetate 50:1 to 10:1) to yield 294 mg of the title compound (0.391 mmol, 78%) as a yellow solid.

R$_f$ (dichloromethane) = 0.15

¹H NMR (300 MHz, Chloroform-*d*) δ 8.76 (d, *J* = 1.6 Hz, 1H), 8.12 (dd, *J* = 7.2, 1.2 Hz, 1H), 8.08 – 7.89 (m, 4H), 7.30 – 7.24 (m, 1H), 7.18 (dd, *J* = 8.5, 5.4 Hz, 2H), 6.94 (d, *J* = 2.9 Hz, 2H), 3.09 (t, *J* = 7.4 Hz, 2H), 2.80 (t, *J* = 7.7 Hz, 2H), 2.27 (s, 3H), 2.13 (s, 3H), 2.09 (s, 3H), 1.90 – 1.63 (m, 4H), 1.50 – 1.13 (m, 30H), 0.93 – 0.80 (m, 6H).

¹³C NMR (101 MHz, CDCl$_3$) δ 200.0, 166.5, 164.5, 143.4, 139.8, 139.5, 136.7, 136.3, 136.1, 134.6, 134.4, 134.3, 130.4, 129.3, 127.6, 126.9, 126.6, 126.4, 124.3, 123.9, 123.4, 121.6, 120.3, 110.0, 109.8, 38.7, 36.1, 32.3, 32.1, 29.8, 29.7, 29.7, 29.5, 29.5, 29.5, 25.0, 22.8, 22.8, 21.2, 18.2, 18.1, 18.1, 18.1, 14.2, 14.2.

IR (ATR) ṽ [cm^{-1}] = 2921 (m), 2851 (m), 1723 (s), 1599 (m), 1486 (s), 1372 (s), 1198 (m), 1114 (m), 1030 (w), 880 (m), 750 (m), 552 (w), 423 (w).

HRMS (ESI, C$_{51}$H$_{64}$N$_2$O$_3$) calc. 753.4990 [M+H]$^+$, found 753.4986 [M+H]$^+$.

4-(3-Acetyl-9*H*-carbazol-9-yl)-2-mesitylisoindoline-1,3-dione

Under inert conditions, 4-(9*H*-carbazol-9-yl)-2-mesitylisoindoline-1,3-dione (344 mg, 0.750 mmol, 1.00 equiv.) and aluminum chloride (120 mg, 0.900 mmol, 1.20 equiv.) were suspended in dry dichloromethane (7.5 mL). Then acetyl chloride (65.0 mg, 0.825 mmol, 1.10 equiv.) was added slowly and the reaction mixture was stirred for 3 hours at room temperature. Water and 1M HCl were added carefully and the layers separated. The aqueous layer was extracted with dichloromethane. The combined organic layers were washed with brine, dried over sodium sulfate and reduced in vacuum. The crude product was purified by column chromatography over SiO_2 (dichloromethane/cyclohexane 1:1 to dichloromethane to dichloromethane/ethyl acetate 50:1 to 20:1) to yield 344 mg of the title compound (0.728 mmol, 97%) as a yellow solid.

R$_f$ (dichloromethane/ethyl acetate 50:1) = 0.41

¹H NMR (400 MHz, Chloroform-*d*) δ 8.78 (s, 1H), 8.23 – 8.16 (m, 1H), 8.14 (d, *J* = 7.3 Hz, 1H), 8.09 – 8.01 (m, 2H), 7.97 (d, *J* = 8.0 Hz, 1H), 7.50 – 7.42 (m, 1H), 7.42 – 7.32 (m, 1H), 7.29 – 7.24 (m, 1H), 7.21 (d, *J* = 8.6 Hz, 1H), 6.93 (d, *J* = 5.3 Hz, 2H), 2.72 (s, 3H), 2.27 (s, 3H), 2.12 (s, 3H), 2.08 (s, 3H).

¹³C NMR (101 MHz, CDCl₃) δ 197.6, 166.4, 164.4, 143.3, 141.4, 139.6, 136.3, 136.2, 136.2, 134.6, 134.5, 133.9, 130.6, 129.4, 129.3, 126.9, 126.8, 126.7, 124.1, 123.9, 123.7, 122.0, 121.8, 120.9, 110.3, 109.8, 26.8, 21.2, 18.2, 18.1.

IR (ATR) $\tilde{v}$ [cm⁻¹] = 3976 (w), 3902 (w), 3835 (w), 3686 (w), 3608 (w), 3472 (w), 3400 (w), 3328 (w), 3235 (w), 3052 (w), 2917 (w), 2322 (w), 2180 (w), 1719 (m), 1593 (m), 1490 (m), 1373 (m), 1251 (m), 1118 (m), 1041 (m), 881 (m), 751 (m), 628 (m), 545 (m), 416 (m).

HRMS (APCI, C₃₁H₂₄N₂O₃) calc. 473.1860 [M+H]⁺, found 473.1858 [M+H]⁺.

2-Mesityl-4-(3-(2,4,6-trimethylbenzoyl)-9*H*-carbazol-9-yl)isoindoline-1,3-dione

Under inert conditions, 4-(9*H*-carbazol-9-yl)-2-mesitylisoindoline-1,3-dione (3.01 g, 7.00 mmol, 1.00 equiv.) and aluminum chloride (1.12 g, 8.40 mmol, 1.20 equiv.) were suspended in dry 1,2-dichlorobenzene (65 mL). Then 2,4,6-trimethylbenzoyl chloride (1.41 g, 7.70 mmol, 1.10 equiv.) was added slowly and the reaction mixture was stirred for 3 hours at 65 °C. Water and 1M HCl were added carefully and the layers separated. The aqueous layer was extracted with dichloromethane. The combined organic layers were washed with brine, dried over sodium sulfate and reduced in vacuum. The crude product was purified by column chromatography over SiO_2 (dichloromethane/cyclohexane 1:1 to dichloromethane to dichloromethane/ethyl acetate 50:1) to yield 2.71 g of the title compound (4.69 mmol, 67%) as a yellow solid.

R_f (dichloromethane/ethyl acetate 50:1) = 0.72

M.p. = 183 °C

^{1}H NMR (400 MHz, DMSO-d_6) δ 8.65 (s, 1H), 8.30 (d, J = 7.7 Hz, 1H), 8.24 – 8.10 (m, 3H), 7.72 (d, J = 8.7 Hz, 1H), 7.52 – 7.38 (m, 2H), 7.38 – 7.28 (m, 2H), 6.97 (d, J = 4.2 Hz, 4H), 2.31 (s, 3H), 2.23 (s, 3H), 2.03 (s, 6H), 2.02 (s, 6H).

^{13}C NMR (101 MHz, DMSO) δ 198.8, 166.1, 164.4, 143.3, 141.1, 138.7, 137.8, 137.3, 137.1, 136.2, 136.2, 134.9, 133.9, 133.4, 132.4, 129.9, 128.8, 128.8, 128.2, 127.1, 126.9, 126.5, 124.0, 123.4, 122.9, 121.5, 121.0, 110.7, 110.7, 20.8, 20.5, 19.0, 17.5, 17.5.

IR (ATR) $\tilde{v}$ [cm^{-1}] = 3933 (w), 3824 (w), 3650 (w), 3481 (w), 3391 (w), 3306 (w), 2918 (w), 2483 (w), 2322 (w), 2260 (w), 2161 (w), 2047 (w), 1714 (m), 1591 (m), 1481 (m), 1369 (m), 1253 (m), 1114 (m), 1028 (w), 880 (m), 746 (m), 540 (m), 421 (m), 0 (vs), 416 (m), 0 (w), 0 (vs), 0 (vs), 0 (vs)

HRMS (APCI, $C_{39}H_{32}N_2O_3$) calc. 577.2486 [M+H]$^+$, found 577.2461 [M+H]$^+$.

EA ($C_{39}H_{32}N_2O_3$) calc. C: 81.23, H: 5.59, N: 4.86; found C: 80.54, H: 5.66, N: 4.81.

4-(3-Dodecanoyl-9*H*-carbazol-9-yl)-2-mesitylisoindoline-1,3-dione

Under inert conditions, 4-(9*H*-carbazol-9-yl)-2-mesitylisoindoline-1,3-dione (323 mg, 0.750 mmol, 1.00 equiv.) and aluminum chloride (120 mg, 0.900 mmol, 1.20 equiv.) were suspended in dry dichloromethane (7.5 mL). Then dodecanoyl chloride (181 mg, 0.825 mmol, 1.10 equiv.) was added slowly and the reaction mixture was stirred for 3 hours at room temperature. Water and 1M HCl were added carefully and the layers separated. The aqueous layer was extracted with dichloromethane. The combined organic layers were washed with brine, dried over sodium sulfate and reduced in vacuum. The crude product was purified by column chromatography over SiO_2 (dichloromethane/cyclohexane 1:1 to dichloromethane to dichloromethane/ethyl acetate 50:1) to yield 320 mg of the title compound (0.525 mmol, 70%) as a yellow solid.

R$_f$ (dichloromethane/ethyl acetate 50:1) = 0.67

^{1}H NMR (400 MHz, Chloroform-*d*) δ 8.79 (d, *J* = 1.6 Hz, 1H), 8.21 (d, *J* = 7.7 Hz, 1H), 8.13 (d, *J* = 7.3 Hz, 1H), 8.10 – 8.00 (m, 2H), 7.96 (d, *J* = 8.0 Hz, 1H), 7.47 – 7.41 (m, 1H), 7.41 – 7.33 (m, 1H), 7.29 – 7.24 (m, 1H), 7.22 (d, *J* = 8.7 Hz, 1H), 6.97 – 6.88 (m, 2H), 3.09 (t, *J* = 7.4 Hz, 2H), 2.27 (s, 3H), 2.13 (s, 3H), 2.09 (s, 3H), 1.81 (p, *J* = 7.5 Hz, 2H), 1.49 – 1.20 (m, 16H), 0.94 – 0.85 (m, 3H).

^{13}C NMR (101 MHz, CDCl$_3$) δ 200.0, 166.4, 164.4, 143.2, 141.3, 139.5, 136.3, 136.2, 136.2, 134.6, 134.5, 133.9, 130.5, 129.3, 129.3, 126.8, 126.7, 126.5, 124.1, 123.9, 123.6, 121.7, 121.7, 120.9, 110.3, 109.8, 38.7, 32.0, 29.8, 29.7, 29.7, 29.6, 29.5, 24.9, 22.8, 21.2, 18.1, 18.1, 14.2.

IR (ATR) $\tilde{v}$ [cm^{-1}] = 3976 (vw), 3843 (vw), 3773 (vw), 3691 (vw), 3479 (vw), 3309 (vw), 3056 (w), 2920 (w), 2851 (w), 2482 (vw), 1717 (m), 1599 (m), 1483 (m), 1371 (m), 1230 (m), 1112 (m), 1002 (w), 881 (m), 747 (m), 617 (w), 548 (m), 417 (w).

HRMS (APCI, C$_{41}$H$_{44}$N$_2$O$_3$) calc. 613.3425 [M+H]$^+$, found 613.3424 [M+H]$^+$.

2-Mesityl-4-(3-(2,2,2-trifluoroacetyl)-9*H*-carbazol-9-yl)isoindoline-1,3-dione

Under inert conditions, 4-(9*H*-carbazol-9-yl)-2-mesitylisoindoline-1,3-dione (431 mg, 1.00 mmol, 1.00 equiv.) and aluminum chloride (320 mg, 2.40 mmol, 2.40 equiv.) were suspended in dry 1,2-dichlorobenzene (10 mL). Then 2,2,2-trifluoroacetic anhydride (231 mg, 1.10 mmol, 1.10 equiv.) was added slowly and the reaction mixture was stirred for 3 hours at 65 °C. Water and 1M HCl were added carefully and the layers separated. The aqueous layer was extracted with dichloromethane. The combined organic layers were washed with brine, dried over sodium sulfate and reduced in vacuum. The crude product was purified by column chromatography over SiO_2 (dichloromethane/cyclohexane 1:1 to dichloromethane) to yield 428 mg of the title compound (0.812 mmol, 81%) as an off-white solid.

R_f (dichloromethane) = 0.56

^{1}H NMR (400 MHz, Chloroform-*d*) δ 8.89 (t, *J* = 1.4 Hz, 1H), 8.26 – 8.17 (m, 2H), 8.15 (d, *J* = 8.9 Hz, 1H), 8.09 (t, *J* = 7.7 Hz, 1H), 7.98 (d, *J* = 8.0 Hz, 1H), 7.50 (ddd, *J* = 8.3, 7.3, 1.3 Hz, 1H), 7.42 (td, *J* = 7.5, 1.0 Hz, 1H), 7.30 – 7.23 (m, 2H), 6.94 (d, *J* = 6.4 Hz, 2H), 2.28 (s, 3H), 2.12 (s, 3H), 2.08 (s, 3H).

^{13}C NMR (101 MHz, CDCl$_3$) δ 180.0, 179.6, 166.3, 164.4, 144.6, 141.6, 139.7, 136.4, 136.2, 136.2, 134.7, 134.5, 133.3, 129.5, 129.4, 129.4, 128.3, 128.3, 127.6, 127.0, 126.6, 124.3, 124.3, 124.2, 123.7, 122.9, 122.4, 121.2, 118.8, 115.9, 110.5, 110.5, 29.8, 21.2, 18.2, 18.1.

^{19}F NMR (376 MHz, CDCl$_3$) δ -74.5.

IR (ATR) $\tilde{v}$ [cm^{-1}] = 3995 (w), 3792 (vw), 3481 (vw), 3368 (vw), 3295 (vw), 3205 (vw), 3062 (w), 2921 (w), 2733 (vw), 1714 (m), 1587 (m), 1483 (m), 1374 (s), 1141 (s), 960 (m), 883 (m), 745 (m), 548 (m), 417 (m).

HRMS (APCI, C$_{31}$H$_{21}$F$_3$N$_2$O$_3$) calc. 527.1577 [M+H]$^+$, found 527.1573 [M+H]$^+$.

4-(3,6-Bis(2,2,2-trifluoroacetyl)-9*H*-carbazol-9-yl)-2-mesitylisoindoline-1,3-dione

Under inert conditions, 4-(9*H*-carbazol-9-yl)-2-mesitylisoindoline-1,3-dione (108 mg, 0.250 mmol, 1.00 equiv.) and aluminum chloride (334 mg, 2.50 mmol, 10.0 equiv.) were suspended in dry 1,2-dichlorobenzene (5 mL). Then 2,2,2-trifluoroacetic anhydride (527 mg, 2.50 mmol, 10.0 equiv.) was added slowly and the reaction mixture was stirred for 3 hours at room temperature. Water and 1M HCl were added carefully and the layers separated. The aqueous layer was extracted with dichloromethane. The combined organic layers were washed with brine, dried over sodium sulfate and reduced in vacuum. The crude product was purified by column chromatography over SiO_2 (dichloromethane/cyclohexane 1:1 to 3:1 to dichloromethane) to yield 90.5 mg of the title compound (0.173 mmol, 69%) as a white solid.

R_f (dichloromethane) = 0.36

¹H NMR (400 MHz, Chloroform-*d*) δ 8.91 (d, *J* = 1.8 Hz, 2H), 8.26 (d, *J* = 7.5 Hz, 1H), 8.22 – 8.17 (m, 2H), 8.15 (d, *J* = 7.6 Hz, 1H), 7.99 (dd, *J* = 8.0, 0.9 Hz, 1H), 7.33 (d, *J* = 8.8 Hz, 2H), 6.94 (s, 2H), 2.27 (s, 3H), 2.09 (s, 6H).

¹³C NMR (101 MHz, Chloroform-*d*) δ 179.7 (q, *J* = 34.9 Hz), 165.9, 164.3, 145.5, 139.9, 136.8, 136.1, 134.8, 134.3, 131.9, 129.5, 127.3, 126.4, 125.2, 124.6 (q, *J* = 2.4 Hz), 124.1, 123.8, 117.1 (q, *J* = 291.4 Hz), 111.1, 21.2, 18.1.

¹⁹F NMR (376 MHz, CDCl₃) δ -74.8.

6-Acetyl-9-(2-mesityl-1,3-dioxoisoindolin-4-yl)-9*H*-carbazole-3-carbonitrile

Under inert conditions, 9-(2-mesityl-1,3-dioxoisoindolin-4-yl)-9*H*-carbazole-3-carbonitrile (130 mg, 0.285 mmol, 1.00 equiv.) and aluminum chloride (45.7 mg, 0.342 mmol, 1.20 equiv.) were suspended in dry 1,2-dichlorobenzene (3 mL). Then acetyl chloride (25.0 mg, 0.314 mmol, 1.10 equiv.) was added slowly and the reaction mixture was stirred for 3 hours at room temperature. Water and 1M HCl were added carefully and the layers separated. The aqueous layer was extracted with dichloromethane. The combined organic layers were washed with brine, dried over sodium sulfate and reduced in vacuum. The crude product was purified by column chromatography over SiO_2 (dichloromethane/cyclohexane 1:1 to dichloromethane) to yield 57.6 mg of the title compound (0.117 mmol, 41%) as a white solid.

**R*f* (dichloromethane) = 0.32

^{1}H NMR (400 MHz, Chloroform-*d*) δ 8.70 (d, *J* = 1.6 Hz, 1H), 8.42 (d, *J* = 1.5 Hz, 1H), 8.21 (dd, *J* = 7.5, 1.0 Hz, 1H), 8.14 – 8.08 (m, 2H), 7.97 (dd, *J* = 8.0, 1.0 Hz, 1H), 7.66 (dd, *J* = 8.5, 1.6 Hz, 1H), 7.31 – 7.21 (m, 2H), 6.93 (s, 2H), 2.68 (s, 3H), 2.26 (s, 3H), 2.09 (s, 6H).

^{13}C NMR (101 MHz, CDCl₃) δ 197.1, 166.1, 164.4, 144.0, 143.2, 139.7, 136.6, 136.1, 134.7, 134.4, 132.5, 131.5, 130.1, 129.4, 127.9, 127.1, 126.5, 125.7, 124.7, 124.2, 122.5, 122.4, 119.7, 111.1, 110.2, 104.8, 26.8, 21.1, 18.1.

IR (ATR) $\tilde{v}$ [cm⁻¹] = 3963 (w), 3886 (w), 3746 (w), 3677 (w), 3466 (w), 3308 (w), 3063 (w), 2918 (w), 2312 (w), 2224 (w), 1730 (m), 1592 (m), 1486 (m), 1373 (m), 1239 (m), 1114 (m), 1029 (w), 818 (m), 743 (m), 681 (m), 605 (m), 423 (w).

HRMS (APCI, $C_{32}H_{23}N_3O_3$) calc. 498.1812 [M+H]⁺, found 498.1806 [M+H]⁺.

4-(3-Acetyl-6-bromo-9*H*-carbazol-9-yl)-2-mesitylisoindoline-1,3-dione

Under inert conditions, 4-(3-bromo-9*H*-carbazol-9-yl)-2-mesitylisoindoline-1,3-dione (510 mg, 1.00 mmol, 1.00 equiv.) and aluminum chloride (160 mg, 1.20 mmol, 1.20 equiv.) were suspended in dry 1,2-dichlorobenzene (10 mL). Then acetyl chloride (86.0 mg, 1.10 mmol, 1.10 equiv.) was added slowly and the reaction mixture was stirred for 3 hours at room temperature. Water and 1M HCl were added carefully and the layers separated. The aqueous layer was extracted with dichloromethane. The combined organic layers were washed with brine, dried over sodium sulfate and reduced in vacuum. The crude product was purified by column chromatography over SiO$_2$ (dichloromethane/cyclohexane 1:2 to 1:1 to dichloromethane to dichloromethane/ethyl acetate 50:1 to 20:1) to yield 393 mg of the title compound (0.711 mmol, 71%) as an off-white solid.

R*f* (dichloromethane/ethyl acetate 50:1) = 0.51

¹H NMR (300 MHz, Chloroform-*d*) δ 8.71 (d, *J* = 1.6 Hz, 1H), 8.31 (d, *J* = 1.9 Hz, 1H), 8.16 (dd, *J* = 7.4, 1.1 Hz, 1H), 8.12 – 8.01 (m, 2H), 7.95 (dd, *J* = 7.9, 1.1 Hz, 1H), 7.52 (dd, *J* = 8.7, 1.9 Hz, 1H), 7.21 (d, *J* = 8.7 Hz, 1H), 7.11 (d, *J* = 8.7 Hz, 1H), 6.93 (s, 2H), 2.71 (s, 3H), 2.27 (s, 3H), 2.09 (s, 3H), 2.08 (s, 3H).

¹³C NMR (101 MHz, CDCl$_3$) δ 197.3, 166.3, 164.4, 143.7, 140.2, 139.7, 136.4, 136.2, 134.7, 134.4, 133.4, 131.0, 129.7, 129.4, 127.3, 126.9, 126.7, 125.8, 124.1, 123.8, 122.8, 122.3, 114.7, 111.8, 110.0, 29.8, 26.8, 21.2, 18.1.

IR (ATR) $\tilde{v}$ [cm^{-1}] = 3470 (vw), 3074 (vw), 2917 (vw), 2734 (vw), 2489 (vw), 2416 (vw), 2314 (vw), 2167 (vw), 1906 (vw), 1663 (m), 1592 (w), 1486 (m), 1376 (m), 1279 (m), 1119 (m), 1027 (w), 953 (w), 817 (m), 751 (m), 555 (w), 423 (w).

HRMS (APCI, C$_{31}$H$_{24}$^{79}BrN$_2$O$_3$) calc. 551.0965 [M+H]$^+$, found 551.0938 [M+H]$^+$.

4-(3-Bromo-6-dodecanoyl-9*H*-carbazol-9-yl)-2-mesitylisoindoline-1,3-dione

Under inert conditions, 4-(3-bromo-9*H*-carbazol-9-yl)-2-mesitylisoindoline-1,3-dione (1.02 g, 2.00 mmol, 1.00 equiv.) and aluminum chloride (320 mg, 2.40 mmol, 1.20 equiv.) were suspended in dry 1,2-dichlorobenzene (20 mL). Then dodecanoyl chloride (481 mg, 2.20 mmol, 1.10 equiv.) was added slowly and the reaction mixture was stirred for 4 hours at 65 °C. Water and 1M HCl were added carefully and the layers separated. The aqueous layer was extracted with dichloromethane. The combined organic layers were washed with brine, dried over sodium sulfate and reduced in vacuum. The crude product was purified by column chromatography over SiO_2 (dichloromethane/cyclohexane 1:1 to 2:1 to 4:1 to dichloromethane) to yield 842 mg of the title compound (1.22 mmol, 61%) as an off-white solid.

R_f (dichloromethane) = 0.36

^{1}H NMR (400 MHz, Chloroform-*d*) δ 8.71 (d, *J* = 1.6 Hz, 1H), 8.31 (d, *J* = 1.9 Hz, 1H), 8.15 (dd, *J* = 7.5, 1.0 Hz, 1H), 8.09 – 8.03 (m, 2H), 7.94 (dd, *J* = 8.0, 1.0 Hz, 1H), 7.51 (dd, *J* = 8.7, 1.9 Hz, 1H), 7.21 (d, *J* = 8.7 Hz, 1H), 7.11 (d, *J* = 8.7 Hz, 1H), 6.93 (s, 2H), 3.07 (t, *J* = 7.4 Hz, 2H), 2.27 (s, 3H), 2.09 (s, 3H), 2.08 (s, 3H), 1.79 (p, *J* = 7.4 Hz, 2H), 1.55 – 1.13 (m, 16H), 0.99 – 0.80 (m, 3H).

^{13}C NMR (101 MHz, CDCl$_3$) δ 199.8, 166.3, 164.4, 143.5, 140.1, 139.6, 136.3, 136.2, 134.7, 134.4, 133.4, 130.9, 129.6, 129.4, 127.2, 126.8, 126.6, 125.8, 124.0, 123.7, 122.7, 121.9, 114.6, 111.8, 110.0, 38.7, 32.0, 29.8, 29.8, 29.7, 29.6, 29.5, 24.9, 22.8, 21.2, 18.2, 18.1, 14.3.

IR (ATR) $\tilde{v}$ [cm^{-1}] = 3061 (vw), 2921 (w), 2851 (w), 2321 (vw), 2171 (vw), 1719 (w), 1592 (w), 1487 (w), 1373 (w), 1281 (w), 1115 (w), 1021 (vw), 882 (w), 750 (w), 550 (w), 420 (w).

HRMS (APCI, C$_{41}$H$_{43}$^{79}BrN$_2$O$_3$) calc. 691.2530 [M+H]$^+$, found 691.2495 [M+H]$^+$.

2-Mesityl-4-(3-(3-methylbutanoyl)-9*H*-carbazol-9-yl)isoindoline-1,3-dione

Under inert conditions, 4-(9*H*-carbazol-9-yl)-2-mesitylisoindoline-1,3-dione (1.08 g, 2.50 mmol, 1.00 equiv.) and aluminum chloride (400 mg, 3.00 mmol, 1.20 equiv.) were suspended in dry 1,2-dichlorobenzene (15 mL). Then 3-methylbutanoyl chloride (331 mg, 2.75 mmol, 1.10 equiv.) was added slowly and the reaction mixture was stirred for 3 hours at room temperature. Water and 1M HCl were added carefully and the layers separated. The aqueous layer was extracted with dichloromethane. The combined organic layers were washed with brine, dried over sodium sulfate and reduced in vacuum. The crude product was purified by column chromatography over SiO_2 (dichloromethane/cyclohexane 1:1 to 2:1 to dichloromethane) to yield 847 mg of the title compound (1.65 mmol, 66%) as a yellow solid.

R_f (dichloromethane) = 0.37

^{1}H NMR (400 MHz, Chloroform-*d*) δ 8.78 (d, *J* = 1.7 Hz, 1H), 8.21 (d, *J* = 7.6 Hz, 1H), 8.14 (dd, *J* = 7.4, 1.0 Hz, 1H), 8.09 – 8.00 (m, 2H), 7.97 (dd, *J* = 8.0, 1.0 Hz, 1H), 7.44 (ddd, *J* = 8.2, 7.3, 1.3 Hz, 1H), 7.37 (td, *J* = 7.5, 1.0 Hz, 1H), 7.28 – 7.24 (m, 1H), 7.21 (d, *J* = 8.6 Hz, 1H), 6.93 (d, *J* = 4.5 Hz, 2H), 2.96 (d, *J* = 6.9, 1.1 Hz, 2H), 2.45 – 2.32 (m, 1H), 2.27 (s, 3H), 2.12 (s, 3H), 2.08 (s, 3H), 1.05 (d, *J* = 6.6 Hz, 6H).

^{13}C NMR (101 MHz, CDCl$_3$) δ 199.7, 166.4, 164.4, 143.2, 141.4, 139.6, 136.3, 136.2, 136.2, 134.6, 134.5, 134.0, 130.9, 129.4, 129.4, 126.8, 126.8, 126.8, 126.6, 124.2, 123.9, 123.7, 121.8, 121.0, 110.3, 109.8, 47.6, 25.6, 23.1, 23.0, 21.2, 18.2, 18.1.

IR (ATR) $\tilde{v}$ [cm^{-1}] = 3048 (vw), 2922 (w), 1719 (m), 1595 (w), 1483 (m), 1371 (m), 1197 (m), 1113 (m), 1005 (w), 882 (m), 747 (m), 550 (w), 421 (w).

HRMS (APCI, C$_{34}$H$_{30}$N$_2$O$_3$) calc. 515.2329 [M+H]$^+$, found 515.2297 [M+H]$^+$.

EA (C$_{34}$H$_{30}$N$_2$O$_3$) calc. C: 79.35, H: 5.88, N: 5.44; found C: 79.18, H: 6.25, N: 5.13.

6-Acetyl-9-(2-(*tert*-butyl)-1,3-dioxoisoindolin-4-yl)-9*H*-carbazole-3-carbonitrile

Under inert conditions, 9-(2-(*tert*-butyl)-1,3-dioxoisoindolin-4-yl)-9*H*-carbazole-3-carbonitrile (197 mg, 0.500 mmol, 1.00 equiv.) and aluminum chloride (80.1 mg, 0.600 mmol, 1.20 equiv.) were suspended in dry 1,2-dichlorobenzene (5 mL). Then acetyl chloride (43.0 mg, 0.550 mmol, 1.10 equiv.) was added slowly and the reaction mixture was stirred for 3 hours at room temperature. Water and 1M HCl were added carefully and the layers separated. The aqueous layer was extracted with dichloromethane. The combined organic layers were washed with brine, dried over sodium sulfate and reduced in vacuum. The crude product was purified by column chromatography over SiO_2 (dichloromethane/cyclohexane 1:1 to 2:1 to dichloromethane) to yield 124 mg of the title compound (0.285 mmol, 57%) as a white solid.

R_f (dichloromethane) = 0.27

¹H NMR (400 MHz, Chloroform-*d*) δ 8.74 – 8.69 (m, 1H), 8.44 (dd, *J* = 1.6, 0.7 Hz, 1H), 8.10 (dd, *J* = 8.7, 1.7 Hz, 1H), 8.04 (dd, *J* = 7.5, 1.0 Hz, 1H), 7.98 (t, *J* = 7.6 Hz, 1H), 7.80 (dd, *J* = 7.8, 1.0 Hz, 1H), 7.65 (dd, *J* = 8.5, 1.6 Hz, 1H), 7.22 – 7.14 (m, 2H), 2.69 (s, 3H), 1.59 (s, 9H).

¹³C NMR (101 MHz, CDCl₃) δ 197.2, 168.2, 166.5, 144.1, 143.4, 135.8, 135.0, 134.2, 131.6, 131.3, 130.0, 127.9, 127.4, 125.7, 124.0, 123.8, 122.4, 122.4, 119.8, 111.2, 110.3, 104.5, 58.6, 29.0, 26.8.

IR (ATR) ṽ [cm⁻¹] = 3820 (vw), 3631 (vw), 3456 (vw), 3370 (vw), 3307 (vw), 3074 (vw), 2976 (vw), 2674 (vw), 2325 (vw), 2221 (w), 2106 (vw), 1884 (vw), 1701 (m), 1595 (w), 1486 (w), 1360 (w), 1239 (m), 1104 (w), 1011 (w), 820 (w), 736 (w), 606 (w), 417 (vw).

HRMS (APCI, $C_{27}H_{21}N_3O_3$) calc. 436.1656 [M+H]⁺, found 436.1650 [M+H]⁺.

EA ($C_{27}H_{21}N_3O_3$) calc. C: 74.47, H: 4.86, N: 9.65; found C: 72.94, H: 5.04, N: 8.97.

1,1'-(9H-Carbazole-3,6-diyl)bis(ethan-1-one)

Under inert conditions, 9H-carbazole (5.02 g, 30.0 mmol, 1.00 equiv.) and aluminum chloride (11.4 g, 85.2 mmol, 2.84 equiv.) were suspended in dry dichloromethane (160 mL). Then acetyl chloride (6.38 g, 81.3 mmol, 2.71 equiv.) was added slowly and the reaction mixture was stirred for 16 hours at room temperature. Water and 1M HCl were added carefully and the layers separated. The aqueous layer was extracted with dichloromethane. The combined organic layers were washed with brine, dried over sodium sulfate and reduced in vacuum. The crude product was purified by recrystallization in ethanol to yield 5.77 g of the title compound (20.7 mmol, 69%) as an off-white solid.

¹H NMR (400 MHz, DMSO-d_6) δ 12.08 (s, 1H), 9.02 (d, J = 1.8 Hz, 2H), 8.07 (dd, J = 8.6, 1.8 Hz, 2H), 7.66 – 7.54 (m, 2H), 2.69 (s, 6H).

¹³C NMR (101 MHz, DMSO) δ 197.1, 143.3, 129.2, 126.4, 122.8, 122.5, 111.3, 26.7.

IR (ATR) $\tilde{v}$ [cm^{-1}] = 3269 (w), 3041 (w), 2920 (w), 2839 (w), 2646 (w), 2492 (w), 2324 (w), 2164 (w), 2052 (w), 1897 (w), 1660 (m), 1353 (m), 1228 (m), 954 (w), 881 (w), 801 (m), 635 (m), 418 (w).

HRMS (APCI, $C_{16}H_{13}NO_2$) calc. 252.1019 [M+H]$^+$, found 252.1007 [M+H]$^+$.

1-(9H-Carbazol-3-yl)ethan-1-one

Under inert conditions, 9H-carbazole (480 mg, 2.50 mmol, 1.00 equiv.) was dissolved in dry THF (10 mL) and cooled to -78 °C. A 1.6 M solution of methyllithium in diethyl ether (3.91 mL, 6.26 mmol, 2.50 equiv.) was added slowly. The reaction mixture was stirred at room temperature for one hour. Then, 2 M hydrochloric acid was added and the extracted with dichloromethane. The organic layers were combined, washed with brine, dried over sodium sulfate and reduced in vacuum to yield 518 mg of the title compound (2.48 mmol, 99%) as an off-white solid.

¹H NMR (300 MHz, DMSO-d_6) δ 11.73 (s, 1H), 8.85 (d, J = 1.7 Hz, 1H), 8.27 (d, J = 7.8 Hz, 1H), 8.03 (dd, J = 8.6, 1.7 Hz, 1H), 7.55 (d, J = 8.4 Hz, 2H), 7.50 – 7.37 (m, 1H), 7.24 (t, J = 7.5, 7.5 Hz, 1H), 2.68 (s, 3H).

IR (ATR) $\tilde{v}$ [cm⁻¹] = 3985 (m), 3164 (m), 3038 (m), 2922 (m), 2667 (m), 1894 (m), 1649 (s), 1497 (m), 1331 (s), 1244 (s), 1126 (s), 1019 (m), 954 (m), 884 (s), 807 (s), 727 (s), 631 (s), 417 (s).

HRMS (APCI, $C_{14}H_{12}NO$) calc. 210.0913 [M+H]⁺, found 210.0913 [M+H]⁺.

5.1.3.3   Tristriazolotriazine (TTT) Project

General Procedure for the Synthesis of Tetrazoles from Aromatic Nitriles (GP 7)[194]

In a sealable vial, a nitrile derivative (1.00 equiv.), sodium azide (3.00 equiv.) and ammonium chloride (3.00 equiv.) were evacuated and backfilled with argon. Dry DMF was added and the resulting mixture heated at 130 °C until completion (usually 16 h, monitored with TLC). After cooling to room temperature, the reaction mixture was poured into an excess of 1M HCl and mixed thoroughly. The white solid was filtered off, washed several times with water and thoroughly dried. The product was purified by silica column chromatography or used without further purification.

4-(3,6-Di-*tert*-butyl-9*H*-carbazol-9-yl)benzonitrile

In a sealable vial, 4-fluorobenzonitrile (1.21 g, 10.0 mmol, 1.00 equiv.), 3,6-di-*tert*-butyl-9*H*-carbazole (2.79 g, 10.0 mmol, 1.00 equiv.) and tripotassium phosphate (4.25 g, 20.0 mmol, 2.00 equiv.) were evacuated and backfilled with argon. Dry DMSO (40 ml) was added and the resulting mixture heated at 110 °C for 16 h. After cooling to room temperature, the reaction mixture was poured into an excess of water and extracted with dichloromethane three times. The combined organic layers were washed with brine, dried over sodium sulfate and reduced in vacuum. The crude product was purified by column chromatography over SiO_2 (dichloromethane/pentane 1:2) to yield 3.74 g of the title compound (9.83 mmol, 98%) as a white solid.

R_f (dichloromethane/pentane 1:2) = 0.42

^{1}H NMR (400 MHz, Chloroform-*d*) δ 8.18 – 8.13 (m, 2H), 7.90 – 7.84 (m, 2H), 7.75 – 7.69 (m, 2H), 7.54 – 7.47 (m, 2H), 7.42 (d, *J* = 8.7 Hz, 2H), 1.48 (s, 18H).

^{13}C NMR (101 MHz, CDCl$_3$) δ 144.2, 142.7, 138.3, 134.0, 126.6, 124.2, 124.1, 118.7, 116.7, 109.9, 109.2, 34.9, 32.1.

IR (ATR) $\tilde{v}$ [cm^{-1}] = 3061 (vw), 2959 (w), 2561 (vw), 2484 (vw), 2380 (vw), 2226 (w), 2108 (vw), 1883 (vw), 1753 (vw), 1603 (w), 1469 (m), 1366 (w), 1261 (w), 1107 (w), 1035 (w), 810 (m), 610 (w), 551 (w), 421 (w), 0 (vs), 414 (vw).

HRMS (APCI, C$_{27}$H$_{28}$N$_2$) calc. 381.2325 [M+H]$^+$, found 381.2310 [M+H]$^+$.

4-(9*H*-Carbazol-9-yl)benzonitrile

In a sealable vial, 4-fluorobenzonitrile (606 mg, 5.00 mmol, 1.00 equiv.), carbazole (837 mg, 5.00 mmol, 1.00 equiv.) and tripotassium phosphate (2.12 g, 10.0 mmol, 2.00 equiv.) were evacuated and backfilled with argon. Dry DMSO (16 ml) was added and the resulting mixture heated at 110 °C for 16 h. After cooling to room temperature, the reaction mixture was poured into an excess of water and extracted with dichloromethane three times. The combined organic layers were washed with brine, dried over sodium sulfate and reduced in vacuum. The crude product was purified by column chromatography over SiO_2 (dichloromethane/pentane 1:1 to 2:1) to yield 1.21 g of the title compound (4.51 mmol, 90%) as a white solid.

R_f (dichloromethane/pentane 1:1) = 0.27

1H NMR (400 MHz, Chloroform-*d*) δ 8.15 (dd, *J* = 7.6, 1.2 Hz, 2H), 7.94 – 7.86 (m, 2H), 7.79 – 7.71 (m, 2H), 7.45 (dd, *J* = 5.9, 1.4 Hz, 4H), 7.34 (ddd, *J* = 8.0, 5.9, 2.1 Hz, 2H).

13C NMR (101 MHz, CDCl3) δ 142.2, 140.0, 134.1, 127.2, 126.5, 124.1, 121.1, 120.7, 118.5, 110.6, 109.7.

IR (ATR) ṽ [cm⁻¹] = 3984 (w), 3851 (w), 3743 (w), 3532 (w), 3472 (w), 3054 (w), 2925 (w), 2850 (w), 2677 (w), 2558 (w), 2322 (w), 2228 (w), 2101 (w), 1915 (w), 1798 (w), 1600 (m), 1448 (m), 1336 (m), 1224 (m), 1022 (w), 914 (w), 839 (m), 751 (m), 622 (m), 548 (m), 423 (w).

HRMS (APCI, $C_{19}H_{12}N_2$) calc. 269.1073 [M+H]⁺, found 269.1069 [M+H]⁺.

4-(Diphenylamino)benzonitrile

 In a sealable vial, 4-fluorobenzonitrile (0.484 g, 4.00 mmol, 1.00 equiv.), diphenylamine (677 mg, 4.00 mmol, 1.00 equiv.) and tripotassium phosphate (1.70 g, 8.00 mmol, 2.00 equiv.) were evacuated and backfilled with argon. Dry DMSO (16 ml) was added and the resulting mixture heated at 110 °C for 16 h. After cooling to room temperature, the reaction mixture was poured into an excess of water and extracted with dichloromethane three times. The combined organic layers were washed with brine, dried over sodium sulfate and reduced in vacuum. The crude product was purified by column chromatography over SiO_2 (dichloromethane/pentane 1:2) to yield 530 mg of the title compound (1.96 mmol, 49%) as a white solid.

R_f (dichloromethane/pentane 1:2) = 0.42

^{1}H NMR (400 MHz, Chloroform-d) δ 7.45 – 7.39 (m, 2H), 7.37 – 7.30 (m, 4H), 7.17 (tt, J = 8.3, 1.2 Hz, 6H), 6.99 – 6.93 (m, 2H).

^{13}C NMR (101 MHz, CDCl$_3$) δ 151.7, 146.0, 133.3, 129.9, 126.2, 125.2, 119.8, 119.8, 102.5.

IR (ATR) $\tilde{v}$ [cm^{-1}] = 3037 (vw), 2915 (vw), 2851 (vw), 2755 (vw), 2616 (vw), 2504 (vw), 2324 (vw), 2211 (m), 2026 (vw), 1910 (vw), 1584 (m), 1486 (m), 1320 (m), 1171 (m), 1072 (w), 830 (m), 696 (m), 544 (m), 407 (w).

HRMS (APCI, C$_{19}$H$_{14}$N$_2$) calc. 271.1230 [M+H]$^+$, found 271.1224 [M+H]$^+$.

EA (C$_{19}$H$_{14}$N$_2$) calc. C: 84.42, H: 5.22, N: 10.36; found C: 84.38, H: 5.20, N: 10.31.

2,6-Bis(3,6-di-*tert*-butyl-9*H*-carbazol-9-yl)benzonitrile

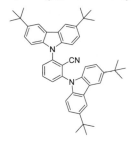 In a sealable vial, 2,6-difluorobenzonitrile (626 mg, 4.50 mmol, 1.00 equiv.), 3,6-di-*tert*-butyl-9*H*-carbazole (2.64 g, 9.45 mmol, 2.10 equiv.) and tripotassium phosphate (3.82 g, 18.0 mmol, 4.00 equiv.) were evacuated and backfilled with argon. Dry DMSO (36 ml) was added and the resulting mixture heated at 110 °C for 16 h. After cooling to room temperature, the reaction mixture was poured into an excess of water and extracted with dichloromethane three times. The combined organic layers were washed with brine, dried over sodium sulfate and reduced in vacuum. The crude product was purified by column chromatography over SiO$_2$ (dichloromethane/pentane 1:4) to yield 2.78 g of the title compound (4.23 mmol, 94%) as a white solid.

R$_f$ (dichloromethane/pentane 1:4) = 0.38

^{1}H NMR (400 MHz, Chloroform-*d*) δ 8.17 (d, *J* = 1.8 Hz, 4H), 7.93 (t, *J* = 8.1 Hz, 1H), 7.70 (d, *J* = 8.1 Hz, 2H), 7.56 (dd, *J* = 8.6, 1.9 Hz, 4H), 7.32 (d, *J* = 8.6 Hz, 4H), 1.50 (s, 36H).

^{13}C NMR (101 MHz, CDCl$_3$) δ 144.1, 143.5, 139.2, 134.9, 128.3, 124.3, 124.2, 116.9, 113.6, 112.7, 109.3, 35.0, 32.1.

IR (ATR) $\tilde{v}$ [cm^{-1}] = 3957 (vw), 3839 (vw), 3625 (vw), 3552 (vw), 3432 (vw), 3293 (vw), 3071 (vw), 2951 (w), 2231 (vw), 1571 (w), 1486 (m), 1357 (w), 1262 (w), 1033 (w), 819 (w), 613 (w), 422 (w).

HRMS (APCI, C$_{47}$H$_{51}$N$_3$) calc. 658.4156 [M+H]$^+$, found 658.4142 [M+H]$^+$.

3,5-Bis(3,6-di-*tert*-butyl-9*H*-carbazol-9-yl)benzonitrile

In a sealable vial, 3,5-difluorobenzonitrile (626 mg, 4.50 mmol, 1.00 equiv.), 3,6-di-*tert*-butyl-9*H*-carbazole (2.64 g, 9.45 mmol, 2.10 equiv.) and tripotassium phosphate (3.82 g, 18.0 mmol, 4.00 equiv.) were evacuated and backfilled with argon. Dry DMSO (36 ml) was added and the resulting mixture heated at 110 °C for 16 h. After cooling to room temperature, the reaction mixture was poured into an excess of water and extracted with dichloromethane three times. The combined organic layers were washed with brine, dried over sodium sulfate and reduced in vacuum. The crude product was purified by column chromatography over SiO$_2$ (dichloromethane/pentane 1:4 to dichloromethane) to yield 1.82 g of the title compound (2.77 mmol, 61%) as a white solid.

R$_f$ (dichloromethane/pentane 1:4) = 0.14

¹H NMR (400 MHz, Chloroform-*d*) δ 8.18 (d, *J* = 1.8 Hz, 4H), 8.12 (d, *J* = 2.0 Hz, 1H), 7.96 (d, *J* = 1.9 Hz, 2H), 7.54 (dd, *J* = 8.7, 1.9 Hz, 4H), 7.48 (d, *J* = 8.6 Hz, 4H), 1.50 (s, 36H).

¹³C NMR (101 MHz, CDCl$_3$) δ 144.4, 141.2, 138.4, 128.4, 127.4, 124.3, 124.2, 117.7, 116.8, 115.5, 109.0, 35.0, 32.1.

IR (ATR) ṽ [cm^{-1}] = 3976 (vw), 3872 (vw), 3773 (vw), 3678 (vw), 3526 (vw), 3413 (vw), 3247 (vw), 3079 (vw), 2923 (w), 2659 (vw), 2439 (vw), 2324 (vw), 2233 (vw), 1993 (vw), 1873 (vw), 1737 (vw), 1589 (w), 1473 (w), 1295 (w), 1107 (vw), 1035 (vw), 891 (w), 805 (w), 740 (w), 612 (w), 422 (w).

HRMS (ESI, C$_{47}$H$_{51}$N$_3$) calc. 658.4156 [M+H]⁺, found 658.4122 [M+H]⁺.

4-(9,9-Dimethylacridin-10(9)-yl)benzonitrile

In a sealable vial, 4-fluorobenzonitrile (303 mg, 2.50 mmol, 1.00 equiv.), 9,9-dimethyl-9,10-dihydroacridine (524 mg, 2.50 mmol, 1.00 equiv.) and tripotassium phosphate (1.06 g, 5.00 mmol, 2.00 equiv.) were evacuated and backfilled with argon. Dry DMSO (10 ml) was added and the resulting mixture heated at 110 °C for 16 h. After cooling to room temperature, the reaction mixture was poured into an excess of water and extracted with dichloromethane three times. The combined organic layers were washed with brine, dried over sodium sulfate and reduced in vacuum. The crude product was purified by column chromatography over SiO$_2$ (dichloromethane/pentane 1:1) to yield 520 mg of the title compound (1.68 mmol, 67%) as a colorless solid.

R$_f$ (dichloromethane/pentane 1:1) = 0.49

1H NMR (400 MHz, DMSO-d_6) δ 8.17 – 8.09 (m, 2H), 7.66 – 7.56 (m, 2H), 7.52 (dd, J = 7.6, 1.7 Hz, 2H), 7.05 – 6.92 (m, 4H), 6.21 (dd, J = 8.0, 1.4 Hz, 2H), 1.61 (s, 6H).

13C NMR (101 MHz, DMSO) δ 145.4, 139.7, 135.3, 131.2, 130.8, 126.6, 125.6, 121.4, 118.5, 114.5, 110.6, 35.7, 31.0.

IR (ATR) $\tilde{v}$ [cm^{-1}] = 3973 (w), 3878 (w), 3765 (w), 3699 (w), 3620 (w), 3503 (w), 3410 (w), 3053 (w), 2973 (w), 2595 (w), 2387 (w), 2324 (w), 2229 (w), 2112 (w), 1924 (w), 1815 (w), 1700 (w), 1586 (m), 1445 (m), 1321 (m), 1045 (w), 924 (w), 837 (w), 744 (m), 622 (m), 558 (m), 433 (m).

HRMS (APCI, C$_{22}$H$_{18}$N$_2$) calc. 311.1543 [M+H]$^+$, found 311.1532 [M+H]$^+$.

3,4,5-Tris(3,6-di-*tert*-butyl-9*H*-carbazol-9-yl)benzonitrile

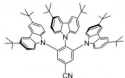 In a sealable vial, 3,4,5-trifluorobenzonitrile (393 mg, 2.50 mmol, 1.00 equiv.), 3,6-di-*tert*-butyl-9*H*-carbazole (2.20 g, 7.88 mmol, 3.15 equiv.) and tripotassium phosphate (3.19 g, 15.0 mmol, 6.00 equiv.) were evacuated and backfilled with argon. Dry DMSO (32 ml) was added and the resulting mixture heated at 110 °C for 16 h. After cooling to room temperature, the reaction mixture was poured into an excess of water and extracted with dichloromethane three times. The combined organic layers were washed with brine, dried over sodium sulfate and reduced in vacuum. The crude product was purified by column chromatography over SiO$_2$ (dichloromethane/pentane 1:2 to 1:1) to yield 1.17 g of the title compound (1.25 mmol, 50%) as a colorless solid.

R$_f$ (dichloromethane/pentane 1:1) = 0.61

¹H NMR (400 MHz, Chloroform-*d*) δ 8.07 (s, 2H), 7.79 – 7.70 (m, 4H), 7.30 – 7.27 (m, 2H), 7.06 (dd, *J* = 8.6, 1.9 Hz, 4H), 6.98 (d, *J* = 8.6 Hz, 4H), 6.69 (d, *J* = 8.6 Hz, 2H), 6.60 (ddd, *J* = 8.6, 2.0, 0.9 Hz, 2H), 1.36 (s, 36H), 1.23 (s, 18H).

¹³C NMR (101 MHz, CDCl$_3$) δ 143.7, 143.2, 139.2, 138.3, 137.1, 136.8, 132.4, 124.6, 124.3, 123.4, 122.7, 117.3, 116.1, 115.2, 112.5, 109.7, 109.2, 34.7, 34.4, 32.0, 31.9.

IR (ATR) $\tilde{v}$ [cm⁻¹] = 3986 (vw), 3883 (vw), 3693 (vw), 3621 (vw), 3506 (vw), 3442 (vw), 2954 (w), 2361 (vw), 2232 (w), 1747 (vw), 1485 (m), 1362 (w), 1262 (w), 1135 (w), 1034 (w), 876 (w), 809 (m), 731 (w), 612 (w), 428 (w).

HRMS (APCI, C$_{67}$H$_{74}$N$_4$) calc. 935.5986 [M+H]⁺, found 935.5972 [M+H]⁺.

6-(3,6-Di-*tert*-butyl-9*H*-carbazol-9-yl)nicotinonitrile

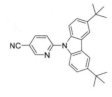

In a sealable vial, 6-fluoronicotinonitrile (0.488 g, 4.00 mmol, 1.00 equiv.), 3,6-di-*tert*-butyl-9*H*-carbazole (1.12 g, 4.00 mmol, 1.00 equiv.) and tripotassium phosphate (1.70 g, 8.00 mmol, 2.00 equiv.) were evacuated and backfilled with argon. Dry DMSO (16 ml) was added and the resulting mixture heated at 110 °C for 16 h. After cooling to room temperature, the reaction mixture was poured into an excess of water and extracted with dichloromethane three times. The combined organic layers were washed with brine, dried over sodium sulfate and reduced in vacuum. The crude product was purified by column chromatography over SiO_2 (dichloromethane/pentane 1:2 to 1:1) to yield 900 mg of the title compound (2.36 mmol, 59%) as a white solid.

R_f (dichloromethane/pentane 1:1) = 0.44

¹H NMR (400 MHz, Chloroform-*d*) δ 8.85 (dd, *J* = 2.3, 0.8 Hz, 1H), 8.11 (d, *J* = 2.0 Hz, 2H), 7.99 (dd, *J* = 8.6, 2.4 Hz, 1H), 7.92 (d, *J* = 8.7 Hz, 2H), 7.73 (d, *J* = 8.5 Hz, 1H), 7.52 (dd, *J* = 8.7, 2.0 Hz, 2H), 1.48 (s, 18H).

¹³C NMR (101 MHz, CDCl₃) δ 154.8, 152.7, 145.7, 141.1, 137.1, 125.5, 124.4, 117.0, 116.6, 116.5, 111.8, 104.9, 34.9, 32.0.

IR (ATR) $\tilde{v}$ [cm⁻¹] = 2955 (w), 2865 (w), 2229 (vw), 1593 (w), 1486 (m), 1390 (m), 1260 (w), 1037 (w), 806 (m), 613 (w), 550 (w), 410 (w).

HRMS (APCI, $C_{26}H_{27}N_3$) calc. 382.2278 [M+H]⁺, found 382.2272 [M+H]⁺.

EA ($C_{26}H_{27}N_3$) calc. C: 81.85, H: 7.13, N: 11.01; found C: 81.73, H: 7.09, N: 10.93.

5-(3,6-Di-*tert*-butyl-9*H*-carbazol-9-yl)picolinonitrile

In a sealable vial, 5-fluoropicolinonitrile (0.488 g, 4.00 mmol, 1.00 equiv.), 3,6-di-*tert*-butyl-9*H*-carbazole (1.12 g, 4.00 mmol, 1.00 equiv.) and tripotassium phosphate (1.70 g, 8.00 mmol, 2.00 equiv.) were evacuated and backfilled with argon. Dry DMSO (16 ml) was added and the resulting mixture heated at 110 °C for 16 h. After cooling to room temperature, the reaction mixture was poured into an excess of water and extracted with dichloromethane three times. The combined organic layers were washed with brine, dried over sodium sulfate and reduced in vacuum. The crude product was purified by column chromatography over SiO$_2$ (dichloromethane/pentane 1:1) to yield 1.18 g of the title compound (3.08 mmol, 77%) as a white solid.

R$_f$ (dichloromethane/pentane 1:1) = 0.10

¹H NMR (300 MHz, Chloroform-*d*) δ 9.05 (dd, *J* = 2.6, 0.8 Hz, 1H), 8.17 – 8.13 (m, 2H), 8.08 (dd, *J* = 8.4, 2.5 Hz, 1H), 7.93 (dd, *J* = 8.3, 0.8 Hz, 1H), 7.51 (dd, *J* = 8.7, 1.9 Hz, 2H), 7.46 – 7.36 (m, 2H), 1.48 (s, 18H).

¹³C NMR (101 MHz, CDCl$_3$) δ 148.6, 145.0, 138.4, 137.9, 133.2, 130.4, 129.5, 124.6, 124.6, 124.5, 117.2, 116.9, 108.8, 35.0, 35.0, 32.0.

IR (ATR) $\tilde{v}$ [cm^{-1}] = 3041 (w), 2951 (w), 2861 (w), 2231 (w), 1863 (vw), 1579 (w), 1483 (m), 1372 (m), 1227 (m), 1082 (w), 1022 (w), 801 (m), 738 (w), 658 (w), 409 (w).

HRMS (APCI, C$_{26}$H$_{27}$N$_3$) calc. 382.2278 [M+H]$^+$, found 382.2271 [M+H]$^+$.

EA (C$_{26}$H$_{27}$N$_3$) calc. C: 81.85, H: 7.13, N: 11.01; found C: 80.75, H: 6.99, N: 10.81.

4-(3,6-Di-*tert*-butyl-9*H*-carbazol-9-yl)-3,5-dimethylbenzonitrile

 In a sealable vial, 4-fluoro-3,5-dimethylbenzonitrile (1.49 g, 10.0 mmol, 1.00 equiv.), 3,6-di-*tert*-butyl-9*H*-carbazole (2.79 g, 10.0 mmol, 1.00 equiv.) and tripotassium phosphate (4.24 g, 20.0 mmol, 2.00 equiv.) were evacuated and backfilled with argon. Dry DMSO (40 ml) was added and the resulting mixture heated at 110 °C for 16 h. After cooling to room temperature, the reaction mixture was poured into an excess of water and extracted with dichloromethane three times. The combined organic layers were washed with brine, dried over sodium sulfate and reduced in vacuum. The crude product was purified by column chromatography over SiO_2 (dichloromethane/pentane 1:2) to yield 3.67 g of the title compound (8.98 mmol, 90%) as a white solid.

R_f (dichloromethane/pentane 1:2) = 0.44

¹H NMR (400 MHz, Chloroform-*d*) δ 8.17 (d, *J* = 1.8 Hz, 2H), 7.58 (s, 2H), 7.43 (dd, *J* = 8.6, 1.9 Hz, 2H), 6.78 (d, *J* = 8.5 Hz, 2H), 1.92 (s, 6H), 1.47 (s, 18H).

¹³C NMR (101 MHz, CDCl₃) δ 143.0, 140.2, 139.9, 138.2, 132.4, 124.1, 123.2, 118.8, 116.7, 112.5, 108.7, 34.9, 32.2, 17.8.

IR (ATR) $\tilde{v}$ [cm⁻¹] = 2960 (w), 2665 (vw), 2229 (vw), 1603 (vw), 1485 (w), 1361 (w), 1262 (w), 1146 (w), 1031 (w), 870 (w), 807 (w), 741 (w), 602 (w), 464 (vw).

HRMS (APCI, C₂₉H₃₂N₂) calc. 409.2638 [M+H]⁺, found 409.2628 [M+H]⁺.

EA (C₂₉H₃₂N₂) calc. C: 85.25, H: 7.89, N: 6.86; found C: 85.18, H: 7.96, N: 6.71.

2-Bromo-*N*-(2,4-dimethylphenyl)-4,6-dimethylaniline[268]

 To a round-bottomed flask equipped with a reflux condenser, 2-bromo-4,6-dimethylaniline (3.00 g, 15.0 mmol, 1.00 equiv.), 1-iodo-2,4-dimethylbenzene (3.83 g, 16.5 mmol, 1.10 equiv.), sodium *tert*-butoxide (2.16 g, 22.5 mmol, 1.50 equiv.), $Pd_2(dba)_3$ (137 mg, 0.150 mmol, 0.01 equiv.) and 1,1'-bis(diphenylphosphino)ferrocene (dppf, 166 mg, 0.300 mmol, 0.02 equiv.) were added, evacuated and backfilled with argon. Dry toluene (41 mL) were added and the mixture refluxed for 16 hours. After cooling to room temperature, an excess of water was added and extracted with dichloromethane three times. The combined organic layers were washed with brine, dried over sodium sulfate and reduced in vacuum. The crude product was purified by column chromatography over SiO_2 (dichloromethane/pentane 1:8) to yield 4.10 g of the title compound (13.5 mmol, 90%) as a colorless viscous oil which solidified over night to a white solid.

R_f (dichloromethane/pentane 1:8) = 0.50

¹H NMR (400 MHz, Chloroform-*d*) δ 7.31 (d, *J* = 1.9 Hz, 1H), 7.01 – 6.97 (m, 2H), 6.79 (dd, *J* = 8.1, 2.0 Hz, 1H), 6.14 (d, *J* = 8.0 Hz, 1H), 5.18 (s, 1H), 2.33 (s, 3H), 2.31 (s, 3H), 2.24 (s, 3H), 2.13 (s, 3H).

¹³C NMR (101 MHz, CDCl₃) δ 140.9, 136.9, 136.3, 135.7, 131.3, 131.2, 131.0, 128.9, 127.1, 124.6, 121.5, 113.6, 20.7, 20.7, 19.3, 17.9.

IR (ATR) $\tilde{v}$ [cm⁻¹] = 3385 (vw), 2917 (w), 2730 (vw), 1736 (vw), 1617 (w), 1503 (m), 1283 (m), 1133 (w), 1035 (w), 802 (m), 716 (w), 609 (w), 536 (w), 436 (w).

HRMS (APCI, $C_{16}H_{18}{}^{79}BrN$) calc. 304.0695 [M+H]⁺, found 304.0693 [M+H]⁺.

The analytical data is consistent with literature.

1,3,6,8-Tetramethyl-9*H*-carbazole[269]

To a round-bottomed flask equipped with a reflux condenser, 2-bromo-*N*-(2,4-dimethylphenyl)-4,6-dimethylaniline (4.05 g, 13.3 mmol, 1.00 equiv.), potassium carbonate (3.67 g, 26.6 mmol, 2.00 equiv.), palladium (II) acetate (134 mg, 0.600 mmol, 0.05 equiv.) and tri-*tert*-butylphosphonium tetrafluoroborate (386 mg, 1.33 mmol, 0.01 equiv.) were added, evacuated and backfilled with argon. Dry toluene (27 mL) and dry *N,N*-dimethylformamide (27 mL) were added and the mixture refluxed for 7 hours. After cooling to room temperature, an excess of water was added and extracted with dichloromethane three times. The combined organic layers were washed with brine, dried over sodium sulfate and reduced in vacuum. The crude product was purified by column chromatography over SiO_2 (dichloromethane/pentane 1:3 to 1:2) to yield 2.56 g of the title compound (11.4 mmol, 86%) as a white solid.

R_f (dichloromethane/pentane 1:2) = 0.29

M.p. = 148 °C

1H NMR (400 MHz, DMSO-d_6) δ 10.45 (s, 1H), 7.62 (d, *J* = 1.6 Hz, 2H), 7.00 – 6.92 (m, 2H), 2.54 (s, 6H), 2.41 (s, 6H).

13C NMR (101 MHz, DMSO) δ 137.6, 127.3, 127.0, 122.5, 120.0, 117.2, 21.1, 17.2.

IR (ATR) $\tilde{v}$ [cm^{-1}] = 3471 (w), 2905 (w), 2729 (w), 2404 (vw), 2338 (vw), 2172 (vw), 2105 (vw), 1881 (vw), 1713 (w), 1597 (w), 1416 (w), 1296 (m), 1222 (m), 1043 (w), 844 (m), 739 (w), 581 (m), 496 (w), 394 (w).

HRMS (ESI, $C_{16}H_{17}N$) calc. 224.1434 [M+H]$^+$, found 224.1431 [M+H]$^+$.

The analytical data is consistent with literature.

4-(1,3,6,8-Tetramethyl-9*H*-carbazol-9-yl)benzonitrile

 In a sealable vial, 4-bromobenzonitrile (546 mg, 3.00 mmol, 1.00 equiv.), 1,3,6,8-tetramethyl-9*H*-carbazole (703 mg, 3.15 mmol, 1.05 equiv.), sodium *tert*-butoxide (706 mg, 7.35 mmol, 2.45 equiv.), palladium (II) acetate (34.0 mg, 0.150 mmol, 0.05 equiv.) and tri-*tert*-butylphosphonium tetrafluoroborate (131 mg, 0.450 mmol, 0.015 equiv.) were added, evacuated and backfilled with argon. Dry toluene (33 mL) was added and the mixture heated at 80 °C for 18 hours. After cooling to room temperature, an excess of water was added and extracted with dichloromethane three times. The combined organic layers were washed with brine, dried over sodium sulfate and reduced in vacuum. The crude product was purified by column chromatography over SiO_2 (dichloromethane/pentane 2:3) to yield 433 mg of the title compound (1.35 mmol, 45%) as a white solid.

R*f* (dichloromethane/pentane 2:3) = 0.31

¹H NMR (400 MHz, Chloroform-*d*) δ 7.78 – 7.70 (m, 4H), 7.61 – 7.52 (m, 2H), 6.92 (s, 2H), 2.48 (s, 6H), 1.84 (s, 6H).

¹³C NMR (101 MHz, CDCl₃) δ 147.4, 139.6, 132.3, 132.3, 130.6, 130.0, 124.8, 121.1, 118.3, 118.1, 112.7, 29.9, 21.2, 19.8.

IR (ATR) $\tilde{v}$ [cm⁻¹] = 3408 (vw), 3088 (vw), 2919 (w), 2532 (vw), 2349 (vw), 2227 (w), 1918 (vw), 1739 (vw), 1597 (w), 1506 (w), 1413 (m), 1280 (w), 1209 (m), 1052 (w), 855 (m), 730 (w), 643 (w), 562 (m), 432 (w).

HRMS (ESI, C₂₃H₂₀N₂) calc. 325.1699 [M+H]⁺, found 325.1694 [M+H]⁺.

9-(4-(2*H*-Tetrazol-5-yl)phenyl)-3,6-di-*tert*-butyl-9*H*-carbazole

According to the general procedure (GP 7), 4-(3,6-di-*tert*-butyl-9*H*-carbazol-9-yl)benzonitrile (951 mg, 2.50 mmol, 1.00 equiv.), sodium azide (487 mg, 7.50 mmol, 3.00 equiv.) and ammonium chloride (401 mg, 7.50 mmol, 3.00 equiv.) were reacted in dry DMF (16 mL) for 16 h at 130 °C. After cooling to room temperature, the reaction mixture was poured into an excess of 1M HCl and mixed thoroughly. The white solid was filtered off, washed several times with water and thoroughly dried to yield 1005 mg of the title compound (2.48 mmol, 99%) as a white solid.

^{1}H NMR (400 MHz, DMSO-d_6) δ 8.34 – 8.27 (m, 4H), 7.89 (d, *J* = 8.3 Hz, 2H), 7.50 (dd, *J* = 8.7, 1.9 Hz, 2H), 7.43 (d, *J* = 8.6 Hz, 2H), 1.42 (s, 18H).

^{13}C NMR (101 MHz, DMSO) δ 143.0, 139.8, 138.0, 128.8, 126.8, 123.9, 123.2, 116.8, 109.2, 34.6, 31.8.

IR (ATR) $\tilde{v}$ [cm^{-1}] = 3400 (vw), 3061 (vw), 2954 (w), 2863 (w), 2715 (vw), 2618 (vw), 2445 (vw), 2105 (vw), 1878 (vw), 1612 (w), 1473 (w), 1364 (w), 1262 (w), 1016 (w), 945 (w), 877 (w), 814 (w), 741 (w), 612 (w), 518 (vw), 426 (w), 0 (vs), 837 (w), 744 (m), 622 (m), 558 (m), 433 (m).

HRMS (APCI, C$_{27}$H$_{29}$N$_5$) calc. 424.2496 [M+H]$^+$, found 424.2481 [M+H]$^+$.

9-(4-(2*H*-Tetrazol-5-yl)phenyl)-9*H*-carbazole

According to the general procedure (GP 7), 4-(9*H*-carbazol-9-yl)benzonitrile (805 mg, 3.00 mmol, 1.00 equiv.), sodium azide (585 mg, 9.00 mmol, 3.00 equiv.) and ammonium chloride (482 mg, 9.00 mmol, 3.00 equiv.) were reacted in dry DMF (16 mL) for 16 h at 130 °C. After cooling to room temperature, the reaction mixture was poured into an excess of 1M HCl and mixed thoroughly. The white solid was filtered off, washed several times with water and thoroughly dried to yield 915 mg of the title compound (2.94 mmol, 98%) as a white solid.

¹H NMR (300 MHz, DMSO-*d*₆) δ 8.36 (d, *J* = 8.2 Hz, 2H), 8.28 (d, *J* = 7.7 Hz, 2H), 7.91 (d, *J* = 7.5 Hz, 2H), 7.57 – 7.40 (m, 4H), 7.33 (ddd, *J* = 8.0, 6.6, 1.5 Hz, 2H).

IR (ATR) ṽ [cm⁻¹] = 3984 (w), 3017 (m), 2860 (m), 2720 (m), 2649 (m), 1928 (w), 1786 (w), 1602 (m), 1441 (m), 1315 (m), 1223 (s), 1051 (m), 991 (m), 915 (m), 841 (m), 749 (s), 621 (m), 530 (m), 421 (m).

HRMS (APCI, C₁₉H₁₃N₅) calc. 312.1244 [M+H]⁺, found 312.1238 [M+H]⁺.

9,9'-(2-(2*H*-Tetrazol-5-yl)-1,3-phenylene)bis(3,6-di-*tert*-butyl-9*H*-carbazole)

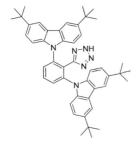

In a sealable vial, 2,6-bis(3,6-di-*tert*-butyl-9*H*-carbazol-9-yl)benzonitrile (658 mg, 1.00 mmol, 1.00 equiv.), sodium azide (195 mg, 3.00 mmol, 3.00 equiv.) and ammonium chloride (161 mg, 3.00 mmol, 3.00 equiv.) were evacuated and backfilled with argon. Dry DMF (6 ml) was added and the resulting mixture heated at 130 °C for 16 h. After cooling to room temperature, the reaction mixture was poured into an excess of 1M HCl and mixed thoroughly. The white solid was filtered off, washed several times with water and thoroughly dried. The crude product was purified by column chromatography over SiO$_2$ (dichloromethane) to yield 650 mg of the title compound (0.927 mmol, 93%) as a white solid.

R$_f$ (dichloromethane) = 0.56

^{1}H NMR (400 MHz, Chloroform-*d*) δ 11.41 (s, 1H), 8.09 (d, *J* = 1.8 Hz, 4H), 7.96 (dd, *J* = 8.4, 7.5 Hz, 1H), 7.77 (d, *J* = 8.0 Hz, 2H), 7.40 (dd, *J* = 8.6, 1.9 Hz, 4H), 7.06 (d, *J* = 8.6 Hz, 4H), 1.43 (s, 36H).

^{13}C NMR (101 MHz, CDCl$_3$) δ 143.6, 139.6, 139.4, 133.7, 130.4, 124.2, 123.7, 116.9, 108.6, 34.9, 32.1.

IR (ATR) ṽ [cm^{-1}] = 3995 (vw), 3889 (vw), 3751 (vw), 3512 (vw), 3308 (vw), 2954 (w), 2561 (vw), 2322 (vw), 2168 (vw), 2105 (vw), 1945 (vw), 1864 (vw), 1746 (vw), 1579 (vw), 1482 (w), 1362 (w), 1262 (w), 1140 (w), 1034 (w), 874 (w), 804 (w), 739 (w), 612 (w), 396 (w).

HRMS (APCI, C$_{47}$H$_{52}$N$_6$) calc. 701.4326 [M+H]$^+$, found 701.4314 [M+H]$^+$.

9,9'-(5-(2*H*-Tetrazol-5-yl)-1,3-phenylene)bis(3,6-di-*tert*-butyl-9*H*-carbazole)

In a sealable vial, 3,5-bis(3,6-di-*tert*-butyl-9*H*-carbazol-9-yl)benzonitrile (1.32 g, 2.00 mmol, 1.00 equiv.), sodium azide (390 mg, 6.00 mmol, 3.00 equiv.) and ammonium chloride (321 mg, 6.00 mmol, 3.00 equiv.) were evacuated and backfilled with argon. Dry DMF (15 ml) was added and the resulting mixture heated at 130 °C for 16 h. After cooling to room temperature, the reaction mixture was poured into an excess of 1M HCl and mixed thoroughly. The white solid was filtered off, washed several times with water and thoroughly dried. The crude product was purified by column chromatography over SiO2 (dichloromethane to dichloromethane/ethyl acetate 50:1 to 20:1) to yield 975 mg of the title compound (1.39 mmol, 70%) as a white solid.

R*f* (dichloromethane/ethyl acetate 20:1) = 0.71

¹H NMR (400 MHz, Chloroform-*d*) δ 8.41 (d, *J* = 1.9 Hz, 2H), 8.15 (d, *J* = 1.8 Hz, 4H), 7.98 (t, *J* = 1.9 Hz, 1H), 7.50 (d, *J* = 8.6 Hz, 4H), 7.45 (dd, *J* = 8.7, 1.9 Hz, 4H), 1.46 (s, 36H).

¹³C NMR (101 MHz, CDCl3) δ 143.9, 141.0, 138.7, 126.4, 124.1, 124.0, 123.0, 116.6, 109.3, 34.9, 32.1.

IR (ATR) $\tilde{v}$ [cm⁻¹] = 3980 (vw), 3872 (vw), 3590 (vw), 3368 (vw), 3155 (vw), 3058 (vw), 2953 (w), 1697 (w), 1597 (w), 1478 (m), 1363 (w), 1295 (m), 1035 (w), 819 (w), 686 (w), 616 (w), 422 (w).

HRMS (APCI, C47H52N6) calc. 701.4326 [M+H]⁺, found 701.4299 [M+H]⁺.

10-(4-(2*H*-Tetrazol-5-yl)phenyl)-9,9-dimethyl-9,10-dihydroacridine

In a sealable vial, 4-(9,9-dimethylacridin-10(9)-yl)benzonitrile (776 mg, 2.50 mmol, 1.00 equiv.), sodium azide (487 mg, 7.50 mmol, 3.00 equiv.) and ammonium chloride (401 mg, 7.50 mmol, 3.00 equiv.) were evacuated and backfilled with argon. Dry DMF (16 ml) was added and the resulting mixture heated at 130 °C for 16 h. After cooling to room temperature, the reaction mixture was poured into an excess of 1M HCl and mixed thoroughly. The white solid was filtered off, washed several times with water and thoroughly dried to yield 867 mg of the title compound (2.45 mmol, 98%) as a white solid.

¹H NMR (400 MHz, DMSO-d_6) δ 8.37 – 8.31 (m, 2H), 7.66 – 7.59 (m, 2H), 7.51 (dd, J = 7.7, 1.6 Hz, 2H), 6.99 (ddd, J = 8.3, 7.3, 1.6 Hz, 2H), 6.93 (td, J = 7.4, 1.4 Hz, 2H), 6.21 (dd, J = 8.2, 1.3 Hz, 2H), 1.63 (s, 6H).

¹³C NMR (101 MHz, DMSO) δ 143.3, 140.0, 132.2, 129.9, 129.8, 126.6, 125.5, 120.9, 113.7, 35.6, 31.2.

IR (ATR) $\tilde{v}$ [cm⁻¹] = 3980 (w), 3842 (vw), 3678 (vw), 2967 (w), 2858 (w), 2726 (w), 2603 (w), 1594 (m), 1484 (m), 1334 (m), 1268 (m), 1155 (w), 992 (w), 928 (w), 853 (w), 745 (m), 623 (m), 522 (w), 382 (w).

HRMS (APCI, $C_{22}H_{19}N_5$) calc. 354.1713 [M+H]⁺, found 354.1707 [M+H]⁺.

9,9',9''-(5-(2*H*-Tetrazol-5-yl)benzene-1,2,3-triyl)tris(3,6-di-*tert*-butyl-9*H*-carbazole)

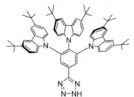

In a sealable vial, 3,4,5-tris(3,6-di-*tert*-butyl-9*H*-carbazol-9-yl)benzonitrile (935 mg, 1.00 mmol, 1.00 equiv.), sodium azide (195 mg, 3.00 mmol, 3.00 equiv.) and ammonium chloride (160 mg, 3.00 mmol, 3.00 equiv.) were evacuated and backfilled with argon. Dry DMF (10 ml) was added and the resulting mixture heated at 130 °C for 16 h. After cooling to room temperature, the reaction mixture was poured into an excess of 1M HCl and mixed thoroughly. The white solid was filtered off, washed several times with water and thoroughly dried. The crude product was purified by column chromatography over SiO_2 (dichloromethane/ethyl acetate 20:1 to 1:1) to yield 850 mg of the title compound (0.871 mmol, 87%) as a white solid.

R*f* (dichloromethane/ethyl acetate 1:1) = 0.82

¹H NMR (400 MHz, Chloroform-*d*) δ 14.09 (s, 1H), 8.56 (s, 2H), 7.75 (d, *J* = 1.6 Hz, 4H), 7.28 (d, *J* = 1.9 Hz, 2H), 7.08 – 6.97 (m, 8H), 6.74 (d, *J* = 8.6 Hz, 2H), 6.60 (dd, *J* = 8.6, 1.9 Hz, 2H), 1.35 (s, 36H), 1.23 (s, 18H).

¹³C NMR (101 MHz, CDCl₃) δ 143.2, 142.7, 138.9, 138.6, 137.3, 127.7, 124.4, 124.1, 123.3, 122.5, 115.9, 115.1, 109.8, 109.5, 34.7, 34.4, 32.0, 31.9.

IR (ATR) $\tilde{v}$ [cm⁻¹] = 3991 (vw), 3820 (vw), 3733 (vw), 3605 (vw), 3270 (vw), 3190 (vw), 3043 (vw), 2952 (w), 2324 (vw), 2162 (vw), 2027 (vw), 1862 (vw), 1747 (vw), 1560 (w), 1472 (m), 1362 (w), 1262 (w), 1034 (w), 875 (w), 806 (m), 608 (w), 423 (w).

HRMS (APCI, C₆₇H₇₅N₇) calc. 978.6157 [M+H]⁺, found 978.6150 [M+H]⁺.

9-(5-(2*H*-Tetrazol-5-yl)pyridin-2-yl)-3,6-di-*tert*-butyl-9*H*-carbazole

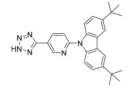 According to the general procedure (GP 7), 6-(3,6-di-*tert*-butyl-9*H*-carbazol-9-yl)nicotinonitrile (763 mg, 2.00 mmol, 1.00 equiv.), sodium azide (391 mg, 6.00 mmol, 3.00 equiv.) and ammonium chloride (321 mg, 6.00 mmol, 3.00 equiv.) were reacted in dry DMF (12 mL) for 16 h at 130 °C. After cooling to room temperature, the reaction mixture was poured into an excess of 1M HCl and mixed thoroughly. The white solid was filtered off, washed several times with water and thoroughly dried to yield 848 mg of the title compound (2.00 mmol, 100%) as a white solid.

¹H NMR (400 MHz, DMSO-d_6) δ 9.30 (dd, J = 2.4, 0.7 Hz, 1H), 8.62 (dd, J = 8.6, 2.4 Hz, 1H), 8.31 (d, J = 1.9 Hz, 2H), 8.03 (dd, J = 8.6, 0.8 Hz, 1H), 7.91 (d, J = 8.8 Hz, 2H), 7.52 (dd, J = 8.8, 2.0 Hz, 2H), 1.42 (s, 18H).

¹³C NMR (101 MHz, DMSO) δ 153.0, 147.5, 144.2, 137.5, 136.9, 124.2, 124.0, 118.0, 116.7, 111.5, 34.6, 31.7.

IR (ATR) $\tilde{v}$ [cm⁻¹] = 3056 (vw), 2955 (w), 2770 (vw), 2215 (vw), 1883 (vw), 1753 (vw), 1609 (w), 1470 (m), 1373 (w), 1297 (w), 991 (w), 831 (w), 741 (w), 616 (w), 424 (w).

HRMS (APCI, $C_{26}H_{28}N_6$) calc. 425.2448 [M+H]⁺, found 425.2440 [M+H]⁺.

9-(6-(2*H*-Tetrazol-5-yl)pyridin-3-yl)-3,6-di-*tert*-butyl-9*H*-carbazole

According to the general procedure (GP 7), 5-(3,6-di-*tert*-butyl-9*H*-carbazol-9-yl)picolinonitrile (763 mg, 2.00 mmol, 1.00 equiv.), sodium azide (391 mg, 6.00 mmol, 3.00 equiv.) and ammonium chloride (321 mg, 6.00 mmol, 3.00 equiv.) were reacted in dry DMF (12 mL) for 16 h at 130 °C. After cooling to room temperature, the reaction mixture was poured into an excess of 1M HCl and mixed thoroughly. The white solid was filtered off, washed several times with water and thoroughly dried to yield 827 mg of the title compound (1.94 mmol, 97%) as a white solid.

^{1}H NMR (400 MHz, DMSO-d_6) δ 9.13 (dd, J = 2.5, 0.8 Hz, 1H), 8.48 (dd, J = 8.4, 0.7 Hz, 1H), 8.39 (dd, J = 8.4, 2.5 Hz, 1H), 8.35 (d, J = 1.9 Hz, 2H), 7.52 (dd, J = 8.7, 1.9 Hz, 2H), 7.45 (d, J = 8.6 Hz, 2H), 1.42 (s, 18H).

^{13}C NMR (101 MHz, DMSO) δ 147.5, 143.5, 141.6, 138.0, 136.1, 135.4, 124.1, 123.8, 123.5, 117.0, 109.1, 34.6, 31.8.

IR (ATR) $\tilde{v}$ [cm^{-1}] = 3047 (vw), 2952 (w), 2866 (vw), 2323 (vw), 2185 (vw), 1583 (vw), 1490 (w), 1364 (w), 1259 (w), 1062 (w), 823 (w), 727 (w), 612 (w), 488 (w), 418 (w).

HRMS (APCI, C$_{26}$H$_{28}$N$_6$) calc. 425.2448 [M+H]$^+$, found 425.2438 [M+H]$^+$.

EA (C$_{26}$H$_{28}$N$_6$) calc. C: 73.56, H: 6.65, N: 19.80; found C: 73.55, H: 6.58, N: 19.44.

N,N-Diphenyl-4-(2*H*-tetrazol-5-yl)aniline

According to the general procedure (GP 7), 4-(diphenylamino)benzonitrile (406 mg, 1.50 mmol, 1.00 equiv.), sodium azide (293 mg, 4.50 mmol, 3.00 equiv.) and ammonium chloride (241 mg, 4.50 mmol, 3.00 equiv.) were reacted in dry DMF (9 mL) for 16 h at 130 °C. After cooling to room temperature, the reaction mixture was poured into an excess of 1M HCl and mixed thoroughly. The white solid was filtered off, washed several times with water and thoroughly dried to yield 430 mg of the title compound (1.37 mmol, 91%) as a white solid.

^{1}H NMR (400 MHz, DMSO-d_6) δ 7.93 – 7.84 (m, 2H), 7.40 – 7.33 (m, 4H), 7.27 – 7.20 (m, 1H), 7.16 (dt, *J* = 7.5, 1.3 Hz, 2H), 7.12 (dt, *J* = 8.5, 1.3 Hz, 4H), 7.06 – 7.01 (m, 2H).

^{13}C NMR (101 MHz, DMSO) δ 187.5, 183.9, 175.0, 167.5, 166.6, 165.9, 165.9, 163.0, 163.0, 162.1, 158.7, 68.4, 58.7.

IR (ATR) $\tilde{v}$ [cm^{-1}] = 3056 (vw), 2849 (vw), 2729 (vw), 2618 (vw), 2323 (vw), 1589 (w), 1484 (m), 1281 (m), 1156 (w), 1056 (w), 996 (w), 844 (w), 693 (m), 621 (w), 514 (m), 406 (w).

HRMS (APCI, C$_{19}$H$_{15}$N$_5$) calc. 314.1400 [M+H]$^+$, found 314.1395 [M+H]$^+$.

EA (C$_{19}$H$_{15}$N$_5$) calc. C: 72.93, H: 4.83, N: 22.35; found C: 72.76, H: 4.74, N: 22.21.

3,6-Di-*tert*-butyl-9-(2,6-dimethyl-4-(2*H*-tetrazol-5-yl)phenyl)-9*H*-carbazole

According to the general procedure (GP 7), 4-(3,6-di-*tert*-butyl-9*H*-carbazol-9-yl)-3,5-dimethylbenzonitrile (3.06 g, 7.50 mmol, 1.00 equiv.), sodium azide (1.46 g, 22.5 mmol, 3.00 equiv.) and ammonium chloride (1.20 g, 22.5 mmol, 3.00 equiv.) were reacted in dry DMF (45 mL) for 16 h at 130 °C. After cooling to room temperature, the reaction mixture was poured into an excess of 1M HCl and mixed thoroughly. The white solid was filtered off, washed several times with water and thoroughly dried to yield 3.34 g of the title compound (7.43 mmol, 99%) as a white solid.

^{1}H NMR (400 MHz, DMSO-d_6) δ 8.33 (d, J = 1.8 Hz, 2H), 8.04 (s, 2H), 7.44 (dd, J = 8.6, 1.9 Hz, 2H), 6.82 (d, J = 8.5 Hz, 2H), 1.87 (s, 6H), 1.41 (s, 18H).

^{13}C NMR (101 MHz, DMSO) δ 142.1, 138.8, 137.9, 136.8, 128.9, 128.2, 127.3, 123.9, 122.5, 117.1, 108.4, 34.5, 31.9, 17.3.

IR (ATR) $\tilde{v}$ [cm^{-1}] = 3041 (vw), 2955 (w), 2864 (vw), 2605 (vw), 2492 (vw), 1854 (vw), 1615 (vw), 1488 (w), 1362 (w), 1262 (w), 1034 (w), 947 (w), 882 (w), 804 (w), 727 (w), 619 (w), 465 (vw), 398 (vw).

HRMS (APCI, C$_{29}$H$_{33}$N$_5$) calc. 452.2809 [M+H]$^+$, found 452.2801 [M+H]$^+$.

EA (C$_{29}$H$_{33}$N$_5$) calc. C: 77.13, H: 7.37, N: 15.51; found C: 77.99, H: 7.50, N: 14.58.

9-(4-(2*H*-Tetrazol-5-yl)phenyl)-1,3,6,8-tetramethyl-9*H*-carbazole

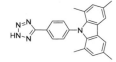 According to the general procedure (GP 7), 4-(1,3,6,8-tetramethyl-9*H*-carbazol-9-yl)benzonitrile (406 mg, 1.25 mmol, 1.00 equiv.), sodium azide (244 mg, 3.75 mmol, 3.00 equiv.) and ammonium chloride (201 mg, 3.75 mmol, 3.00 equiv.) were reacted in dry DMF (5 mL) for 16 h at 130 °C. After cooling to room temperature, the reaction mixture was poured into an excess of 1M HCl and mixed thoroughly. The white solid was filtered off, washed several times with water and thoroughly dried to yield 456 mg of the title compound (1.24 mmol, 99%) as a white solid.

¹H NMR (400 MHz, DMSO-d_6) δ 8.17 (d, J = 8.4 Hz, 2H), 7.80 (d, J = 1.7 Hz, 2H), 7.75 (d, J = 8.4 Hz, 2H), 6.92 (d, J = 1.6 Hz, 2H), 2.41 (s, 6H), 1.84 (s, 6H).

¹³C NMR (101 MHz, DMSO) δ 144.1, 139.0, 132.1, 130.2, 128.7, 127.1, 123.7, 120.7, 117.9, 20.8, 19.0.

IR (ATR) $\tilde{v}$ [cm⁻¹] = 3554 (vw), 2920 (w), 2735 (w), 1711 (vw), 1615 (w), 1499 (m), 1415 (w), 1278 (w), 1207 (m), 1058 (w), 843 (m), 728 (w), 647 (w), 526 (w).

HRMS (ESI, $C_{23}H_{21}N_5$) calc. 368.1870 [M+H]⁺, found 368.1822 [M+H]⁺.

3,7,11-Tris(4-(9H-carbazol-9-yl)phenyl)tris([1,2,4]triazolo)[4,3-a:4',3'-c:4'',3''-e][1,3,5]triazine

In a sealable vial, 9-(4-(2H-tetrazol-5-yl)phenyl)-9H-carbazole (856 mg, 2.75 mmol, 1.00 equiv.) and cyanuric chloride (152 mg, 0.825 mmol, 0.30 equiv.) were evacuated and backfilled with argon. Dry toluene (55 mL) and lutidine (589 mg, 5.50 mmol, 2.00 equiv.) were added and the resulting mixture heated at 80 °C for 4 hours. After cooling to room temperature, the reaction mixture was poured into an excess of water, acidified with 1M HCl and filtered. The residue was washed with water and dichloromethane to yield 578 mg of the title compound (0.625 mmol, 78%) as an off-white solid.

^{1}H NMR (400 MHz, DMSO-d_6) δ 8.41 (d, J = 8.6 Hz, 6H), 8.32 (d, J = 7.5 Hz, 6H), 8.03 (d, J = 8.6 Hz, 6H), 7.62 (d, J = 8.3 Hz, 6H), 7.53 (ddd, J = 8.3, 7.1, 1.3 Hz, 6H), 7.40 – 7.33 (m, 6H).

^{13}C NMR (101 MHz, DMSO) δ 148.8, 142.2, 139.7, 131.8, 128.9, 128.2, 126.5, 126.5, 123.2, 123.1, 120.7, 120.6, 109.9, 99.5.

IR (ATR) $\tilde{v}$ [cm^{-1}] = 3855 (vw), 3358 (vw), 3253 (vw), 3056 (vw), 2928 (vw), 2800 (vw), 2655 (vw), 2516 (vw), 2375 (vw), 2312 (vw), 2171 (vw), 2052 (vw), 1930 (vw), 1779 (vw), 1592 (w), 1448 (m), 1335 (w), 1225 (w), 1012 (w), 914 (vw), 838 (w), 743 (m), 622 (w), 426 (w).

HRMS (FAB-MS, 3-NBA, $C_{60}H_{37}N_{12}$) calc. 925.3259, found 925.3263.

3,7,11-Tris(4-(3,6-di-*tert*-butyl-9*H*-carbazol-9-yl)phenyl)tris([1,2,4]triazolo)[4,3-a:4',3'-c:4'',3''-e][1,3,5]triazine

In a sealable vial, 9-(4-(2*H*-tetrazol-5-yl)phenyl)-3,6-di-*tert*-butyl-9*H*-carbazole (847 mg, 2.00 mmol, 1.00 equiv.) and cyanuric chloride (111 mg, 0.600 mmol, 0.300 equiv.) were evacuated and backfilled with argon. Dry toluene (40 mL) and lutidine (107 mg, 4.00 mmol, 2.00 equiv.) were added and the resulting mixture heated at 80 °C for 4 hours. After cooling to room temperature, the reaction mixture was poured into an excess of water, acidified with 1M HCl and extracted with dichloromethane three times. The combined organic layers were washed with brine, dried over sodium sulfate, reduced in vacuum and purified by silica column chromatography (dichloromethane) to yield 530 mg of the title compound (0.421 mmol, 70%) as a white solid.

R_f (dichloromethane) = 0.65

^{1}H NMR (400 MHz, Chloroform-*d*) δ 8.52 – 8.44 (m, 6H), 8.18 – 8.13 (m, 6H), 7.89 (d, J = 8.7 Hz, 6H), 7.57 (d, J = 8.7 Hz, 6H), 7.51 (dd, J = 8.6, 1.9 Hz, 6H), 1.49 (s, 54H).

^{13}C NMR (101 MHz, CDCl$_3$) δ 150.8, 143.8, 142.0, 140.9, 138.7, 132.0, 126.3, 124.0, 121.7, 116.5, 109.6, 34.9, 32.1.

IR (ATR) $\tilde{v}$ [cm^{-1}] = 3049 (vw), 2951 (vw), 2656 (vw), 2560 (vw), 2474 (vw), 2243 (vw), 2166 (vw), 2087 (vw), 1980 (vw), 1864 (vw), 1744 (vw), 1592 (w), 1481 (w), 1363 (w), 1295 (w), 1035 (vw), 808 (w), 735 (w), 611 (w), 487 (vw), 414 (vw).

HRMS (ESI, C$_{84}$H$_{84}$N$_{12}$) calc. 1261.7015 [M+H]$^+$, found 1261.6928 [M+H]$^+$.

EA (C$_{84}$H$_{84}$N$_{12}$) calc. C: 79.97, H: 6.71, N: 13.32; found C: 79.64, H: 6.68, N: 13.18.

4,4',4''-(Tris([1,2,4]triazolo)[4,3-a:4',3'-c:4'',3''-e][1,3,5]triazine-3,7,11-triyl)tris(N,N-diphenylaniline)

In a sealable vial, N,N-diphenyl-4-(2H-tetrazol-5-yl)aniline (313 mg, 1.00 mmol, 1.00 equiv.) and cyanuric chloride (55.3 mg, 0.300 mmol, 0.30 equiv.) were evacuated and backfilled with argon. Dry toluene (20 mL) and lutidine (53.5 mg, 2.00 mmol, 2.00 equiv.) were added and the resulting mixture heated at 80 °C for 4 hours. After cooling to room temperature, the reaction mixture was poured into an excess of water, acidified with 1M HCl and extracted with dichloromethane three times. The combined organic layers were washed with brine, dried over sodium sulfate, reduced in vacuum and purified by silica column chromatography (dichloromethane/ethyl acetate 99:1 to 98:2) to yield 238 mg of the title compound (0.255 mmol, 85%) as a white solid.

R_f (dichloromethane/ethyl acetate 98:2) = 0.73

1H NMR (400 MHz, Chloroform-*d*) δ 8.02 – 7.94 (m, 6H), 7.32 (dd, *J* = 8.4, 7.3 Hz, 12H), 7.24 – 7.17 (m, 12H), 7.17 – 7.08 (m, 12H).

13C NMR (101 MHz, CDCl3) δ 151.0, 151.0, 146.7, 140.5, 131.2, 129.7, 126.0, 124.5, 120.2, 115.7.

IR (ATR) $\tilde{v}$ [cm⁻¹] = 3035 (vw), 2775 (vw), 2613 (vw), 2320 (vw), 2245 (vw), 2173 (vw), 2086 (vw), 1938 (vw), 1734 (vw), 1582 (m), 1472 (m), 1285 (m), 1196 (m), 1075 (w), 834 (w), 694 (m), 619 (w), 503 (m), 410 (w).

HRMS (ESI, C60H42N12) calc. 931.3728 [M+H]⁺, found 931.3722 [M+H]⁺.

EA (C60H42N12) calc. C: 77.40, H: 4.55, N: 18.05; found C: 76.28, H: 4.57, N: 17.52.

3,7,11-Tris(3,5-bis(3,6-di-*tert*-butyl-9*H*-carbazol-9-yl)phenyl)tris([1,2,4]triazolo)[4,3-a:4',3'-c:4'',3''-e][1,3,5]triazine

In a sealable vial, 9,9'-(5-(2*H*-tetrazol-5-yl)-1,3-phenylene)bis(3,6-di-*tert*-butyl-9*H*-carbazole) (350 mg, 0.500 mmol, 1.00 equiv.) and cyanuric chloride (27.7 mg, 0.150 mmol, 0.30 equiv.) were evacuated and backfilled with argon. Dry toluene (10 mL) and lutidine (26.8 mg, 1.00 mmol, 2.00 equiv.) were added and the resulting mixture heated at 80 °C for 4 hours. After cooling to room temperature, the reaction mixture was poured into an excess of water, acidified with 1M HCl and extracted with dichloromethane three times. The combined organic layers were washed with brine, dried over sodium sulfate, reduced in vacuum and purified by silica column chromatography (dichloromethane/pentane 1:1) to yield 211 mg of the title compound (0.101 mmol, 67%) as an off-white solid.

R_f (dichloromethane/pentane 1:1) = 0.80

1H NMR (400 MHz, Chloroform-*d*) δ 8.45 (d, *J* = 2.0 Hz, 6H), 8.21 – 8.14 (m, 15H), 7.81 (d, *J* = 8.7 Hz, 12H), 7.58 (dd, *J* = 8.7, 1.9 Hz, 12H), 1.48 (s, 108H).

13C NMR (101 MHz, CDCl3) δ 150.1, 143.9, 140.8, 140.5, 138.8, 127.1, 125.9, 124.2, 124.1, 116.6, 109.6, 34.9, 32.1.

IR (ATR) $\tilde{v}$ [cm^{-1}] = 2953 (w), 2664 (vw), 2284 (vw), 2193 (vw), 2111 (vw), 2042 (vw), 1709 (vw), 1588 (w), 1472 (w), 1362 (w), 1294 (w), 1106 (vw), 1034 (vw), 876 (vw), 807 (w), 688 (vw), 612 (w), 469 (vw).

HRMS (ESI, C144H153N15) calc. 2093.2506 [M+H]$^+$, found 2093.2454 [M+H]$^+$.

EA (C144H153N15) calc. C: 82.60, H: 7.37, N: 10.03; found C: 81.50 H: 7.38, N: 9.50.

3,7,11-Tris(4-(9,9-dimethylacridin-10(9*H*)-yl)phenyl)tris([1,2,4]triazolo)[4,3-a:4',3'-c:4'',3''-e][1,3,5]triazine

In a sealable vial, 10-(4-(2*H*-tetrazol-5-yl)phenyl)-9,9-dimethyl-9,10-dihydroacridine (566 mg, 1.60 mmol, 1.00 equiv.) and cyanuric chloride (88.6 mg, 0.480 mmol, 0.30 equiv.) were evacuated and backfilled with argon. Dry toluene (32 mL) and lutidine (85.8 mg, 3.20 mmol, 2.00 equiv.) were added and the resulting mixture heated at 80 °C for 4 hours. After cooling to room temperature, the reaction mixture was poured into an excess of water, acidified with 1M HCl and extracted with dichloromethane three times. The combined organic layers were washed with brine, dried over sodium sulfate, reduced in vacuum and purified by silica column chromatography (dichloromethane/ethyl acetate 98:2 to 90:10) to yield 449 mg of the title compound (0.427 mmol, 89%) as an off-white solid.

R_f (dichloromethane/ethyl acetate 98:2) = 0.23

¹H NMR (400 MHz, DMSO-d_6) δ 8.41 (d, J = 8.3 Hz, 6H), 7.74 (d, J = 8.3 Hz, 6H), 7.55 (dd, J = 7.7, 1.6 Hz, 6H), 7.08 (td, J = 7.7, 1.6 Hz, 6H), 6.98 (td, J = 7.5, 1.3 Hz, 6H), 6.34 (dd, J = 8.1, 1.3 Hz, 6H), 1.67 (s, 18H).

¹³C NMR (101 MHz, DMSO) δ 148.8, 143.6, 142.2, 140.0, 132.8, 131.2, 130.1, 126.6, 125.6, 124.4, 121.0, 113.9, 35.7, 31.2.

IR (ATR) ṽ [cm⁻¹] = 3974 (vw), 3857 (vw), 3630 (vw), 3420 (vw), 3304 (vw), 2968 (w), 2621 (vw), 2376 (vw), 2217 (vw), 2103 (vw), 1911 (vw), 1591 (w), 1477 (w), 1324 (w), 1046 (w), 923 (w), 844 (w), 742 (m), 623 (w), 532 (w), 421 (vw).

HRMS (ESI, $C_{69}H_{54}N_{12}$) calc. 1051.4667 [M+H]⁺, found 1051.4610 [M+H]⁺.

EA ($C_{69}H_{54}N_{12}$) calc. C: 78.83, H: 5.18, N: 15.99; found C: 78.75 H: 5.46, N: 15.17.

3,7,11-Tris(3,4,5-tris(3,6-di-*tert*-butyl-9*H*-carbazol-9-yl)phenyl)tris([1,2,4]triazolo)[4,3-a:4',3'-c:4'',3''-e][1,3,5]triazine

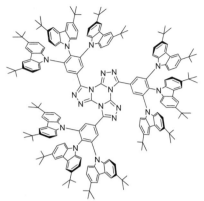

In a sealable vial, 9,9',9''-(5-(2*H*-tetrazol-5-yl)benzene-1,2,3-triyl)tris(3,6-di-*tert*-butyl-9*H*-carbazole) (649 mg, 0.663 mmol, 1.00 equiv.) and cyanuric chloride (36.7 mg, 0.199 mmol, 0.300 equiv.) were evacuated and backfilled with argon. Dry toluene (14 mL) and lutidine (140 mg, 1.33 mmol, 2.00 equiv.) were added and the resulting mixture heated at 80 °C for 4 hours. After cooling to room temperature, the reaction mixture was poured into an excess of water, acidified with 1M HCl and extracted with dichloromethane three times. The combined organic layers were washed with brine, dried over sodium sulfate, reduced in vacuum and purified by silica column chromatography (dichloromethane/pentane 1:2 to 1:1) to yield 390 mg of the title compound (0.133 mmol, 67%) as a yellow solid.

R$_f$ (dichloromethane/pentane 1:1) = 0.45

¹H NMR (400 MHz, Chloroform-*d*) δ 8.57 (s, 6H), 7.78 – 7.69 (m, 12H), 7.35 (d, *J* = 8.6 Hz, 12H), 7.33 – 7.28 (m, 6H), 7.14 (d, *J* = 8.7 Hz, 12H), 6.86 (d, *J* = 8.5 Hz, 6H), 6.66 (d, *J* = 8.6 Hz, 6H), 1.36 (s, 108H), 1.27 (s, 54H).

¹³C NMR (101 MHz, CDCl₃) δ 149.5, 143.2, 142.8, 140.8, 138.6, 138.3, 137.2, 135.6, 130.6, 124.5, 124.2, 124.1, 123.4, 122.6, 115.8, 115.0, 110.0, 109.9, 34.7, 34.5, 32.1, 31.9.

IR (ATR) ṽ [cm⁻¹] = 3928 (vw), 3856 (vw), 3546 (vw), 3414 (vw), 3050 (vw), 2953 (w), 2659 (vw), 2382 (vw), 2281 (vw), 2165 (vw), 2107 (vw), 1862 (vw), 1747 (vw), 1591 (w), 1472 (m), 1362 (w), 1262 (m), 1137 (w), 1034 (w), 875 (w), 806 (m), 609 (w), 422 (w).

HRMS (ESI, C₂₀₄H₂₂₂N₁₈) calc. 2924.7998 [M+H]⁺, found 2924.7997 [M+H]⁺.

EA (C₂₀₄H₂₂₂N₁₈) calc. C: 83.74, H: 7.65, N: 8.62; found C: 83.37 H: 7.92, N: 8.15.

3,7,11-Tris(6-(3,6-di-*tert*-butyl-9*H*-carbazol-9-yl)pyridin-3-yl)tris([1,2,4]triazolo)[4,3-a:4',3'-c:4'',3''-e][1,3,5]triazine

In a sealable vial, 9-(5-(2*H*-tetrazol-5-yl)pyridin-2-yl)-3,6-di-*tert*-butyl-9*H*-carbazole (425 mg, 1.00 mmol, 1.00 equiv.) and cyanuric chloride (55.3 mg, 0.300 mmol, 0.30 equiv.) were evacuated and backfilled with argon. Dry toluene (20 mL) and lutidine (53.5 mg, 2.00 mmol, 2.00 equiv.) were added and the resulting mixture heated at 80 °C for 4 hours. After cooling to room temperature, the reaction mixture was poured into an excess of water, acidified with 1M HCl and extracted with dichloromethane three times. The combined organic layers were washed with brine, dried over sodium sulfate, reduced in vacuum and purified by silica column chromatography (dichloromethane/ethyl acetate 97:3) to yield 279 mg of the title compound (0.222 mmol, 74%) as an off-white solid.

R_f (dichloromethane/ethyl acetate 97:3) = 0.61

¹H NMR (400 MHz, Chloroform-*d*) δ 9.48 (d, *J* = 2.4 Hz, 3H), 8.78 (dd, *J* = 8.7, 2.5 Hz, 3H), 8.11 (d, *J* = 2.0 Hz, 6H), 8.07 (d, *J* = 8.7 Hz, 6H), 7.95 (d, *J* = 8.6 Hz, 3H), 7.55 (dd, *J* = 8.8, 2.0 Hz, 6H), 1.48 (s, 54H).

¹³C NMR (101 MHz, CDCl₃) δ 150.7, 148.7, 145.2, 139.7, 137.5, 125.3, 124.4, 116.5, 116.4, 111.9, 35.0, 32.0, 29.9.

IR (ATR) $\tilde{v}$ [cm⁻¹] = 2955 (w), 1587 (m), 1461 (m), 1388 (m), 1298 (m), 1036 (w), 951 (vw), 809 (w), 613 (w), 499 (vw), 410 (w).

HRMS (ESI, C₈₁H₈₁N₁₅) calc. 1264.6872 [M+H]⁺, found 1264.6829 [M+H]⁺.

EA (C₈₁H₈₁N₁₅) calc. C: 76.93, H: 6.46, N: 16.61; found C: 76.28, H: 6.42, N: 16.31.

3,7,11-Tris(5-(3,6-di-*tert*-butyl-9*H*-carbazol-9-yl)pyridin-2-yl)tris([1,2,4]triazolo)[4,3-a',3'-c:4'',3''-e][1,3,5]triazine

In a sealable vial, 9-(6-(2*H*-tetrazol-5-yl)pyridin-3-yl)-3,6-di-*tert*-butyl-9*H*-carbazole (425 mg, 1.00 mmol, 1.00 equiv.) and cyanuric chloride (55.3 mg, 0.300 mmol, 0.300 equiv.) were evacuated and backfilled with argon. Dry toluene (20 mL) and lutidine (53.5 mg, 2.00 mmol, 2.00 equiv.) were added and the resulting mixture heated at 80 °C for 4 hours. After cooling to room temperature, the reaction mixture was poured into an excess of water, acidified with 1M HCl and extracted with dichloromethane three times. The combined organic layers were washed with brine, dried over sodium sulfate, reduced in vacuum and purified by silica column chromatography (dichloromethane/ethyl acetate 97:3) to yield 87.0 mg of the title compound (0.069 mmol, 23%) as an off-white solid.

R_f (dichloromethane/ethyl acetate 97:3) = 0.58

^{1}H NMR (400 MHz, Chloroform-*d*) δ 9.22 (d, *J* = 2.4 Hz, 3H), 8.40 (d, *J* = 8.4 Hz, 3H), 8.23 (dd, *J* = 8.4, 2.4 Hz, 3H), 8.17 (dd, *J* = 1.7, 0.9 Hz, 6H), 7.59 – 7.48 (m, 12H), 1.49 (s, 54H).

^{13}C NMR (101 MHz, CDCl$_3$) δ 149.6, 147.7, 144.5, 141.3, 141.0, 138.4, 137.6, 133.8, 126.4, 124.4, 124.4, 116.7, 109.2, 35.0, 32.1.

IR (ATR) $\tilde{v}$ [cm^{-1}] = 3052 (vw), 2954 (w), 1585 (w), 1475 (m), 1363 (w), 1261 (w), 1107 (w), 1034 (w), 945 (vw), 877 (w), 808 (w), 611 (w), 407 (w).

HRMS (ESI, C$_{81}$H$_{81}$N$_{15}$) calc. 1264.6872 [M+H]$^+$, found 1264.6864 [M+H]$^+$.

3,7,11-Tris(4-(3,6-di-*tert*-butyl-9*H*-carbazol-9-yl)-3,5-dimethylphenyl)tris([1,2,4]triazolo)[4,3-a:4',3'-c:4'',3''-e][1,3,5]triazine

In a sealable vial, 3,6-di-*tert*-butyl-9-(2,6-dimethyl-4-(2*H*-tetrazol-5-yl)phenyl)-9*H*-carbazole (452 mg, 1.00 mmol, 1.00 equiv.) and cyanuric chloride (55.3 mg, 0.300 mmol, 0.300 equiv.) were evacuated and backfilled with argon. Dry toluene (20 mL) and lutidine (53.5 mg, 2.00 mmol, 2.00 equiv.) were added and the resulting mixture heated at 80 °C for 4 hours. After cooling to room temperature, the reaction mixture was poured into an excess of water, acidified with 1M HCl and extracted with

dichloromethane three times. The combined organic layers were washed with brine, dried over sodium sulfate, reduced in vacuum and purified by silica column chromatography (dichloromethane/ethyl acetate 99:1 to 98:2) to yield 314 mg of the title compound (0.234 mmol, 78%) as a white solid.

R_f (dichloromethane/ethyl acetate 98:2) = 0.70

^{1}H NMR (400 MHz, Chloroform-*d*) δ 8.21 (d, *J* = 1.8 Hz, 6H), 8.15 (s, 6H), 7.48 (dd, *J* = 8.6, 1.9 Hz, 6H), 6.97 (d, *J* = 8.5 Hz, 6H), 2.06 (s, 18H), 1.50 (s, 54H).

^{13}C NMR (101 MHz, CDCl$_3$) δ 151.0, 142.8, 140.9, 139.2, 139.0, 138.5, 130.4, 124.0, 123.9, 123.2, 116.6, 109.1, 34.9, 32.2, 18.2.

IR (ATR) $\tilde{v}$ [cm^{-1}] = 3052 (vw), 2956 (w), 1577 (w), 1478 (m), 1362 (w), 1294 (w), 1032 (vw), 942 (w), 877 (w), 812 (w), 620 (w), 469 (w).

HRMS (ESI, C$_{90}$H$_{96}$N$_{12}$) calc. 1345.7954 [M+H]$^+$, found 1345.7911 [M+H]$^+$.

EA (C$_{90}$H$_{96}$N$_{12}$) calc. C: 80.32, H: 7.19, N: 12.49; found C: 80.51, H: 7.24, N: 12.07.

3,7,11-Tris(4-(1,3,6,8-tetramethyl-9*H*-carbazol-9-yl)phenyl)tris([1,2,4]triazolo)[4,3-a:4',3'-c:4'',3''-e][1,3,5]triazine

In a sealable vial, 9-(4-(2*H*-tetrazol-5-yl)phenyl)-1,3,6,8-tetramethyl-9*H*-carbazole (393 mg, 1.07 mmol, 1.00 equiv.) and cyanuric chloride (59.2 mg, 0.321 mmol, 0.300 equiv.) were evacuated and backfilled with argon. Dry toluene (22 mL) and lutidine (57.4 mg, 2.14 mmol, 2.00 equiv.) were added and the resulting mixture heated at 80 °C for 4 hours. After cooling to room temperature, the reaction mixture was poured into an excess of water, acidified with 1M HCl and extracted with dichloromethane three times. The combined organic layers were washed with brine, dried over sodium sulfate, reduced in vacuum and purified by silica column chromatography (dichloromethane/ethyl acetate 95:5 to 85:15) to yield 298 mg of the title compound (0.273 mmol, 85%) as a white solid.

R_f (dichloromethane/ethyl acetate 95:5) = 0.44

¹H NMR (300 MHz, DMSO-d_6) δ 8.18 (d, *J* = 8.4 Hz, 6H), 7.87 – 7.77 (m, 12H), 6.97 (d, *J* = 1.7 Hz, 6H), 2.44 (s, 18H), 1.96 (s, 18H).

¹³C NMR (101 MHz, DMSO) δ 148.6, 144.5, 142.1, 139.1, 131.1, 130.2, 130.2, 128.8, 124.9, 123.8, 120.8, 117.9, 20.8, 19.1.

IR (ATR) ṽ [cm⁻¹] = 3379 (vw), 3299 (vw), 3227 (vw), 2914 (vw), 2389 (vw), 2323 (vw), 2234 (vw), 2072 (vw), 1913 (vw), 1733 (vw), 1586 (w), 1484 (w), 1410 (w), 1276 (w), 1205 (w), 1013 (w), 846 (w), 644 (w), 576 (w), 437 (vw).

HRMS (ESI, $C_{72}H_{60}N_{12}$) calc. 1092.5064 [M]⁺, found 1092.5047 [M]⁺.

3,6-Di-*tert*-butyl-9-(4-(5,7-diphenyl-[1,2,4]triazolo[4,3-a][1,3,5]triazin-3-yl)phenyl)-9*H*-carbazole

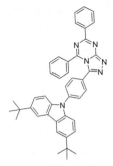

In a sealable vial, 9-(4-(2*H*-tetrazol-5-yl)phenyl)-3,6-di-*tert*-butyl-9*H*-carbazole (339 mg, 0.800 mmol, 1.00 equiv.) and 2-chloro-4,6-diphenyl-1,3,5-triazine (214 mg, 0.800 mmol, 1.00 equiv.) were evacuated and backfilled with argon. Dry toluene (16 mL) and lutidine (86.0 mg, 0.800 mmol, 1.00 equiv.) were added and the resulting mixture heated at 80 °C for 4 hours. After cooling to room temperature, the reaction mixture was poured into an excess of water, acidified with 1M HCl and extracted with dichloromethane three times. The combined organic layers were washed with brine, dried over sodium sulfate, reduced in vacuum and purified by silica column chromatography (dichloromethane/pentane 1:1 to 5:1) to yield 62.0 mg of the title compound (0.0961 mmol, 12%) as a yellow solid.

R$_f$ (dichloromethane/pentane 1:1) = 0.27

^{1}H NMR (400 MHz, Chloroform-*d*) δ 9.18 – 9.10 (m, 2H), 8.81 – 8.74 (m, 2H), 8.71 – 8.64 (m, 2H), 8.16 (t, J = 1.3, 1.3 Hz, 2H), 7.82 – 7.69 (m, 5H), 7.65 – 7.54 (m, 3H), 7.50 (d, J = 1.4 Hz, 4H), 1.48 (s, 18H).

^{13}C NMR (101 MHz, CDCl$_3$) δ 166.8, 163.7, 159.9, 154.4, 143.5, 141.2, 138.9, 135.5, 134.2, 132.7, 131.7, 129.8, 129.6, 129.6, 129.0, 129.0, 128.1, 126.7, 123.9, 123.9, 116.5, 109.5, 34.9, 32.2.

IR (ATR) $\tilde{v}$ [cm^{-1}] = 3060 (vw), 2958 (w), 1603 (w), 1471 (m), 1365 (w), 1296 (w), 1219 (w), 1034 (w), 768 (m), 691 (w), 612 (w), 532 (vw), 417 (vw).

HRMS (ESI, C$_{42}$H$_{38}$N$_6$) calc. 627.3231 [M+H]$^+$, found 627.3212 [M+H]$^+$.

EA (C$_{42}$H$_{38}$N$_6$) calc. C: 80.48, H: 6.11, N: 13.41; found C: 79.69 H: 6.07, N: 13.01.

9,9'-((5-Phenylbis([1,2,4]triazolo)[4,3-a:4',3'-c][1,3,5]triazine-3,9-diyl)bis(4,1-phenylene))bis(3,6-di-*tert*-butyl-9*H*-carbazole)

In a sealable vial, 9-(4-(2*H*-tetrazol-5-yl)phenyl)-3,6-di-*tert*-butyl-9*H*-carbazole (318 mg, 0.750 mmol, 1.00 equiv.) and 2,4-dichloro-6-phenyl-1,3,5-triazine (76.3 mg, 0.338 mmol, 0.450 equiv.) were evacuated and backfilled with argon. Dry toluene (15 mL) and lutidine (80.0 mg, 0.750 mmol, 1.00 equiv.) were added and the resulting mixture heated at 80 °C for 4 hours. After cooling to room temperature, the reaction mixture was poured into an excess of water, acidified with 1M HCl and extracted with dichloromethane three times. The combined organic layers were washed with brine, dried over sodium sulfate, reduced in vacuum and purified by silica column chromatography (dichloromethane/ethyl acetate 98:2 to 90:10) to yield 226 mg of the title compound (0.240 mmol, 71%) as a yellow solid.

R_f (dichloromethane/ethyl acetate 98:2) = 0.34

¹H NMR (400 MHz, Chloroform-*d*) δ 8.86 – 8.79 (m, 1H), 8.70 (dd, *J* = 8.3, 6.1 Hz, 2H), 8.41 (d, *J* = 8.5 Hz, 1H), 8.17 – 8.11 (m, 5H), 7.91 (d, *J* = 8.4 Hz, 1H), 7.86 (d, *J* = 8.2 Hz, 1H), 7.82 – 7.68 (m, 2H), 7.62 – 7.43 (m, 8H), 7.37 (d, *J* = 21.8 Hz, 3H), 7.28 (d, *J* = 8.7 Hz, 1H), 1.49 (d, *J* = 0.9 Hz, 18H), 1.46 (d, *J* = 3.8 Hz, 18H).

¹³C NMR (101 MHz, CDCl₃) δ 161.7, 151.4, 150.9, 148.8, 148.1, 144.4, 143.8, 143.7, 143.7, 143.6, 141.4, 141.3, 141.0, 140.5, 138.7, 138.7, 138.6, 132.7, 131.5, 131.5, 131.4, 131.1, 129.9, 129.7, 129.2, 129.0, 128.4, 126.6, 126.4, 126.3, 126.2, 126.0, 124.1, 124.0, 123.9, 123.5, 123.2, 116.6, 116.5, 116.4, 109.6, 109.5, 109.3, 109.2, 34.9, 34.9, 34.9, 32.1, 32.1.

IR (ATR) ṽ [cm⁻¹] = 3856 (vw), 3761 (vw), 3630 (vw), 3509 (vw), 3414 (vw), 3206 (vw), 3055 (w), 2952 (w), 2565 (vw), 2379 (vw), 2294 (vw), 1735 (w), 1607 (w), 1471 (m), 1362 (w), 1293 (w), 1104 (w), 1034 (w), 808 (w), 740 (w), 611 (w), 422 (w).

HRMS (ESI, C₆₃H₆₁N₉) calc. 944.5123 [M+H]⁺, found 944.5108 [M+H]⁺.

EA (C₆₃H₆₁N₉) calc. C: 80.14, H: 6.51, N: 13.35; found C: 79.41, H: 6.56, N: 13.00.

3,6-Di-*tert*-butyl-9-(4-(5,7-diphenyl-[1,2,4]triazolo[4,3-a][1,3,5]triazin-3-yl)-2,6-dimethylphenyl)-9*H*-carbazole

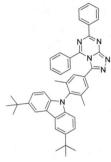

In a sealable vial, 3,6-di-*tert*-butyl-9-(2,6-dimethyl-4-(2*H*-tetrazol-5-yl)phenyl)-9*H*-carbazole (361 mg, 0.800 mmol, 1.00 equiv.) and 2-chloro-4,6-diphenyl-1,3,5-triazine (214 mg, 0.800 mmol, 1.00 equiv.) were evacuated and backfilled with argon. Dry toluene (16 mL) and lutidine (86.0 mg, 0.800 mmol, 1.00 equiv.) were added and the resulting mixture heated at 80 °C for 4 hours. After cooling to room temperature, the reaction mixture was poured into an excess of water, acidified with 1M HCl and extracted with dichloromethane three times. The combined organic layers were washed with brine, dried over sodium sulfate, reduced in vacuum and purified by silica column chromatography (dichloromethane/pentane 1:1 to 2:1) to yield 101 mg of the title compound (0.155 mmol, 19%) as a yellow solid.

R_f (dichloromethane) = 0.65

^{1}H NMR (400 MHz, Chloroform-*d*) δ 8.87 – 8.81 (m, 4H), 8.32 (s, 2H), 8.20 (dd, *J* = 1.9, 0.6 Hz, 2H), 7.73 – 7.67 (m, 2H), 7.67 – 7.60 (m, 4H), 7.46 (ddd, *J* = 8.6, 4.5, 1.9 Hz, 2H), 6.91 (dd, *J* = 8.5, 0.6 Hz, 2H), 2.03 (s, 6H), 1.49 (s, 18H).

^{13}C NMR (101 MHz, CDCl$_3$) δ 174.7, 165.9, 162.0, 142.6, 139.5, 138.7, 138.6, 138.0, 134.6, 134.1, 129.9, 129.8, 129.7, 129.1, 128.0, 126.4, 123.9, 123.1, 116.6, 108.9, 34.9, 32.2, 18.0.

IR (ATR) $\tilde{v}$ [cm^{-1}] = 3981 (vw), 3914 (vw), 3818 (vw), 3647 (vw), 3546 (vw), 3440 (vw), 3290 (vw), 3056 (vw), 2956 (w), 2498 (vw), 2322 (vw), 2247 (vw), 2051 (vw), 1769 (vw), 1595 (w), 1476 (m), 1378 (w), 1266 (w), 1012 (w), 911 (w), 729 (w), 617 (w), 474 (vw), 415 (w).

HRMS (ESI, C$_{44}$H$_{42}$N$_6$) calc. 655.3544 [M+H]$^+$, found 655.3534 [M+H]$^+$.

9,9'-((5-Phenylbis([1,2,4]triazolo)[4,3-a:4',3'-c][1,3,5]triazine-3,9-diyl)bis(2,6-dimethyl-4,1-phenylene))bis(3,6-di-*tert*-butyl-9*H*-carbazole)

In a sealable vial, 3,6-di-*tert*-butyl-9-(2,6-dimethyl-4-(2*H*-tetrazol-5-yl)phenyl)-9*H*-carbazole (361 mg, 0.800 mmol, 1.00 equiv.) and 2,4-dichloro-6-phenyl-1,3,5-triazine (81.4 mg, 0.360 mmol, 0.450 equiv.) were evacuated and backfilled with argon. Dry toluene (16 mL) and lutidine (86.0 mg, 0.800 mmol, 1.00 equiv.) were added and the resulting mixture heated at 80 °C for 4 hours. After cooling to room temperature, the reaction mixture was poured into an excess of water, acidified with 1M HCl and extracted with dichloromethane three times. The combined organic layers were washed with brine, dried over sodium sulfate, reduced in vacuum and purified by silica column chromatography (dichloromethane/ethyl acetate 98:2 to 92:8) to yield 308 mg of the title compound (0.387 mmol, 86%) as a yellow solid.

R_f (dichloromethane/ethyl acetate 98:2) = 0.31

1H NMR (400 MHz, Chloroform-*d*) δ 8.35 (s, 2H), 8.19 (dd, *J* = 11.1, 1.9 Hz, 4H), 7.63 (dd, *J* = 19.1, 7.7 Hz, 3H), 7.54 – 7.36 (m, 6H), 7.17 (s, 2H), 6.95 (d, *J* = 8.5 Hz, 2H), 6.74 (d, *J* = 8.5 Hz, 2H), 2.04 (s, 6H), 1.72 (s, 6H), 1.48 (s, 36H).

13C NMR (101 MHz, CDCl₃) δ 151.0, 149.2, 149.1, 149.1, 144.4, 142.8, 142.6, 139.1, 138.9, 138.6, 138.4, 138.3, 137.4, 132.6, 130.4, 130.0, 129.9, 129.5, 128.4, 126.0, 125.2, 123.9, 123.8, 123.1, 116.7, 116.5, 109.2, 108.7, 34.9, 34.9, 32.2, 32.2, 32.2, 29.9, 18.2, 17.7.

IR (ATR) ṽ [cm⁻¹] = 3963 (vw), 3856 (vw), 3771 (vw), 3649 (vw), 3525 (vw), 3429 (vw), 3234 (vw), 3168 (vw), 3054 (vw), 2953 (w), 2659 (vw), 2323 (vw), 2204 (vw), 1739 (vw), 1625 (w), 1476 (w), 1362 (w), 1293 (w), 1032 (w), 876 (w), 808 (w), 744 (w), 620 (w), 470 (w).

HRMS (ESI, C₆₇H₆₉N₉) calc. 1000.5749 [M+H]⁺, found 1000.5695 [M+H]⁺.

5.2   Crystal Structures

5.2.1   Crystallographic Data Solved by Dr. Martin Nieger

Crystal structures in this section were measured and solved by Dr. Martin Nieger at the University of Helsinki.

Name in this thesis	Number in this thesis	Code used by Dr. Nieger
4CzCNTl	43	SB899_HY
4CzCNTlallyl	47d	SB987_HY
4CzCNTlOMePh	47k	SB906_HY
4CzCNTlpgyl	47e	SB908_HY
4CzdTl	45	SB990_HY
4CzdTlpgyl	48b	SB1023_HY
4CzdTlPh	48c	SB988_HY
4CzCNOXDMe	71b	SB1001_HY
4CzCNOXDCH$_2$Cl	71f	SB1054_HY
4CzCNOXD2,6OMePh	71i	SB1047_HY
4CzCNOXDMes	71h	SB1049_HY
4CzCNOXDC$_6$F$_5$	71n	SB1046_HY

4CzCNTl (43) SB899_HY

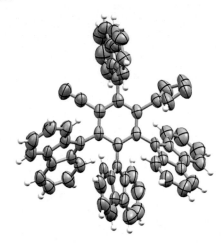

Crystal data

$C_{60.50}H_{41}Cl_9N_9$	$Z = 8$
$M_r = 1213.07$	$F(000) = 4960$
Monoclinic, $C2/c$	$D_x = 1.413$ Mg m^{-3}
$a = 46.5638\ (16)$ Å	Cu $K\alpha$ radiation, $\lambda = 1.54178$ Å
$b = 10.0857\ (3)$ Å	$\mu = 4.43$ mm^{-1}
$c = 24.2863\ (8)$ Å	$T = 123$ K
$\beta = 90.876\ (2)°$	$0.36 \times 0.24 \times 0.16$ mm
$V = 11404.2\ (6)$ Å^3	

Data collection

39614 measured reflections	$\theta_{max} = 69.8°,\ \theta_{min} = 3.6°$
10685 independent reflections	$h = -56 \rightarrow 52$
6919 reflections with $I > 2\sigma(I)$	$k = -10 \rightarrow 12$
$R_{int} = 0.032$	$l = -29 \rightarrow 29$

4CzCNTlallyl (**47d**) SB987_HY

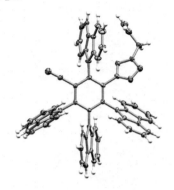

Crystal data

$C_{59}H_{37}N_9 \cdot 5(C_2D_6OS)$	$F(000) = 2656$
$M_r = 1292.80$	$D_x = 1.330$ Mg m^{-3}
Monoclinic, Cc (no.9)	Cu $K\alpha$ radiation, $\lambda = 1.54178$ Å
$a = 20.2281$ (8) Å	Cell parameters from 9318 reflections
$b = 19.3476$ (7) Å	$\theta = 3.2–72.2°$
$c = 17.3533$ (6) Å	$\mu = 2.12$ mm^{-1}
$\beta = 108.119$ (2)°	$T = 123$ K
$V = 6454.7$ (4) Å^3	Blocks, yellow
$Z = 4$	$0.32 \times 0.20 \times 0.20$ mm

Data collection

Bruker D8 VENTURE diffractometer with Photon100 detector	12395 independent reflections
Radiation source: INCOATEC microfocus sealed tube	11731 reflections with $I > 2\sigma(I)$
Detector resolution: 10.4167 pixels mm^{-1}	$R_{int} = 0.045$
rotation in ϕ and ω, 1°, shutterless scans	$\theta_{max} = 72.3°$, $\theta_{min} = 3.2°$
Absorption correction: multi-scan SADABS (Sheldrick, 2014)	$h = -24\rightarrow24$
$T_{min} = 0.569$, $T_{max} = 0.706$	$k = -23\rightarrow23$
37497 measured reflections	$l = -21\rightarrow21$

Refinement

Refinement on F^2	Secondary atom site location: difference Fourier map
Least-squares matrix: full	Hydrogen site location: mixed
$R[F^2 > 2\sigma(F^2)] = 0.062$	H-atom parameters constrained
$wR(F^2) = 0.164$	$w = 1/[\sigma^2(F_o^2) + (0.0926P)^2 + 15.5946P]$ where $P = (F_o^2 + 2F_c^2)/3$
$S = 1.03$	$(\Delta/\sigma)_{max} < 0.001$
12395 reflections	$\Delta\rangle_{max} = 1.02$ e Å^{-3}
779 parameters	$\Delta\rangle_{min} = -1.03$ e Å^{-3}
226 restraints	Absolute structure: Flack x determined using 5256 quotients [(I+)-(I-)]/[(I+)+(I-)] (Parsons, Flack and Wagner, Acta Cryst. B69 (2013) 249-259).
Primary atom site location: structure-invariant direct methods	Absolute structure parameter: 0.035 (7)

4CzCNTlOMePh (47k) SB906_HY

Crystal data

$C_{63}H_{39}N_9O \cdot 2(C_2H_6OS)$	$F(000) = 2288$
$M_r = 1094.29$	$D_x = 1.324$ Mg m^{-3}
Monoclinic, $P2_1/c$ (no.14)	Cu $K\alpha$ radiation, $\lambda = 1.54178$ Å
$a = 13.3713$ (4) Å	Cell parameters from 9967 reflections
$b = 18.2748$ (5) Å	$\theta = 3.1–71.8°$
$c = 22.5593$ (6) Å	$\mu = 1.35$ mm^{-1}
$\beta = 94.989$ (1)°	$T = 123$ K
$V = 5491.7$ (3) Å^3	Blocks, orange
$Z = 4$	$0.18 \times 0.14 \times 0.06$ mm

Data collection

Bruker D8 VENTURE diffractometer with Photon100 detector	10769 independent reflections
Radiation source: INCOATEC microfocus sealed tube	8347 reflections with $I > 2\sigma(I)$
Detector resolution: 10.4167 pixels mm^{-1}	$R_{int} = 0.044$
rotation in ϕ and ω, 1°, shutterless scans	$\theta_{max} = 72.2°, \theta_{min} = 3.1°$
Absorption correction: multi-scan SADABS (Sheldrick, 2014)	$h = -16 \rightarrow 16$
$T_{min} = 0.818, T_{max} = 0.915$	$k = -22 \rightarrow 21$
49642 measured reflections	$l = -27 \rightarrow 27$

Refinement

Refinement on F^2	Primary atom site location: structure-invariant direct methods
Least-squares matrix: full	Secondary atom site location: difference Fourier map
$R[F^2 > 2\sigma(F^2)] = 0.067$	Hydrogen site location: difference Fourier map
$wR(F^2) = 0.195$	H-atom parameters constrained
$S = 1.03$	$w = 1/[\sigma^2(F_o^2) + (0.1045P)^2 + 5.0689P]$ where $P = (F_o^2 + 2F_c^2)/3$
10769 reflections	$(\Delta/\sigma)_{max} < 0.001$
724 parameters	$\Delta\rangle_{max} = 0.88$ e Å^{-3}
36 restraints	$\Delta\rangle_{min} = -0.59$ e Å^{-3}

4CzCNTlpgyl (47e) SB908_HY

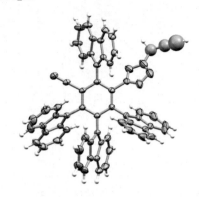

Crystal data

$C_{59}H_{35}N_9$	$V = 2347.39$ (16) $Å^3$
$M_r = 869.96$	$Z = 2$
Triclinic, P-1 (no.2)	$F(000) = 904$
$a = 12.7632$ (5) Å	$D_x = 1.231$ Mg m^{-3}
$b = 13.7682$ (5) Å	Cu $K\alpha$ radiation, $\lambda = 1.54178$ Å
$c = 15.4497$ (6) Å	$\mu = 0.59$ mm^{-1}
$\alpha = 81.329$ (2)°	$T = 123$ K
$\beta = 66.579$ (2)°	$0.22 \times 0.20 \times 0.08$ mm
$\gamma = 70.469$ (2)°	

Data collection

26059 measured reflections	$\theta_{max} = 72.2°$, $\theta_{min} = 3.1°$
9090 independent reflections	$h = -15 \rightarrow 15$
7460 reflections with $I > 2\sigma(I)$	$k = -16 \rightarrow 16$
$R_{int} = 0.023$	$l = -18 \rightarrow 19$

4CzdTl (45) SB990_HY

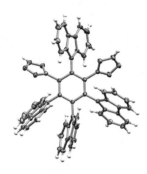

Crystal data

$C_{56}H_{34}N_{12}·3(C_3D_6O)$	$Z = 2$
$M_r = 1067.29$	$F(000) = 1100$
Triclinic, $P\text{-}1$ (no.2)	$D_x = 1.309$ Mg m^{-3}
$a = 14.1106$ (3) Å	Cu $K\alpha$ radiation, $\lambda = 1.54178$ Å
$b = 14.1728$ (3) Å	Cell parameters from 9954 reflections
$c = 15.7942$ (4) Å	$\theta = 3.0–72.2°$
$\alpha = 68.274$ (1)°	$\mu = 0.66$ mm^{-1}
$\beta = 81.264$ (1)°	$T = 123$ K
$\gamma = 67.405$ (1)°	Blocks, colourless
$V = 2708.86$ (11) Å^3	$0.18 \times 0.12 \times 0.04$ mm

Data collection

Bruker D8 VENTURE diffractometer with Photon100 detector	10599 independent reflections
Radiation source: INCOATEC microfocus sealed tube	9000 reflections with $I > 2\sigma(I)$
Detector resolution: 10.4167 pixels mm^{-1}	$R_{int} = 0.033$
rotation in ϕ and ω, 1°, shutterless scans	$\theta_{max} = 72.3°$, $\theta_{min} = 3.0°$
Absorption correction: multi-scan *SADABS* (Sheldrick, 2014)	$h = -17 \rightarrow 17$
$T_{min} = 0.890$, $T_{max} = 0.971$	$k = -17 \rightarrow 17$
41293 measured reflections	$l = -18 \rightarrow 19$

Refinement

Refinement on F^2	Secondary atom site location: difference Fourier map
Least-squares matrix: full	Hydrogen site location: mixed
$R[F^2 > 2\sigma(F^2)] = 0.041$	H atoms treated by a mixture of independent and constrained refinement
$wR(F^2) = 0.105$	$w = 1/[\sigma^2(F_o^2) + (0.0484P)^2 + 1.119P]$ where $P = (F_o^2 + 2F_c^2)/3$
$S = 1.03$	$(\Delta/\sigma)_{max} = 0.001$
10599 reflections	$\Delta\rangle_{max} = 0.29$ e Å^{-3}
734 parameters	$\Delta\rangle_{min} = -0.27$ e Å^{-3}
2 restraints	Extinction correction: *SHELXL2014/7* (Sheldrick 2014), Fc*=kFc[1+0.001xFc$^2\lambda^3$/sin(2θ)]$^{-1/4}$
Primary atom site location: structure-invariant direct methods	Extinction coefficient: 0.00105 (11)

4CzdTlpgyl (48b) SB1023_HY

Crystal data

$C_{62}H_{38}N_{12} \cdot CD_2Cl_2$	$Z = 2$
$M_r = 1037.98$	$F(000) = 1072$
Triclinic, P-1 (no.2)	$D_x = 1.360$ Mg m^{-3}
$a = 12.3140$ (4) Å	Cu $K\alpha$ radiation, $\lambda = 1.54178$ Å
$b = 13.7068$ (4) Å	Cell parameters from 9718 reflections
$c = 17.4497$ (5) Å	$\theta = 2.7$–72.0°
$\alpha = 68.562$ (1)°	$\mu = 1.60$ mm^{-1}
$\beta = 87.449$ (1)°	$T = 123$ K
$\gamma = 68.378$ (1)°	Blocks, yellow
$V = 2534.71$ (13) Å^3	$0.24 \times 0.20 \times 0.16$ mm

Data collection

Bruker D8 VENTURE diffractometer with Photon100 detector	9903 independent reflections
Radiation source: INCOATEC microfocus sealed tube	8944 reflections with $I > 2\sigma(I)$
Detector resolution: 10.4167 pixels mm^{-1}	$R_{int} = 0.028$
rotation in ϕ and ω, 1°, shutterless scans	$\theta_{max} = 72.1°$, $\theta_{min} = 2.7°$
Absorption correction: multi-scan *SADABS* (Sheldrick, 2014)	$h = -14 \rightarrow 15$
$T_{min} = 0.664$, $T_{max} = 0.795$	$k = -16 \rightarrow 16$
46641 measured reflections	$l = -21 \rightarrow 21$

Refinement

Refinement on F^2	Secondary atom site location: difference Fourier map
Least-squares matrix: full	Hydrogen site location: difference Fourier map
$R[F^2 > 2\sigma(F^2)] = 0.058$	H-atom parameters constrained
$wR(F^2) = 0.156$	$w = 1/[\sigma^2(F_o^2) + (0.0751P)^2 + 3.4733P]$ where $P = (F_o^2 + 2F_c^2)/3$
$S = 1.04$	$(\Delta/\sigma)_{max} < 0.001$
9903 reflections	$\Delta\rangle_{max} = 1.63$ e Å^{-3}
690 parameters	$\Delta\rangle_{min} = -1.23$ e Å^{-3}
7 restraints	Extinction correction: *SHELXL2014/7* (Sheldrick 2014, $Fc^* = kFc[1+0.001xFc^2\lambda^3/\sin(2\theta)]^{-1/4}$
Primary atom site location: structure-invariant direct methods	Extinction coefficient: 0.00060 (13)

4CzdTlPh (48c) SB988_HY

Crystal data

$C_{68}H_{42}N_{12} \cdot CDCl_3$	$Z = 2$
$M_r = 1147.51$	$F(000) = 1184$
Triclinic, $P\text{-}1$ (no.2)	$D_x = 1.363$ Mg m^{-3}
$a = 14.0357$ (10) Å	Cu $K\alpha$ radiation, $\lambda = 1.54178$ Å
$b = 14.434$ (1) Å	Cell parameters from 9555 reflections
$c = 15.3815$ (11) Å	$\theta = 3.1\text{–}72.1°$
$\alpha = 100.878$ (2)°	$\mu = 1.93$ mm^{-1}
$\beta = 112.090$ (2)°	$T = 123$ K
$\gamma = 94.568$ (2)°	Blocks, yellow
$V = 2796.4$ (3) Å^3	$0.32 \times 0.28 \times 0.16$ mm

Data collection

Bruker D8 VENTURE diffractometer with Photon100 detector	10957 independent reflections
Radiation source: INCOATEC microfocus sealed tube	10267 reflections with $I > 2\sigma(I)$
Detector resolution: 10.4167 pixels mm^{-1}	$R_{int} = 0.025$
rotation in ϕ and ω, 1°, shutterless scans	$\theta_{max} = 72.1°$, $\theta_{min} = 3.2°$
Absorption correction: multi-scan *SADABS* (Sheldrick, 2014)	$h = -17 \rightarrow 16$
$T_{min} = 0.618$, $T_{max} = 0.724$	$k = -17 \rightarrow 17$
50883 measured reflections	$l = -18 \rightarrow 18$

Refinement

Refinement on F^2	Secondary atom site location: difference Fourier map
Least-squares matrix: full	Hydrogen site location: difference Fourier map
$R[F^2 > 2\sigma(F^2)] = 0.039$	H-atom parameters constrained
$wR(F^2) = 0.104$	$w = 1/[\sigma^2(F_o^2) + (0.0481P)^2 + 1.6388P]$ where $P = (F_o^2 + 2F_c^2)/3$
$S = 1.05$	$(\Delta/\sigma)_{max} = 0.001$
10957 reflections	$\Delta\rangle_{max} = 0.97$ e Å^{-3}
758 parameters	$\Delta\rangle_{min} = -0.95$ e Å^{-3}
0 restraints	Extinction correction: *SHELXL2014*/7 (Sheldrick 2014), Fc*=kFc[1+0.001xFc$^2\lambda^3$/sin(2θ)]$^{-1/4}$
Primary atom site location: structure-invariant direct methods	Extinction coefficient: 0.00131 (11)

4CzCNOXDMe (71b) SB1001_HY

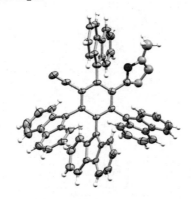

Crystal data

$C_{58}H_{35}N_7O\cdot CD_2Cl_2\cdot H_2O$	$Z = 2$
$M_r = 950.88$	$F(000) = 984$
Triclinic, $P\text{-}1$ (no.2)	$D_x = 1.318$ Mg m^{-3}
$a = 13.2369$ (6) Å	Cu $K\alpha$ radiation, $\lambda = 1.54178$ Å
$b = 13.9041$ (6) Å	Cell parameters from 9863 reflections
$c = 15.3862$ (7) Å	$\theta = 3.1$–71.6°
$\alpha = 81.926$ (3)°	$\mu = 1.64$ mm^{-1}
$\beta = 64.225$ (3)°	$T = 123$ K
$\gamma = 70.048$ (3)°	Rods, yellow
$V = 2396.9$ (2) Å^3	$0.36 \times 0.10 \times 0.08$ mm

Data collection

Bruker D8 VENTURE diffractometer with Photon100 detector	9195 independent reflections
Radiation source: INCOATEC microfocus sealed tube	7178 reflections with $I > 2\sigma(I)$
Detector resolution: 10.4167 pixels mm^{-1}	$R_{int} = 0.045$
rotation in ϕ and ω, 1°, shutterless scans	$\theta_{max} = 71.7°$, $\theta_{min} = 3.2°$
Absorption correction: multi-scan SADABS (Sheldrick, 2014)	$h = -16 \rightarrow 16$
$T_{min} = 0.612$, $T_{max} = 0.864$	$k = -16 \rightarrow 17$
29704 measured reflections	$l = -18 \rightarrow 16$

Refinement

Refinement on F^2	Primary atom site location: structure-invariant direct methods
Least-squares matrix: full	Secondary atom site location: difference Fourier map
$R[F^2 > 2\sigma(F^2)] = 0.084$	Hydrogen site location: inferred from neighbouring sites
$wR(F^2) = 0.239$	H-atom parameters constrained
$S = 1.03$	$w = 1/[\sigma^2(F_o^2) + (0.1105P)^2 + 3.4008P]$ where $P = (F_o^2 + 2F_c^2)/3$
9195 reflections	$(\Delta/\sigma)_{max} < 0.001$
588 parameters	$\Delta\rangle_{max} = 1.04$ e Å^{-3}
120 restraints	$\Delta\rangle_{min} = -0.36$ e Å^{-3}

4CzCNOXDCH₂Cl (71f) SB1054_HY

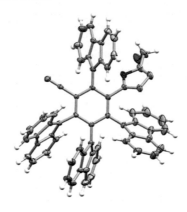

Crystal data

C₅₈H₃₄ClN₇O·1.5(CDCl₃)	$Z = 6$
$M_r = 1060.93$	$F(000) = 3258$
Triclinic, $P\text{-}1$ (no.2)	$D_x = 1.405$ Mg m⁻³
$a = 13.5049$ (3) Å	Cu $K\alpha$ radiation, $\lambda = 1.54178$ Å
$b = 15.7884$ (4) Å	Cell parameters from 9198 reflections
$c = 36.3880$ (8) Å	$\theta = 3.1\text{–}72.1°$
$\alpha = 92.931$ (1)°	$\mu = 3.29$ mm⁻¹
$\beta = 91.593$ (1)°	$T = 123$ K
$\gamma = 103.613$ (1)°	Blocks, yellow
$V = 7524.5$ (3) Å³	0.35 × 0.25 × 0.20 mm

Data collection

Bruker D8 VENTURE diffractometer with Photon100 detector	24357 reflections with $I > 2\sigma(I)$
Radiation source: INCOATEC microfocus sealed tube	$R_{int} = 0.041$
rotation in ϕ and ω, 0.5°, shutterless scans	$\theta_{max} = 72.2°$, $\theta_{min} = 2.4°$
Absorption correction: multi-scan *SADABS* (Sheldrick, 2014)	$h = -16 \rightarrow 16$
$T_{min} = 0.484$, $T_{max} = 0.623$	$k = -19 \rightarrow 19$
111530 measured reflections	$l = -44 \rightarrow 44$
29475 independent reflections	

Refinement

Refinement on F^2	Primary atom site location: dual
Least-squares matrix: full	Secondary atom site location: difference Fourier map
$R[F^2 > 2\sigma(F^2)] = 0.082$	Hydrogen site location: inferred from neighbouring sites
$wR(F^2) = 0.229$	H-atom parameters constrained
$S = 1.09$	$w = 1/[\sigma^2(F_o^2) + (0.115P)^2 + 10.3P]$ where $P = (F_o^2 + 2F_c^2)/3$
29475 reflections	$(\Delta/\sigma)_{max} = 0.001$
1810 parameters	$\Delta\rangle_{max} = 1.65$ e Å⁻³
1809 restraints	$\Delta\rangle_{min} = -0.72$ e Å⁻³

4CzCNOXD2,6OMePh (71i) SB1047_HY

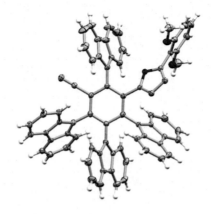

Crystal data

$C_{65}H_{41}N_7O_3 \cdot 2(CH_2Cl_2)$	$Z = 2$
$M_r = 1137.90$	$F(000) = 1176$
Triclinic, $P\text{-}1$ (no.2)	$D_x = 1.425$ Mg m^{-3}
$a = 12.9677$ (4) Å	Cu $K\alpha$ radiation, $\lambda = 1.54178$ Å
$b = 13.0856$ (4) Å	Cell parameters from 8996 reflections
$c = 18.0734$ (6) Å	$\theta = 3.6\text{--}71.8°$
$\alpha = 97.395$ (2)°	$\mu = 2.50$ mm^{-1}
$\beta = 108.972$ (2)°	$T = 123$ K
$\gamma = 108.753$ (2)°	Plates, yellow
$V = 2651.53$ (15) Å^3	$0.20 \times 0.12 \times 0.04$ mm

Data collection

Bruker D8 VENTURE diffractometer with Photon100 detector	10296 independent reflections
Radiation source: INCOATEC microfocus sealed tube	8504 reflections with $I > 2\sigma(I)$
Detector resolution: 10.4167 pixels mm^{-1}	$R_{int} = 0.046$
rotation in ϕ and ω, 1°, shutterless scans	$\theta_{max} = 71.9°$, $\theta_{min} = 2.7°$
Absorption correction: multi-scan SADABS (Sheldrick, 2014)	$h = -15 \rightarrow 13$
$T_{min} = 0.730$, $T_{max} = 0.889$	$k = -15 \rightarrow 16$
34556 measured reflections	$l = -22 \rightarrow 21$

Refinement

Refinement on F^2	Primary atom site location: dual
Least-squares matrix: full	Secondary atom site location: difference Fourier map
$R[F^2 > 2\sigma(F^2)] = 0.065$	Hydrogen site location: inferred from neighbouring sites
$wR(F^2) = 0.143$	H-atom parameters constrained
$S = 1.12$	$w = 1/[\sigma^2(F_o^2) + (0.0263P)^2 + 4.109P]$ where $P = (F_o^2 + 2F_c^2)/3$
10296 reflections	$(\Delta/\sigma)_{max} < 0.001$
678 parameters	$\Delta\rangle_{max} = 0.28$ e Å^{-3}
0 restraints	$\Delta\rangle_{min} = -0.34$ e Å^{-3}

4CzCNOXDMes (71h) SB1049_HY

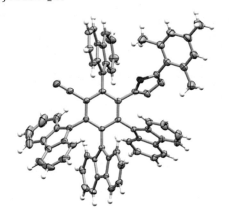

Crystal data

$C_{66}H_{43}N_7O·2(CDCl_3)$	$Z = 2$
$M_r = 1190.82$	$F(000) = 1224$
Triclinic, P-1 (no.2)	$D_x = 1.377$ Mg m^{-3}
$a = 12.7750$ (5) Å	Cu $K\alpha$ radiation, $\lambda = 1.54178$ Å
$b = 15.5618$ (6) Å	Cell parameters from 9853 reflections
$c = 15.6101$ (6) Å	$\theta = 2.0–72.1°$
$\alpha = 73.349$ (2)°	$\mu = 3.14$ mm^{-1}
$\beta = 81.473$ (2)°	$T = 123$ K
$\gamma = 75.841$ (2)°	Blocks, yellow
$V = 2872.6$ (2) Å^3	$0.22 \times 0.16 \times 0.06$ mm

Data collection

Bruker D8 VENTURE diffractometer with Photon100 detector	11273 independent reflections
Radiation source: INCOATEC microfocus sealed tube	9897 reflections with $I > 2\sigma(I)$
Detector resolution: 10.4167 pixels mm^{-1}	$R_{int} = 0.036$
rotation in ϕ and ω, 1°, shutterless scans	$\theta_{max} = 72.3°$, $\theta_{min} = 3.0°$
Absorption correction: multi-scan $SADABS$ (Sheldrick, 2014)	$h = -15 \rightarrow 15$
$T_{min} = 0.618$, $T_{max} = 0.806$	$k = -19 \rightarrow 19$
46498 measured reflections	$l = -19 \rightarrow 19$

Refinement

Refinement on F^2	Primary atom site location: dual
Least-squares matrix: full	Secondary atom site location: difference Fourier map
$R[F^2 > 2\sigma(F^2)] = 0.055$	Hydrogen site location: inferred from neighbouring sites
$wR(F^2) = 0.139$	H-atom parameters constrained
$S = 1.03$	$w = 1/[\sigma^2(F_o^2) + (0.0465P)^2 + 3.3094P]$ where $P = (F_o^2 + 2F_c^2)/3$
11273 reflections	$(\Delta/\sigma)_{max} < 0.001$
670 parameters	$\Delta\rangle_{max} = 0.51$ e Å^{-3}
0 restraints	$\Delta\rangle_{min} = -0.38$ e Å^{-3}

4CzCNOXDC6F5 (71n) SB1046_HY

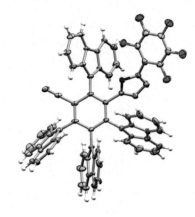

Crystal data

$C_{63}H_{32}F_5N_7O \cdot CDCl_3$	$Z = 4$
$M_r = 1118.33$	$F(000) = 2280$
Triclinic, P-1 (no.2)	$D_x = 1.474$ Mg m^{-3}
$a = 13.6061$ (4) Å	Cu $K\alpha$ radiation, $\lambda = 1.54178$ Å
$b = 14.1874$ (4) Å	Cell parameters from 9779 reflections
$c = 28.1025$ (9) Å	$\theta = 3.4$–71.9°
$\alpha = 89.663$ (2)°	$\mu = 2.26$ mm^{-1}
$\beta = 87.885$ (2)°	$T = 123$ K
$\gamma = 68.354$ (2)°	Blocks, yellow
$V = 5038.6$ (3) Å^3	$0.36 \times 0.16 \times 0.06$ mm

Data collection

Bruker D8 VENTURE diffractometer with Photon100 detector	19562 independent reflections
Radiation source: INCOATEC microfocus sealed tube	16159 reflections with $I > 2\sigma(I)$
Detector resolution: 10.4167 pixels mm^{-1}	$R_{int} = 0.049$
rotation in ϕ and ω, 1°, shutterless scans	$\theta_{max} = 72.0°$, $\theta_{min} = 3.2°$
Absorption correction: multi-scan $SADABS$ (Sheldrick, 2014)	$h = -16 \rightarrow 16$
$T_{min} = 0.711$, $T_{max} = 0.864$	$k = -17 \rightarrow 17$
73037 measured reflections	$l = -34 \rightarrow 31$

Refinement

Refinement on F^2	Primary atom site location: dual
Least-squares matrix: full	Secondary atom site location: difference Fourier map
$R[F^2 > 2\sigma(F^2)] = 0.094$	Hydrogen site location: inferred from neighbouring sites
$wR(F^2) = 0.256$	H-atom parameters constrained
$S = 1.07$	$w = 1/[\sigma^2(F_o^2) + (0.115P)^2 + 18.450P]$ where $P = (F_o^2 + 2F_c^2)/3$
19562 reflections	$(\Delta/\sigma)_{max} < 0.001$
1369 parameters	$\Delta\rangle_{max} = 1.06$ e Å^{-3}
0 restraints	$\Delta\rangle_{min} = -0.76$ e Å^{-3}

5.2.2 Crystallographic Data Solved by Dr. Olaf Fuhr

Crystal structures in this section were measured and solved by Dr. Olaf Fuhr at the Institute of Nanotechnology (INT) at the Karlsruhe Institute of Technology.

Name in this thesis	Number in this thesis	Code used by Dr. Fuhr
85	85	FHU122
FPI4NMe$_2$Ph	101m	FHUcy131
FPI3,5MeOPh	101o	FHUcy135
CzPIMes	102g	FHUcy107
CzPI4CNPh	102j	FHUcy132
CzPI2TolPh	105	FHUcy084
dCOMeCzPIMes	137	FHUcy148_tw
COCF$_3$CzPIMes	123	FHUcy059
tCzBN	149	FHU184
3,5-2tCzBN	152	FHUttO18

85 FHU122

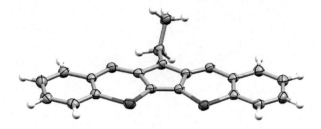

Identification code	FHU122
Empirical formula	$C_{18}H_{13}N_3S_2$
Formula weight	335.43
Temperature/K	140.15
Crystal system	orthorhombic
Space group	Pbca
a/Å	5.8629(2)
b/Å	17.0375(9)
c/Å	29.8197(17)
α/°	90
β/°	90
γ/°	90
Volume/Å^3	2978.7(3)
Z	8
ρ_{calc}g/cm^3	1.496
μ/mm^{-1}	2.112
F(000)	1392.0
Crystal size/mm^3	0.29 × 0.03 × 0.02
Radiation	GaKα (λ = 1.34143)
2Θ range for data collection/°	9.032 to 114.96
Index ranges	-3 ≤ h ≤ 7, -21 ≤ k ≤ 20, -36 ≤ l ≤ 37
Reflections collected	9008
Independent reflections	3011 [R_{int} = 0.0576, R_{sigma} = 0.0492]
Indep. refl. with I>=2σ (I)	2230
Data/restraints/parameters	3011/0/209
Goodness-of-fit on F^2	0.999
Final R indexes [I>=2σ (I)]	R_1 = 0.0529, wR_2 = 0.1285
Final R indexes [all data]	R_1 = 0.0787, wR_2 = 0.1378
Largest diff. peak/hole / e Å^{-3}	0.48/-0.39

FPI4NMe₂Ph (101m) FHUcy131

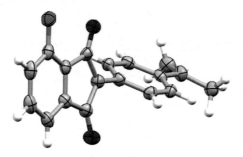

Identification code	FHUcy131
Empirical formula	$C_{16}H_{13}FN_2O_2$
Formula weight	284.28
Temperature/K	180.15
Crystal system	triclinic
Space group	P-1
a/Å	7.6170(4)
b/Å	11.8624(7)
c/Å	14.7436(8)
α/°	101.058(4)
β/°	98.208(4)
γ/°	90.854(4)
Volume/Å³	1292.82(13)
Z	4
ρ_{calc}g/cm³	1.461
μ/mm⁻¹	0.897
F(000)	592.0
Crystal size/mm³	0.36 × 0.19 × 0.18
Radiation	CuKα (λ = 1.54186)
2Θ range for data collection/°	12.372 to 140.912
Index ranges	-9 ≤ h ≤ 9, -7 ≤ k ≤ 14, -18 ≤ l ≤ 16
Reflections collected	9730
Independent reflections	4722 [R_{int} = 0.0277, R_{sigma} = 0.0218]
Indep. refl. with I>=2σ (I)	4482
Data/restraints/parameters	4722/0/401
Goodness-of-fit on F²	1.159
Final R indexes [I>=2σ (I)]	R_1 = 0.0863, wR_2 = 0.2422
Final R indexes [all data]	R_1 = 0.0885, wR_2 = 0.2441
Largest diff. peak/hole / e Å⁻³	0.66/-0.32

FPI3,5MeOPh (101o) FHUcy135

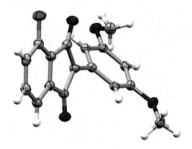

Identification code	FHUCY135
Empirical formula	$C_{16}H_{12}FNO_4$
Formula weight	301.27
Temperature/K	150.15
Crystal system	orthorhombic
Space group	$Pna2_1$
a/Å	20.0504(8)
b/Å	3.83480(10)
c/Å	16.9836(5)
α/°	90
β/°	90
γ/°	90
Volume/Å³	1305.86(7)
Z	4
ρ_{calc}g/cm³	1.532
μ/mm⁻¹	0.654
F(000)	624.0
Crystal size/mm³	0.18 × 0.04 × 0.03
Radiation	GaKα (λ = 1.34143)
2Θ range for data collection/°	22.466 to 114.928
Index ranges	-19 ≤ h ≤ 25, -4 ≤ k ≤ 1, -21 ≤ l ≤ 19
Reflections collected	5318
Independent reflections	2276 [R_{int} = 0.0153, R_{sigma} = 0.0120]
Indep. refl. with I>=2σ (I)	2202
Data/restraints/parameters	2276/1/202
Goodness-of-fit on F²	1.067
Final R indexes [I>=2σ (I)]	R_1 = 0.0260, wR_2 = 0.0688
Final R indexes [all data]	R_1 = 0.0269, wR_2 = 0.0710
Largest diff. peak/hole / e Å⁻³	0.14/-0.17
Flack parameter	0.2(2)

CzPIMes (102g) FHUcy107

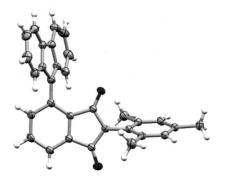

Identification code	FHUcy107
Empirical formula	$C_{29}H_{22}N_2O_2$
Formula weight	430.48
Temperature/K	180.15
Crystal system	triclinic
Space group	P-1
a/Å	7.9174(2)
b/Å	8.0688(2)
c/Å	17.9122(5)
α/°	92.152(2)
β/°	91.382(2)
γ/°	106.489(2)
Volume/Å³	1095.74(5)
Z	2
ρ_{calc}g/cm³	1.305
μ/mm⁻¹	0.419
F(000)	452.0
Crystal size/mm³	0.25 × 0.23 × 0.22
Radiation	GaKα (λ = 1.34143)
2Θ range for data collection/°	10.144 to 124.97
Index ranges	-10 ≤ h ≤ 5, -10 ≤ k ≤ 10, -23 ≤ l ≤ 23
Reflections collected	13312
Independent reflections	5099 [R_{int} = 0.0201, R_{sigma} = 0.0235]
Indep. refl. with I>=2σ (I)	4099
Data/restraints/parameters	5099/0/302
Goodness-of-fit on F²	1.129
Final R indexes [I>=2σ (I)]	R_1 = 0.0654, wR_2 = 0.1612
Final R indexes [all data]	R_1 = 0.0773, wR_2 = 0.1653
Largest diff. peak/hole / e Å⁻³	0.32/-0.24

CzPI4CNPh (102j) FHUcy132

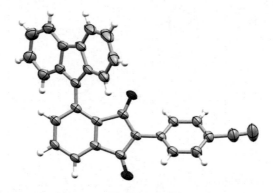

Identification code	FHUcy132
Empirical formula	$C_{27}H_{15}N_3O_2$
Formula weight	413.42
Temperature/K	180.15
Crystal system	orthorhombic
Space group	Pbcn
a/Å	26.3262(10)
b/Å	7.9940(4)
c/Å	19.6706(7)
α/°	90
β/°	90
γ/°	90
Volume/Å³	4139.7(3)
Z	8
ρ_{calc}g/cm³	1.327
μ/mm⁻¹	0.689
F(000)	1712.0
Crystal size/mm³	0.36 × 0.19 × 0.05
Radiation	CuKα (λ = 1.54186)
2Θ range for data collection/°	11.568 to 140.38
Index ranges	-27 ≤ h ≤ 31, -9 ≤ k ≤ 9, -23 ≤ l ≤ 10
Reflections collected	8420
Independent reflections	3800 [R_{int} = 0.0243, R_{sigma} = 0.0208]
Data/restraints/parameters	3800/0/289
Goodness-of-fit on F²	1.044
Final R indexes [I>=2σ (I)]	R_1 = 0.0512, wR_2 = 0.1377
Final R indexes [all data]	R_1 = 0.0583, wR_2 = 0.1474
Largest diff. peak/hole / e Å⁻³	0.19/-0.29

CzPI2TolPh (105) FHUcy084

Identification code	FHUcy084
Empirical formula	$C_{33}H_{22}N_2O_2$
Formula weight	478.52
Temperature/K	150.15
Crystal system	triclinic
Space group	P-1
a/Å	9.5107(2)
b/Å	11.7527(3)
c/Å	12.6891(3)
α/°	113.861(2)
β/°	110.802(2)
γ/°	90.889(2)
Volume/Å³	1191.36(5)
Z	2
ρ_{calc}g/cm³	1.334
μ/mm⁻¹	0.425
F(000)	500.0
Crystal size/mm³	0.28 × 0.21 × 0.17
Radiation	GaKα (λ = 1.34143)
2Θ range for data collection/°	7.214 to 119.976
Index ranges	-12 ≤ h ≤ 10, -15 ≤ k ≤ 15, -9 ≤ l ≤ 16
Reflections collected	13616
Independent reflections	5194 [R_{int} = 0.0123, R_{sigma} = 0.0080]
Indep. refl. with I>=2σ (I)	4997
Data/restraints/parameters	5194/0/335
Goodness-of-fit on F²	1.042
Final R indexes [I>=2σ (I)]	R_1 = 0.0390, wR_2 = 0.0984
Final R indexes [all data]	R_1 = 0.0402, wR_2 = 0.0995
Largest diff. peak/hole / e Å⁻³	0.23/-0.32

dCOMeCzPIMes (137) FHUcy148_tw

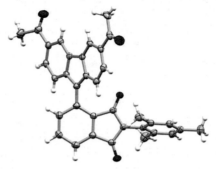

Identification code	FHUcy148_tw
Empirical formula	$C_{33}H_{26}N_2O_4$
Formula weight	514.56
Temperature/K	180.15
Crystal system	monoclinic
Space group	$P2_1/c$
a/Å	10.1380(4)
b/Å	32.1586(11)
c/Å	8.2799(4)
α/°	90
β/°	104.235(3)
γ/°	90
Volume/Å^3	2616.56(19)
Z	4
ρ_{calc}g/cm^3	1.306
μ/mm^{-1}	0.695
F(000)	1080.0
Crystal size/mm^3	0.39 × 0.08 × 0.07
Radiation	CuKα (λ = 1.54186)
2Θ range for data collection/°	9 to 141.586
Index ranges	$-12 \leq h \leq 11, -38 \leq k \leq 38, -5 \leq l \leq 9$
Reflections collected	4823
Independent reflections	4823 [R_{int} = ?, R_{sigma} = 0.0186]
Indep. refl. with I>=2σ (I)	4221
Data/restraints/parameters	4823/0/358
Goodness-of-fit on F^2	1.086
Final R indexes [I>=2σ (I)]	R_1 = 0.0412, wR_2 = 0.1096
Final R indexes [all data]	R_1 = 0.0480, wR_2 = 0.1159
Largest diff. peak/hole / e Å^{-3}	0.21/-0.25

COCF₃CzPIMes (123) FHUcy059

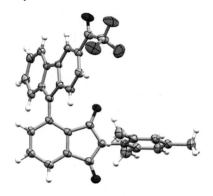

Identification code	FHUcy059
Empirical formula	$C_{31}H_{21}F_3N_2O_3$
Formula weight	526.50
Temperature/K	180.15
Crystal system	triclinic
Space group	P-1
a/Å	8.0911(3)
b/Å	8.6786(4)
c/Å	18.4184(8)
α/°	101.749(3)
β/°	93.434(3)
γ/°	97.139(3)
Volume/Å³	1251.57(9)
Z	2
$\rho_{calc}g/cm^3$	1.397
μ/mm⁻¹	0.887
F(000)	544.0
Crystal size/mm³	0.35 × 0.3 × 0.21
Radiation	CuKα (λ = 1.54186)
2Θ range for data collection/°	10.514 to 140.806
Index ranges	-9 ≤ h ≤ 5, -10 ≤ k ≤ 10, -21 ≤ l ≤ 22
Reflections collected	9468
Independent reflections	4604 [R_{int} = 0.0248, R_{sigma} = 0.0181]
Data/restraints/parameters	4604/0/355
Goodness-of-fit on F²	1.042
Final R indexes [I>=2σ (I)]	R_1 = 0.0486, wR_2 = 0.1322
Final R indexes [all data]	R_1 = 0.0503, wR_2 = 0.1340
Largest diff. peak/hole / e Å⁻³	0.23/-0.29

tCzBN (149) FHU184

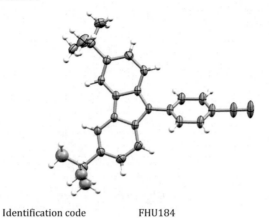

Identification code	FHU184
Empirical formula	$C_{28}H_{29}Cl_3N_2$
Formula weight	499.88
Temperature/K	150.15
Crystal system	triclinic
Space group	P-1
a/Å	10.0742(3)
b/Å	10.9079(3)
c/Å	12.8574(3)
α/°	94.228(2)
β/°	101.808(2)
γ/°	97.560(2)
Volume/Å³	1363.56(6)
Z	2
ρ_{calc}g/cm³	1.218
μ/mm⁻¹	2.070
F(000)	524.0
Crystal size/mm³	0.19 × 0.17 × 0.16
Radiation	GaKα (λ = 1.34143)
2Θ range for data collection/°	6.144 to 121.998
Index ranges	-4 ≤ h ≤ 12, -14 ≤ k ≤ 14, -16 ≤ l ≤ 16
Reflections collected	15144
Independent reflections	6152 [R_{int} = 0.0228, R_{sigma} = 0.0185]
Data/restraints/parameters	6152/0/304
Goodness-of-fit on F²	1.089
Final R indexes [I>=2σ (I)]	R_1 = 0.0957, wR_2 = 0.2816
Final R indexes [all data]	R_1 = 0.1053, wR_2 = 0.2910
Largest diff. peak/hole / e Å⁻³	1.41/-0.96

3,5-2tCzBN (**152**) FHUttO18

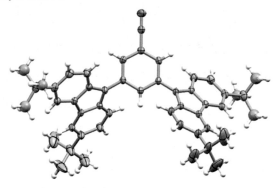

Identification code	FHUTTO18
Empirical formula	$C_{48}H_{53}Cl_2N_3$
Formula weight	742.83
Temperature/K	180.15
Crystal system	monoclinic
Space group	C2/c
a/Å	17.4372(4)
b/Å	17.7918(5)
c/Å	15.7914(3)
α/°	90
β/°	119.3820(10)
γ/°	90
Volume/Å³	4268.93(18)
Z	4
ρ_{calc}g/cm³	1.156
μ/mm⁻¹	1.074
F(000)	1584.0
Crystal size/mm³	0.22 × 0.21 × 0.2
Radiation	GaKα (λ = 1.34143)
2Θ range for data collection/°	6.912 to 125.038
Index ranges	-20 ≤ h ≤ 23, -23 ≤ k ≤ 23, -18 ≤ l ≤ 20
Reflections collected	30180
Independent reflections	5164 [R_{int} = 0.0150, R_{sigma} = 0.0110]
Indep. refl. with I>=2σ (I)	4415
Data/restraints/parameters	5164/0/257
Goodness-of-fit on F²	1.098
Final R indexes [I>=2σ (I)]	R_1 = 0.0729, wR_2 = 0.2110
Final R indexes [all data]	R_1 = 0.0808, wR_2 = 0.2188
Largest diff. peak/hole / e Å⁻³	0.65/-0.55

6 List of Abbreviations

3-NBA	3-nitrobenzyl alcohol
A	acceptor
a.u.	arbitrary unit
AFM	atomic force microscopy
AIE	aggregation-induced emission
Alq$_3$	tris(8-hydroxyquinolinato)aluminium
APCI	atmospheric pressure chemical ionization
ATR	attenuated total reflection
BTT	bistriazolotriazine
calc.	calculated
CBP	4,4'-bis(carbazol-9-yl)biphenyl
cd	candela
CD	circular dichroism
CDCB	carbazolyl dicyanobenzenes
CIE	*commission internationale de l'éclairage*
CPEL	circularly polarized electroluminescence
CRC1176	Collaborative Research Centre 1176
CT	charge-transfer state
CuAAC	copper-catalyzed alkyne-azide cycloaddition
CzSi	9-(4-*tert*-Butylphenyl)-3,6-bis(triphenylsilyl)-9H-carbazole
D	donor
DCB	1,4-di(carba-zol-9-yl)benzene
DCM	dichloromethane
DMAC	9,9-Dimethyl-9,10-dihydroacridine
DMSO	dimethyl sulfoxide
DPEPO	bis[2-(diphenylphosphino)phenyl] ether oxide
dppf	1,1'-bis(diphenylphosphino)ferrocene
DSC	differential scanning calorimetry
EA	elemental analysis
EBL	electron blocking layer

EIL	electron injection layer
EML	emissive layer
EQE	external quantum efficiency
ESI	electrospray ionization
ETL	electron transport layer
FAB	fast atom bombardment
FWHM	full width at half maximum
GmbH	*Gesellschaft mit beschränkter Haftung*
HFC	hyperfine-coupling
HIL	hole injection layer
HOMO	highest occupied molecular orbital
hpf	hours post fertilization
HTL	hole transport layer
IR	Infrared Spectroscopy
Irppy3	Tris[2-phenylpyridinato-C^2,N]iridium(III)
ISC	intersystem crossing
ITO	indium tin oxide
KIT	Karlsruhe Institute of Technology
LE	locally excited state
LED	light-emitting diode
LUMO	lowest unoccupied molecular orbital
mCBP	3,3'-di(carbazol-9-yl)biphenyl
MCF7	Michigan Cancer Foundation-7
mCP	1,3-bis(carbazol-9-yl)benzene
MR	multi-resonance
MS	mass spectrometry
MTT	3-(4,5-dimethylthiazol-2-yl)-2,5-diphenyltetrazolium bromide
MTT	monotriazolotriazine
NIR	near-infrared
NMR	nuclear magnetic resonance spectroscopy
Odots	organic dots
OFET	organic field-effect transistors

OLED	organic light-emitting diode
PEDOT:PSS	poly(3,4-ethylenedioxythiophene) polystyrene sulfonate
PESA	photoelectron spectroscopy in air
PLQY	photoluminescence quantum yield
PMMA	poly(methyl methacrylate)
PPT	2,8-bis(diphenylphosphoryl)dibenzo[b,d]thiophene
RISC	reverse intersystem crossing
S_0	ground state
S_1	first excited singlet state
SCFET	single crystal field-effect transistors
SPhos	2-dicyclohexylphosphino-2′,6′-dimethoxybiphenyl
STA	simultaneous thermal analysis
T_1	first excited triplet state
TADF	thermally activated delayed fluorescence
TCO	transparent conductive oxides
TCSPC	time-correlated single photon counting
TD-DFT	Time-Dependent Density Functional Theory
TFAA	trifluoroacetic anhydride
TGA	thermal gravimetric analysis
THF	tetrahydrofuran
TLC	thin layer chromatography
TPP	triphenylphosphonium
TRES	time-resolved emission spectra
TRFI	time-resolved fluorescence imaging
TTA	triplet-triplet-annihilation
TTT	tristriazolotriazine
TTT	tris[1,2,4]triazolo[1,3,5]triazine
TV	television
UV	ultraviolet
ΔE_{ST}	energy gap between S_1 and T_1

7 Bibliography

[1] Ban Ki Moon, "Ban Ki Moon Message," can be found under www.light2015.org/dam/About/Resources/Ban_Ki_Moon_Message.pdf, **2015**, visited at 01.11.2019.

[2] E. Spuling, *Unpublished M. Sc. Thesis: Synthetic Approaches to Novel [2.2]Paracyclophane-Based Fluorophores Employing the Thermally Activated Delayed Fluorescence Principle*, Karlsruhe Institute of Technology, **2015**.

[3] M. Balter, *Science* **2004**, *304*, 663–665.

[4] "The boy's life of Edison by Meadowcroft, W. H. (William Henry): Harper & Bros, New York Hardcover - Owl & Company Bookshop (Calvello Books)," can be found under https://www.abebooks.com/first-edition/boys-life-Edison-Meadowcroft-William-Henry/30216924494/bd, visited at 01.11.2019

[5] D. Volz, *Zweikernige Kupfer(I)-Komplexe Als OLED-Leuchtstoffe: Synthese, Eigenschaften Und Neue Konzepte*, Logos-Verlag, **2014**.

[6] "Licht: Stromverbrauch, Kosten, Spartipps," can be found under https://www.entega.de/blog/stromverbrauch-licht/, **2018**, visited at 01.11.2019.

[7] P. W. Atkins, J. de Paula, *Atkins: Physikalische Chemie*, Wiley-VCH, Weinheim, **2013**.

[8] "en.lighten - Efficient lighting for ceveloping and emerging countries," can be found under http://www.enlighten-initiative.org/, visited at 01.11.2019.

[9] "How much electricity is used for lighting in the United States? - FAQ - U.S. Energy Information Administration (EIA)," can be found under https://www.eia.gov/tools/faqs/faq.php?id=99&t=3, visited at 01.11.2019.

[10] esa, "Earth from Space: Night lights," can be found under https://www.esa.int/Our_Activities/Observing_the_Earth/Earth_from_Space_Night_lights, visited at 01.11.2019.

[11] "Abent.-Sternede" can be found under https://www.abenteuer-sterne.de/lichtverschmutzung/, visited at 01.11.2019.

[12] M. Singh, H. M. Haverinen, P. Dhagat, G. E. Jabbour, *Adv. Mater.* **2010**, *22*, 673–685.

[13] O. Nuyken, S. Jungermann, V. Wiederhirn, E. Bacher, K. Meerholz, *Monatshefte Für Chem. Chem. Mon.* **2006**, *137*, 811–824.

[14] A. Arjona-Esteban, D. Volz, in *Highly Effic. OLEDs*, John Wiley & Sons, Ltd, **2018**, pp. 543–572.

[15] "ISE 2018: LG puts focus on Transparent OLED," can be found under https://www.installation-international.com/technology/ise-2018-lg-puts-focus-transparent-oled, **2018**, visited at 01.11.2019.

[16] "Samsung shows off the world's first 'stretchable' display," can be found under http://www.dailymail.co.uk/~/article-4535272/index.html, **2017**, visited at 01.11.2019.

[17] "OLEDat InfosNews OLED-Tv Disp", can be found under https://www.oled.at/erstes-video-von-audis-oled-beleuchtungskonzept/, **2012**, visited at 01.11.2019.

[18] "Audi promises OLED lighting in the 2017 A8 Saloon flagship car | OLED-Info," can be found under http://www.oled-info.com/audi-promises-oled-lighting-2017-a8-saloon-flagship-car, visited at 01.11.2019.

[19] "OLED: Audi verlegt Außenspiegel in die Fahrzeugtür - Golem.de," can be found under https://www.golem.de/news/oled-audi-verlegt-aussenspiegel-in-die-fahrzeugtuer-1807-135303.html, visited at 01.11.2019

[20] Henry Joseph Round, *Electr. World* **1907**, *19*, 309.

[21] A. Bernanose, *Br. J. Appl. Phys.* **1955**, *6*, S54–S55.

[22] M. Pope, H. P. Kallmann, P. Magnante, *J. Chem. Phys.* **1963**, *38*, 2042–2043.

[23] M. Sano, M. Pope, H. Kallmann, *J. Chem. Phys.* **1965**, *43*, 2920–2921.

[24] C. W. Tang, S. A. VanSlyke, *Appl. Phys. Lett.* **1987**, *51*, 913–915.

[25] A. J. Heeger, *Angew. Chem.* **2001**, *113*, 2660–2682.

[26] W. Brütting, C. Adachi, *Physics of Organic Semiconductors*, John Wiley & Sons, **2012**.

[27] A. Köhler, H. Bässler, *Electronic Processes in Organic Semiconductors: An Introduction*, John Wiley & Sons, **2015**.

[28] R. A. Marcus, *Rev. Mod. Phys.* **1993**, *65*, 599–610.

[29] A. Troisi, D. L. Cheung, D. Andrienko, *Phys. Rev. Lett.* **2009**, *102*, 116602.

[30] D. Hertel, C. D. Müller, K. Meerholz, *Chem. Unserer Zeit* **2005**, *39*, 336–347.

[31] H. Yersin, *Highly Efficient OLEDs: Materials Based on Thermally Activated Delayed Fluorescence*, John Wiley & Sons, **2018**.

[32] T. D. Schmidt, W. Brütting, in *Highly Effic. OLEDs*, John Wiley & Sons, Ltd, **2018**, pp. 199–228.

[33] H. Bässler, A. Köhler, in *Unimolecular Supramol. Electron. Chem. Phys. Meet Met.-Mol. Interfaces* (Ed.: R.M. Metzger), Springer Berlin Heidelberg, Berlin, Heidelberg, **2012**, pp. 1–65.

[34] S. Kim, K. Hong, K. Kim, I. Lee, J.-L. Lee, *J. Mater. Chem.* **2012**, *22*, 2039–2044.

[35] E. Spuling, *Synthesis of New [2.2]Paracyclophane Derivatives for Application in Material Sciences*, Logos-Verlag, **2019**.

[36] C. Adachi, M. A. Baldo, M. E. Thompson, S. R. Forrest, *J. Appl. Phys.* **2001**, *90*, 5048–5051.

[37] Y. Tao, K. Yuan, T. Chen, P. Xu, H. Li, R. Chen, C. Zheng, L. Zhang, W. Huang, *Adv. Mater.* **2014**, *26*, 7931–7958.

[38] M. A. Baldo, D. F. O'Brien, Y. You, A. Shoustikov, S. Sibley, M. E. Thompson, S. R. Forrest, *Nature* **1998**, *395*, 151–154.

[39] M. A. Baldo, S. Lamansky, P. E. Burrows, M. E. Thompson, S. R. Forrest, *Appl. Phys. Lett.* **1999**, *75*, 4–6.

[40] S. Lamansky, P. Djurovich, D. Murphy, F. Abdel-Razzaq, H.-E. Lee, C. Adachi, P. E. Burrows, S. R. Forrest, M. E. Thompson, *J. Am. Chem. Soc.* **2001**, *123*, 4304–4312.

[41] D. Bruce, M. M. Richter, *Anal. Chem.* **2002**, *74*, 1340–1342.

[42] L. Chen, H. You, C. Yang, X. Zhang, J. Qin, D. Ma, *J. Mater. Chem.* **2006**, *16*, 3332–3339.

[43] F.-I. Wu, P.-I. Shih, Y.-H. Tseng, G.-Y. Chen, C.-H. Chien, C.-F. Shu, Y.-L. Tung, Y. Chi, A. K.-Y. Jen, *J. Phys. Chem. B* **2005**, *109*, 14000–14005.

[44] J. Lu, Y. Tao, Y. Chi, Y. Tung, *Synth. Met.* **2005**, *155*, 56–62.

[45] P. T. Furuta, L. Deng, S. Garon, M. E. Thompson, J. M. J. Fréchet, *J. Am. Chem. Soc.* **2004**, *126*, 15388–15389.

[46] A. S. Ionkin, W. J. Marshall, Y. Wang, *Organometallics* **2005**, *24*, 619–627.

[47] Y.-L. Tung, S.-W. Lee, Y. Chi, L.-S. Chen, C.-F. Shu, F.-I. Wu, A. J. Carty, P.-T. Chou, S.-M. Peng, G.-H. Lee, *Adv. Mater.* **2005**, *17*, 1059–1064.

[48] Y.-L. Tung, L.-S. Chen, Y. Chi, P.-T. Chou, Y.-M. Cheng, E. Y. Li, G.-H. Lee, C.-F. Shu, F.-I. Wu, A. J. Carty, *Adv. Funct. Mater.* **2006**, *16*, 1615–1626.

[49] C. Adachi, M. A. Baldo, S. R. Forrest, M. E. Thompson, *Appl. Phys. Lett.* **2000**, *77*, 904–906.

[50] T. Sajoto, P. I. Djurovich, A. B. Tamayo, J. Oxgaard, W. A. Goddard, M. E. Thompson, *J. Am. Chem. Soc.* **2009**, *131*, 9813–9822.

[51] H. Uoyama, K. Goushi, K. Shizu, H. Nomura, C. Adachi, *Nature* **2012**, *492*, 234–238.

[52] Q. Zhang, B. Li, S. Huang, H. Nomura, H. Tanaka, C. Adachi, *Nat. Photonics* **2014**, *8*, 326–332.

[53] S. Y. Lee, T. Yasuda, Y. S. Yang, Q. Zhang, C. Adachi, *Angew. Chem. Int. Ed.* **2014**, *53*, 6402–6406.

[54] Z. Yang, Z. Mao, Z. Xie, Y. Zhang, S. Liu, J. Zhao, J. Xu, Z. Chi, M. P. Aldred, *Chem. Soc. Rev.* **2017**, *46*, 915–1016.

[55] M. Y. Wong, E. Zysman-Colman, *Adv. Mater.* **2017**, *29*, 1605444.

[56] X. Liang, Z.-L. Tu, Y.-X. Zheng, *Chem. – Eur. J.* **2019**, *25*, 5623–5642.

[57] J. Lee, K. Shizu, H. Tanaka, H. Nakanotani, T. Yasuda, H. Kaji, C. Adachi, *J. Mater. Chem. C* **2015**, *3*, 2175–2181.

[58] X. Cai, X. Li, G. Xie, Z. He, K. Gao, K. Liu, D. Chen, Y. Cao, S.-J. Su, *Chem. Sci.* **2016**, *7*, 4264–4275.

[59] S. Y. Lee, T. Yasuda, H. Komiyama, J. Lee, C. Adachi, *Adv. Mater.* **2016**, *28*, 4019–4024.

[60] A. Kretzschmar, C. Patze, S. T. Schwaebel, U. H. F. Bunz, *J. Org. Chem.* **2015**, *80*, 9126–9131.

[61] I. S. Park, S. Y. Lee, C. Adachi, T. Yasuda, *Adv. Funct. Mater.* **2016**, *26*, 1813–1821.

[62] K. Shizu, Y. Sakai, H. Tanaka, S. Hirata, C. Adachi, H. Kaji, *ITE Trans. Media Technol. Appl.* **2015**, *3*, 108–113.

[63] M. Moral, L. Muccioli, W.-J. Son, Y. Olivier, J. C. Sancho-García, *J. Chem. Theory Comput.* **2015**, *11*, 168–177.

[64] M. A. El-Sayed, *Acc. Chem. Res.* **1968**, *1*, 8–16.

[65] J. C. Deaton, S. C. Switalski, D. Y. Kondakov, R. H. Young, T. D. Pawlik, D. J. Giesen, S. B. Harkins, A. J. M. Miller, S. F. Mickenberg, J. C. Peters, *J. Am. Chem. Soc.* **2010**, *132*, 9499–9508.

[66] Q. Zhang, J. Li, K. Shizu, S. Huang, S. Hirata, H. Miyazaki, C. Adachi, *J. Am. Chem. Soc.* **2012**, *134*, 14706–14709.

[67] P. Rajamalli, D. Chen, W. Li, I. D. W. Samuel, D. B. Cordes, A. M. Z. Slawin, E. Zysman-Colman, *J. Mater. Chem. C* **2019**, *7*, 6664–6671.

[68] D. de S. Pereira, D. Ryun Lee, N. A. Kukhta, K. Hyung Lee, C. Long Kim, A. S. Batsanov, J. Yeob Lee, A. P. Monkman, *J. Mater. Chem. C* **2019**, *7*, 10481–10490.

[69] W. Zhang, J. Jin, Z. Huang, S. Zhuang, L. Wang, *Sci. Rep.* **2016**, *6*, 30178.

[70] M.-K. Hung, K.-W. Tsai, S. Sharma, J.-Y. Wu, S.-A. Chen, *Angew. Chem. Int. Ed.* **2019**, *58*, 11317–11323.

[71] C. A. Parker, C. G. Hatchard, *Trans. Faraday Soc.* **1961**, *57*, 1894–1904.

[72] P. F. Jones, A. R. Calloway, *Chem. Phys. Lett.* **1971**, *10*, 438–443.

[73] M. Sikorski, I. V. Khmelinskii, W. Augustyniak, F. Wilkinson, *J. Chem. Soc. Faraday Trans.* **1996**, *92*, 3487–3490.

[74] B. S. Yamanashi, D. M. Hercules, *Appl. Spectrosc.* **1971**, *25*, 457–460.

[75] F. A. Salazar, A. Fedorov, M. N. Berberan-Santos, *Chem. Phys. Lett.* **1997**, *271*, 361–366.

[76] H. Yersin, U. Monkowius, **2008**, DE 10 2008 033 563 A1.

[77] A. Endo, M. Ogasawara, A. Takahashi, D. Yokoyama, Y. Kato, C. Adachi, *Adv. Mater.* **2009**, *21*, 4802–4806.

[78] M. J. Leitl, V. A. Krylova, P. I. Djurovich, M. E. Thompson, H. Yersin, *J. Am. Chem. Soc.* **2014**, *136*, 16032–16038.

[79] T. Hofbeck, U. Monkowius, H. Yersin, *J. Am. Chem. Soc.* **2015**, *137*, 399–404.

[80] D. M. Zink, D. Volz, T. Baumann, M. Mydlak, H. Flügge, J. Friedrichs, M. Nieger, S. Bräse, *Chem. Mater.* **2013**, *25*, 4471–4486.

[81] D. Volz, D. M. Zink, T. Bocksrocker, J. Friedrichs, M. Nieger, T. Baumann, U. Lemmer, S. Bräse, *Chem. Mater.* **2013**, *25*, 3414–3426.

[82] M. Wallesch, D. Volz, D. M. Zink, U. Schepers, M. Nieger, T. Baumann, S. Bräse, *Chem. – Eur. J.* **2014**, *20*, 6578–6590.

[83] C. Bizzarri, C. Strabler, J. Prock, B. Trettenbrein, M. Ruggenthaler, C.-H. Yang, F. Polo, A. Iordache, P. Brüggeller, L. D. Cola, *Inorg. Chem.* **2014**, *53*, 10944–10951.

[84] Z. Wang, C. Zheng, W. Wang, C. Xu, B. Ji, X. Zhang, *Inorg. Chem.* **2016**, *55*, 2157–2164.

[85] A. Endo, K. Sato, K. Yoshimura, T. Kai, A. Kawada, H. Miyazaki, C. Adachi, *Appl. Phys. Lett.* **2011**, *98*, 083302.

[86] C. Adachi, *Jpn. J. Appl. Phys.* **2014**, *53*, 060101.

[87] X. Cai, S.-J. Su, *Adv. Funct. Mater.* **2018**, *28*, 1802558.

[88] H. Tanaka, K. Shizu, H. Nakanotani, C. Adachi, *Chem. Mater.* **2013**, *25*, 3766–3771.

[89] D. Zhang, M. Cai, Y. Zhang, D. Zhang, L. Duan, *Mater. Horiz.* **2016**, *3*, 145–151.

[90] D. Zhang, X. Cao, Q. Wu, M. Zhang, N. Sun, X. Zhang, Y. Tao, *J. Mater. Chem. C* **2018**, *6*, 3675–3682.

[91] C. S. Oh, D. de S. Pereira, S. H. Han, H.-J. Park, H. F. Higginbotham, A. P. Monkman, J. Y. Lee, *ACS Appl. Mater. Interfaces* **2018**, *10*, 35420–35429.

[92] L.-S. Cui, H. Nomura, Y. Geng, J. U. Kim, H. Nakanotani, C. Adachi, *Angew. Chem. Int. Ed.* **2017**, *56*, 1571–1575.

[93] W. Huang, M. Einzinger, T. Zhu, H. S. Chae, S. Jeon, S.-G. Ihn, M. Sim, S. Kim, M. Su, G. Teverovskiy, et al., *Chem. Mater.* **2018**, *30*, 1462–1466.

[94] N. Sharma, E. Spuling, C. M. Mattern, W. Li, O. Fuhr, Y. Tsuchiya, C. Adachi, S. Bräse, I. D. W. Samuel, E. Zysman-Colman, *Chem. Sci.* **2019**, *10*, 6689–6696.

[95] K. Kawasumi, T. Wu, T. Zhu, H. S. Chae, T. Van Voorhis, M. A. Baldo, T. M. Swager, *J. Am. Chem. Soc.* **2015**, *137*, 11908–11911.

[96] Y. Gao, T. Su, Y. Wu, Y. Geng, M. Zhang, Z.-M. Su, *Chem. Phys. Lett.* **2016**, *666*, 7–12.

[97] G. Méhes, H. Nomura, Q. Zhang, T. Nakagawa, C. Adachi, *Angew. Chem. Int. Ed.* **2012**, *51*, 11311–11315.

[98] K. Nasu, T. Nakagawa, H. Nomura, C.-J. Lin, C.-H. Cheng, M.-R. Tseng, T. Yasuda, C. Adachi, *Chem. Commun.* **2013**, *49*, 10385–10387.

[99] H. Ohkuma, T. Nakagawa, K. Shizu, T. Yasuda, C. Adachi, *Chem. Lett.* **2014**, *43*, 1017–1019.

[100] M. Bian, Y. Wang, X. Guo, F. Lv, Z. Chen, L. Duan, Z. Bian, Z. Liu, H. Geng, L. Xiao, *J. Mater. Chem. C* **2018**, *6*, 10276–10283.

[101] D. J. cram, J. M. Cram, *Acc. Chem. Res.* **1971**, *4*, 204–213.

[102] F. Tuinstra, J. L. Koenig, *J. Chem. Phys.* **1970**, *53*, 1126–1130.

[103] E. Spuling, N. Sharma, I. D. W. Samuel, E. Zysman-Colman, S. Bräse, *Chem. Commun.* **2018**, *54*, 9278–9281.

[104] T. Hatakeyama, K. Shiren, K. Nakajima, S. Nomura, S. Nakatsuka, K. Kinoshita, J. Ni, Y. Ono, T. Ikuta, *Adv. Mater.* **2016**, *28*, 2777–2781.

[105] H. Nakanotani, T. Furukawa, T. Hosokai, T. Hatakeyama, C. Adachi, *Adv. Opt. Mater.* **2017**, *5*, 1700051.

[106] K. Matsui, S. Oda, K. Yoshiura, K. Nakajima, N. Yasuda, T. Hatakeyama, *J. Am. Chem. Soc.* **2018**, *140*, 1195–1198.

[107] Y. Gao, Q.-Q. Pan, L. Zhao, Y. Geng, T. Su, T. Gao, Z.-M. Su, *Chem. Phys. Lett.* **2018**, *701*, 98–102.

[108] X. Liang, Z.-P. Yan, H.-B. Han, Z.-G. Wu, Y.-X. Zheng, H. Meng, J.-L. Zuo, W. Huang, *Angew. Chem. Int. Ed.* **2018**, *57*, 11316–11320.

[109] A. Pershin, D. Hall, V. Lemaur, J.-C. Sancho-Garcia, L. Muccioli, E. Zysman-Colman, D. Beljonne, Y. Olivier, *Nat. Commun.* **2019**, *10*, 1–5.

[110] K. Hong, J.-L. Lee, *Electron. Mater. Lett.* **2011**, *7*, 77–91.

[111] X. Zhang, C. Fuentes-Hernandez, Y. Zhang, M. W. Cooper, S. Barlow, S. R. Marder, B. Kippelen, *J. Appl. Phys.* **2018**, *124*, 055501.

[112] S. Zhang, W. Li, L. Yao, Y. Pan, F. Shen, R. Xiao, B. Yang, Y. Ma, *Chem. Commun.* **2013**, *49*, 11302–11304.

[113] W.-C. Wu, H.-C. Yeh, L.-H. Chan, C.-T. Chen, *Adv. Mater.* **2002**, *14*, 1072–1075.

[114] T.-H. Huang, J. T. Lin, L.-Y. Chen, Y.-T. Lin, C.-C. Wu, *Adv. Mater.* **2006**, *18*, 602–606.

[115] Q. Zhang, D. Tsang, H. Kuwabara, Y. Hatae, B. Li, T. Takahashi, S. Y. Lee, T. Yasuda, C. Adachi, *Adv. Mater.* **2015**, *27*, 2096–2100.

[116] W.-L. Tsai, M.-H. Huang, W.-K. Lee, Y.-J. Hsu, K.-C. Pan, Y.-H. Huang, H.-C. Ting, M. Sarma, Y.-Y. Ho, H.-C. Hu, et al., *Chem. Commun.* **2015**, *51*, 13662–13665.

[117] J. Luo, Z. Xie, J. W. Y. Lam, L. Cheng, H. Chen, C. Qiu, H. S. Kwok, X. Zhan, Y. Liu, D. Zhu, et al., *Chem. Commun.* **2001**, 1740–1741.

[118] B. Z. Tang, X. Zhan, G. Yu, P. P. S. Lee, Y. Liu, D. Zhu, *J. Mater. Chem.* **2001**, *11*, 2974–2978.

[119] J. Mei, Y. Hong, J. W. Y. Lam, A. Qin, Y. Tang, B. Z. Tang, *Adv. Mater.* **2014**, *26*, 5429–5479.

[120] H. Wang, L. Xie, Q. Peng, L. Meng, Y. Wang, Y. Yi, P. Wang, *Adv. Mater.* **2014**, *26*, 5198–5204.

[121] H. Tong, Y. Hong, Y. Dong, M. Häußler, J. W. Y. Lam, Z. Li, Z. Guo, Z. Guo, B. Z. Tang, *Chem. Commun.* **2006**, 3705–3707.

[122] J. Guo, X.-L. Li, H. Nie, W. Luo, S. Gan, S. Hu, R. Hu, A. Qin, Z. Zhao, S.-J. Su, et al., *Adv. Funct. Mater.* **2017**, *27*, 1606458.

[123] J. Guo, X.-L. Li, H. Nie, W. Luo, R. Hu, A. Qin, Z. Zhao, S.-J. Su, B. Z. Tang, *Chem. Mater.* **2017**, *29*, 3623–3631.

[124] J. Huang, H. Nie, J. Zeng, Z. Zhuang, S. Gan, Y. Cai, J. Guo, S.-J. Su, Z. Zhao, B. Z. Tang, *Angew. Chem. Int. Ed.* **2017**, *56*, 12971–12976.

[125] S. Gan, J. Zhou, T. A. Smith, H. Su, W. Luo, Y. Hong, Z. Zhao, B. Zhong Tang, *Mater. Chem. Front.* **2017**, *1*, 2554–2558.

[126] H. Liu, J. Zeng, J. Guo, H. Nie, Z. Zhao, B. Z. Tang, *Angew. Chem.* **2018**, *130*, 9434–9438.

[127] W. H. Weber, C. F. Eagen, *Opt. Lett.* **1979**, *4*, 236–238.

[128] S. Nowy, B. C. Krummacher, J. Frischeisen, N. A. Reinke, W. Brütting, *J. Appl. Phys.* **2008**, *104*, 123109.

[129] J. Frischeisen, D. Yokoyama, A. Endo, C. Adachi, W. Brütting, *Org. Electron.* **2011**, *12*, 809–817.

[130] M. Flämmich, J. Frischeisen, D. S. Setz, D. Michaelis, B. C. Krummacher, T. D. Schmidt, W. Brütting, N. Danz, *Org. Electron.* **2011**, *12*, 1663–1668.

[131] T. D. Schmidt, D. S. Setz, M. Flämmich, J. Frischeisen, D. Michaelis, B. C. Krummacher, N. Danz, W. Brütting, *Appl. Phys. Lett.* **2011**, *99*, 163302.

[132] S. Youn Lee, T. Yasuda, H. Nomura, C. Adachi, *Appl. Phys. Lett.* **2012**, *101*, 093306.

[133] C. Mayr, S. Y. Lee, T. D. Schmidt, T. Yasuda, C. Adachi, W. Brütting, *Adv. Funct. Mater.* **2014**, *24*, 5232–5239.

[134] J. Frischeisen, D. Yokoyama, C. Adachi, W. Brütting, *Appl. Phys. Lett.* **2010**, *96*, 073302.

[135] T. Komino, H. Tanaka, C. Adachi, *Chem. Mater.* **2014**, *26*, 3665–3671.

[136] D. H. Kim, K. Inada, L. Zhao, T. Komino, N. Matsumoto, J. C. Ribierre, C. Adachi, *J. Mater. Chem. C* **2017**, *5*, 1216–1223.

[137] T.-Y. Shang, L.-H. Lu, Z. Cao, Y. Liu, W.-M. He, B. Yu, *Chem. Commun.* **2019**, *55*, 5408–5419.

[138] Y. J. Cho, K. S. Yook, J. Y. Lee, *Adv. Mater.* **2014**, *26*, 6642–6646.

[139] M. Yokoyama, K. Inada, Y. Tsuchiya, H. Nakanotani, C. Adachi, *Chem. Commun.* **2018**, *54*, 8261–8264.

[140] F. L. Vaillant, M. Garreau, S. Nicolai, G. Gryn'ova, C. Corminboeuf, J. Waser, *Chem. Sci.* **2018**, *9*, 5883–5889.

[141] E. Speckmeier, T. Fischer, K. Zeitler, *J. Am. Chem. Soc.* **2018**, DOI 10.1021/jacs.8b08933.

[142] F. Hundemer, *Unpublished M. Sc. Thesis: Modular Acceptor Design of Nitrile-Based TADF Emitters*, Karlsruhe Institute of Technology, **2016**.

[143] B. Akhlaghinia, S. Rezazadeh, *J. Braz. Chem. Soc.* **2012**, *23*, 2197–2203.

[144] Y. Horio, Y. Ootake, S. Sawaki, S. Inukai, M. Agata, M. Umezawa, M. Goto, *Tetrazoleacetic Acid Derivatives Having Aldose Reductase Inhibitory Activity.*, **1992**, EP0506011 (A1).

[145] G. Roman, M. N. Rahman, D. Vukomanovic, Z. Jia, K. Nakatsu, W. A. Szarek, *Chem. Biol. Drug Des.* **2010**, *75*, 68–90.

[146] S. Achamlale, A. Elachqar, A. E. Hallaoui, A. Alami, S. Elhajji, M. L. Roumestant, P. Viallefont, *Phosphorus Sulfur Silicon Relat. Elem.* **1998**, *140*, 103–111.

[147] T. Onaka, H. Umemoto, Y. Miki, A. Nakamura, T. Maegawa, *J. Org. Chem.* **2014**, *79*, 6703–6707.

[148] G. Méhes, K. Goushi, W. J. Potscavage, C. Adachi, *Org. Electron.* **2014**, *15*, 2027–2037.

[149] M. Ferrari, *Nat. Rev. Cancer* **2005**, *5*, 161–171.

[150] N. Rosenfeld, J. W. Young, U. Alon, P. S. Swain, M. B. Elowitz, *Science* **2005**, *307*, 1962–1965.

[151] E. H. Rubinson, A. S. P. Gowda, T. E. Spratt, B. Gold, B. F. Eichman, *Nature* **2010**, *468*, 406–411.

[152] P. Nalbant, L. Hodgson, V. Kraynov, A. Toutchkine, K. M. Hahn, *Science* **2004**, *305*, 1615–1619.

[153] M. Schäferling, *Angew. Chem. Int. Ed.* **2012**, *51*, 3532–3554.

[154] X. Xiong, F. Song, J. Wang, Y. Zhang, Y. Xue, L. Sun, N. Jiang, P. Gao, L. Tian, X. Peng, *J. Am. Chem. Soc.* **2014**, *136*, 9590–9597.

[155] J. Yao, M. Yang, Y. Duan, *Chem. Rev.* **2014**, *114*, 6130–6178.

[156] E. Genin, Z. Gao, J. A. Varela, J. Daniel, T. Bsaibess, I. Gosse, L. Groc, L. Cognet, M. Blanchard-Desce, *Adv. Mater.* **2014**, *26*, 2258–2261.

[157] Z. Guo, S. Park, J. Yoon, I. Shin, *Chem. Soc. Rev.* **2013**, *43*, 16–29.

[158] Y. Gao, G. Feng, T. Jiang, C. Goh, L. Ng, B. Liu, B. Li, L. Yang, J. Hua, H. Tian, *Adv. Funct. Mater.* **2015**, *25*, 2857–2866.

[159] Y. Xiao, R. Zhang, Z. Ye, Z. Dai, H. An, J. Yuan, *Anal. Chem.* **2012**, *84*, 10785–10792.

[160] B. Song, G. Wang, J. Yuan, *Chem. Commun.* **2005**, 3553–3555.

[161] N. Weibel, L. J. Charbonnière, M. Guardigli, A. Roda, R. Ziessel, *J. Am. Chem. Soc.* **2004**, *126*, 4888–4896.

[162] H. Dong, S.-R. Du, X.-Y. Zheng, G.-M. Lyu, L.-D. Sun, L.-D. Li, P.-Z. Zhang, C. Zhang, C.-H. Yan, *Chem. Rev.* **2015**, *115*, 10725–10815.

[163] K. Y. Zhang, Q. Yu, H. Wei, S. Liu, Q. Zhao, W. Huang, *Chem. Rev.* **2018**, *118*, 1770–1839.

[164] M. Sánchez-Navarro, M. Teixidó, E. Giralt, *Acc. Chem. Res.* **2017**, *50*, 1847–1854.

[165] M. Wang, Z. Mao, T.-S. Kang, C.-Y. Wong, J.-L. Mergny, C.-H. Leung, D.-L. Ma, *Chem. Sci.* **2016**, *7*, 2516–2523.

[166] J.-C. G. Bünzli, C. Piguet, *Chem. Soc. Rev.* **2005**, *34*, 1048–1077.

[167] A. Gorman, J. Killoran, C. O'Shea, T. Kenna, W. M. Gallagher, D. F. O'Shea, *J. Am. Chem. Soc.* **2004**, *126*, 10619–10631.

[168] T. Li, D. Yang, L. Zhai, S. Wang, B. Zhao, N. Fu, L. Wang, Y. Tao, W. Huang, *Adv. Sci. Weinh. Baden-Wurtt. Ger.* **2017**, *4*, 1600166.

[169] F. Ni, Z. Zhu, X. Tong, M. Xie, Q. Zhao, C. Zhong, Y. Zou, C. Yang, *Chem. Sci.* **2018**, *9*, 6150–6155.

[170] J. E. Coleman, *Annu. Rev. Biochem.* **1992**, *61*, 897–946.

[171] Q. Zhang, S. Xu, M. Li, Y. Wang, N. Zhang, Y. Guan, M. Chen, C.-F. Chen, H.-Y. Hu, *Chem. Commun.* **2019**, *55*, 5639–5642.

[172] Z. Zhu, D. Tian, P. Gao, K. Wang, Y. Li, X. Shu, J. Zhu, Q. Zhao, *J. Am. Chem. Soc.* **2018**, *140*, 17484–17491.

[173] F. Ni, Z. Zhu, X. Tong, W. Zeng, K. An, D. Wei, S. Gong, Q. Zhao, X. Zhou, C. Yang, *Adv. Sci.* **2019**, *6*, 1801729.

[174] M. Green, P. M. Loewenstein, *Cell* **1988**, *55*, 1179–1188.

[175] T. Schröder, N. Niemeier, S. Afonin, A. S. Ulrich, H. F. Krug, S. Bräse, *J. Med. Chem.* **2008**, *51*, 376–379.

[176] K. Eggenberger, E. Birtalan, T. Schröder, S. Bräse, P. Nick, *ChemBioChem* **2009**, *10*, 2504–2512.

[177] D. K. Kölmel, D. Fürniss, S. Susanto, A. Lauer, C. Grabher, S. Bräse, U. Schepers, *Pharmaceuticals* **2012**, *5*, 1265–1281.

[178] S. W. Münch, *Festphasensynthese Neuartiger Molekularer Transporter Und Funktioneller Peptoide*, Logos-Verlag, **2018**.

[179] I. D. M. Wehl, *High-Throughput Screening and Evaluation of Combinatorial Cell Penetrating Peptoid Libraries to Identify Organelle- and Organ-Specific Drug Delivery Molecules*, Karlsruhe Institute of Technology, **2019**.

[180] F. Hundemer, L. G. von Reventlow, C. Leonhardt, M. Polamo, M. Nieger, S. M. Seifermann, A. Colsmann, S. Bräse, *ChemistryOpen* **2019**, *8*, 1413–1420.

[181] Y. Zhang, C. Zuniga, S.-J. Kim, D. Cai, S. Barlow, S. Salman, V. Coropceanu, J.-L. Brédas, B. Kippelen, S. Marder, *Chem. Mater.* **2011**, *23*, 4002–4015.

[182] Y. Tao, Q. Wang, C. Yang, Q. Wang, Z. Zhang, T. Zou, J. Qin, D. Ma, *Angew. Chem. Int. Ed.* **2008**, *47*, 8104–8107.

[183] Q. Li, L.-S. Cui, C. Zhong, Z.-Q. Jiang, L.-S. Liao, *Org. Lett.* **2014**, *16*, 1622–1625.

[184] S. Gong, Q. Fu, W. Zeng, C. Zhong, C. Yang, D. Ma, J. Qin, *Chem. Mater.* **2012**, *24*, 3120–3127.

[185] S. Chidirala, H. Ulla, A. Valaboju, M. R. Kiran, M. E. Mohanty, M. N. Satyanarayan, G. Umesh, K. Bhanuprakash, V. J. Rao, *J. Org. Chem.* **2016**, *81*, 603–614.

[186] C. Adachi, T. Tsutsui, S. Saito, *Appl. Phys. Lett.* **1989**, *55*, 1489–1491.

[187] J. Kido, C. Ohtaki, K. Hongawa, K. Okuyama, K. Nagai, *Jpn. J. Appl. Phys.* **1993**, *32*, L917.

[188] A. P. Kulkarni, C. J. Tonzola, A. Babel, S. A. Jenekhe, *Chem. Mater.* **2004**, *16*, 4556–4573.

[189] Y. Hamada, C. Adachi, T. Tsutsui, S. Saito, *Jpn. J. Appl. Phys.* **1992**, *31*, 1812.

[190] N. Tamoto, C. Adachi, K. Nagai, *Chem. Mater.* **1997**, *9*, 1077–1085.

[191] Y. Zhu, A. P. Kulkarni, S. A. Jenekhe, *Chem. Mater.* **2005**, *17*, 5225–5227.

[192] J. Lee, K. Shizu, H. Tanaka, H. Nomura, T. Yasuda, C. Adachi, *J. Mater. Chem. C* **2013**, *1*, 4599–4604.

[193] D. Y. Kwon, G. H. Lee, Y. S. Kim, *J. Nanosci. Nanotechnol.* **2015**, 15, 10.

[194] M. Y. Wong, S. Krotkus, G. Copley, W. Li, C. Murawski, D. Hall, G. J. Hedley, M. Jaricot, D. B. Cordes, A. M. Z. Slawin, et al., *ACS Appl. Mater. Interfaces* **2018**, *10*, 33360–33372.

[195] Z. Li, W. Li, C. Keum, E. Archer, B. Zhao, A. M. Z. Slawin, W. Huang, M. C. Gather, I. D. W. Samuel, E. Zysman-Colman, *J. Phys. Chem. C* **2019**, *123*, 24772–24785.

[196] X. Zhang, M. W. Cooper, Y. Zhang, C. Fuentes-Hernandez, S. Barlow, S. R. Marder, B. Kippelen, *ACS Appl. Mater. Interfaces* **2019**, *11*, 12693–12698.

[197] J. Zhang, L. Zhou, H. A. Al-Attar, K. Shao, L. Wang, D. Zhu, Z. Su, M. R. Bryce, A. P. Monkman, *Adv. Funct. Mater.* **2013**, *23*, 4667–4677.

[198] W. Yu, G. Huang, Y. Zhang, H. Liu, L. Dong, X. Yu, Y. Li, J. Chang, *J. Org. Chem.* **2013**, *78*, 10337–10343.

[199] S. Guin, T. Ghosh, S. K. Rout, A. Banerjee, B. K. Patel, *Org. Lett.* **2011**, *13*, 5976–5979.

[200] R. Huisgen, J. Sauer, H. J. Sturm, J. H. Markgraf, *Chem. Ber.* **1960**, *93*, 2106–2124.

[201] Y. Zheng, A. S. Batsanov, V. Jankus, F. B. Dias, M. R. Bryce, A. P. Monkman, *J. Org. Chem.* **2011**, *76*, 8300–8310.

[202] C. Leonhardt, *Unpublished B. Sc. Thesis: Synthese Und Charakterisierung Akzeptor-Modifizierter TADF-Emitter*, Karlsruhe Institute of Technology, **2017**.

[203] M. Montalti (University of Bologna, Italy), A. Credi (University of Bologna, Italy), L. Prodi (University of Bologna, Italy), M. T. Gandolfi (University of Bologna, Italy), *Handbook of Photochemistry*, Taylor & Francis Inc, Boca Raton, **2006**.

[204] B. Li, L. Zhou, H. Cheng, Q. Huang, J. Lan, L. Zhou, J. You, *Chem. Sci.* **2018**, *9*, 1213–1220.

[205] T. Ueno, Y. Urano, H. Kojima, T. Nagano, *J. Am. Chem. Soc.* **2006**, *128*, 10640–10641.

[206] N. I. Rtishchev, D. V. Samoilov, V. P. Martynova, A. V. El'tsov, *Russ. J. Gen. Chem.* **2001**, *71*, 1467–1478.

[207] C. Munkholm, D. R. Parkinson, D. R. Walt, *J. Am. Chem. Soc.* **1990**, *112*, 2608–2612.

[208] P. L. Santos, J. S. Ward, P. Data, A. S. Batsanov, M. R. Bryce, F. B. Dias, A. P. Monkman, *J. Mater. Chem. C* **2016**, *4*, 3815–3824.

[209] R. Ishimatsu, S. Matsunami, K. Shizu, C. Adachi, K. Nakano, T. Imato, *J. Phys. Chem. A* **2013**, *117*, 5607–5612.

[210] H. Yersin, A. P. Monkman, in *Highly Effic. OLEDs*, Wiley, Weinheim, **2018**, pp. 425–463.

[211] P. L. dos Santos, F. B. Dias, A. P. Monkman, *J. Phys. Chem. C* **2016**, *120*, 18259–18267.

[212] A. S. D. Sandanayaka, T. Matsushima, C. Adachi, *J. Phys. Chem. C* **2015**, *119*, 23845–23851.

[213] S. Höfle, M. Pfaff, H. Do, C. Bernhard, D. Gerthsen, U. Lemmer, A. Colsmann, *Org. Electron.* **2014**, *15*, 337–341.

[214] H. Nakanotani, K. Masui, J. Nishide, T. Shibata, C. Adachi, *Sci. Rep.* **2013**, *3*, 2127.

[215] R. V. Schneider, *Synthesis and Characterization of Sequence-Defined Stiff Oligomers Using the Sonogashira Reaction*, Karlsruhe Institute of Technology, **2018**.

[216] O. Franco, M. Jakoby, R. V. Schneider, F. Hundemer, D. Hahn, B. S. Richards, S. Bräse, M. A. R. Meier, U. Lemmer, I. A. Howard, *Front. Chem.* **2020**, *8*, DOI 10.3389/fchem.2020.00126.

[217] R. V. Schneider, K. A. Waibel, A. P. Arndt, M. Lang, R. Seim, D. Busko, S. Bräse, U. Lemmer, M. A. R. Meier, *Sci. Rep.* **2018**, *8*, 1–8.

[218] Y. Sagara, K. Shizu, H. Tanaka, H. Miyazaki, K. Goushi, H. Kaji, C. Adachi, *Chem. Lett.* **2014**, *44*, 360–362.

[219] C. H. Chen, C. W. Tang, *Appl. Phys. Lett.* **2001**, *79*, 3711–3713.

[220] K. Okumoto, H. Kanno, Y. Hamaa, H. Takahashi, K. Shibata, *Appl. Phys. Lett.* **2006**, *89*, 063504.

[221] K. Okumoto, H. Kanno, Y. Hamada, H. Takahashi, K. Shibata, *Appl. Phys. Lett.* **2006**, *89*, 013502.

[222] K. Shizu, J. Lee, H. Tanaka, H. Nomura, T. Yasuda, H. Kaji, C. Adachi, *Pure Appl. Chem.* **2015**, *87*, 627–638.

[223] Y. Sun, H. Jiang, W. Wu, W. Zeng, X. Wu, *Org. Lett.* **2013**, *15*, 1598–1601.

[224] W. Hong, Z. Wei, H. Xi, W. Xu, W. Hu, Q. Wang, D. Zhu, *J. Mater. Chem.* **2008**, *18*, 4814–4820.

[225] Z. Wei, W. Hong, H. Geng, C. Wang, Y. Liu, R. Li, W. Xu, Z. Shuai, W. Hu, Q. Wang, et al., *Adv. Mater.* **2010**, *22*, 2458–2462.

[226] J. Zhang, P. Xiao, F. Dumur, C. Guo, W. Hong, Y. Li, D. Gigmes, B. Graff, J.-P. Fouassier, J. Lalevée, *Macromol. Chem. Phys.* **2016**, *217*, 2145–2153.

[227] P. Dimroth, F. Reicheneder, *Angew. Chem. Int. Ed. Engl.* **1969**, *8*, 751–752.

[228] A. D. Hendsbee, S. M. McAfee, J.-P. Sun, T. M. McCormick, I. G. Hill, G. C. Welch, *J. Mater. Chem. C* **2015**, *3*, 8904–8915.

[229] X. Guo, A. Facchetti, T. J. Marks, *Chem. Rev.* **2014**, *114*, 8943–9021.

[230] Z. Liu, G. Zhang, Z. Cai, X. Chen, H. Luo, Y. Li, J. Wang, D. Zhang, *Adv. Mater.* **2014**, *26*, 6965–6977.

[231] F. Dumur, M. Ibrahim-Ouali, D. Gigmes, *Appl. Sci.* **2018**, *8*, 539.

[232] M. Chapran, R. Lytvyn, C. Begel, G. Wiosna-Salyga, J. Ulanski, M. Vasylieva, D. Volyniuk, P. Data, J. V. Grazulevicius, *Dyes Pigments* **2019**, *162*, 872–882.

[233] M. E. Jang, T. Yasuda, J. Lee, S. Y. Lee, C. Adachi, *Chem. Lett.* **2015**, *44*, 1248–1250.

[234] M. Li, Y. Liu, R. Duan, X. Wei, Y. Yi, Y. Wang, C.-F. Chen, *Angew. Chem. Int. Ed.* **2017**, *56*, 8818–8822.

[235] M. Li, S.-H. Li, D. Zhang, M. Cai, L. Duan, M.-K. Fung, C.-F. Chen, *Angew. Chem. Int. Ed.* **2018**, *57*, 2889–2893.

[236] Y. Danyliv, D. Volyniuk, O. Bezvikonnyi, I. Hladka, K. Ivaniuk, I. Helzhynskyy, P. Stakhira, A. Tomkeviciene, L. Skhirtladze, J. V. Grazulevicius, *Dyes Pigments* **2020**, *172*, 107833.

[237] L. Zhang, M. Li, T.-P. Hu, Y.-F. Wang, Y.-F. Shen, Y.-P. Yi, H.-Y. Lu, Q.-Y. Gao, C.-F. Chen, *Chem. Commun.* **2019**, *55*, 12172–12175.

[238] M. Danz, *Organic Molecules for Use as Emitters*, **2016**, WO2016042070 (A1).

[239] S. Seifermann, M. Danz, G. LIAPTSIS, *Organische moleküle zur verwendung in organischen optoelektronischen vorrichtungen*, **2018**, WO2018024725A1.

[240] S. Seifermann, M. Danz, D. Volz, *Phthalimid und carbazol oder dessen analoge enthaltende verbindungen zur verwendung in organischen optoelektronischen vorrichtungen*, **2018**, WO2018024723A1.

[241] C. Peng, A. Salehi, Y. Chen, M. Danz, G. Liaptsis, F. So, *ACS Appl. Mater. Interfaces* **2017**, *9*, 41421–41427.

[242] P. M. Lorz, F. K. Towae, W. Enke, R. Jäckh, N. Bhargava, W. Hillesheim, in *Ullmanns Encycl. Ind. Chem.*, American Cancer Society, **2007**.

[243] V. N. G. Lindsay, D. Fiset, P. J. Gritsch, S. Azzi, A. B. Charette, *J. Am. Chem. Soc.* **2013**, *135*, 1463–1470.

[244] A. Krasovskiy, P. Knochel, *Angew. Chem. Int. Ed.* **2004**, *43*, 3333–3336.

[245] D. E. Pugachov, T. S. Kostryukova, G. V. Zatonsky, S. Z. Vatsadze, N. V. Vasil'ev, *Chem. Heterocycl. Compd.* **2018**, *54*, 528–534.

356 Bibliography

[246] P.-Y. Wang, H.-S. Fang, W.-B. Shao, J. Zhou, Z. Chen, B.-A. Song, S. Yang, *Bioorg. Med. Chem. Lett.* **2017**, *27*, 4294–4297.

[247] K. A. Hofmann, O. Ehrhart, *Berichte Dtsch. Chem. Ges.* **1911**, *44*, 2713–2717.

[248] K. A. Hofmann, O. Ehrhart, *Berichte Dtsch. Chem. Ges.* **1912**, *45*, 2731–2740.

[249] D. W. Kaiser, G. A. Peters, V. P. Wystrach, *J. Org. Chem.* **1953**, *18*, 1610–1615.

[250] R. Huisgen, H. J. Sturm, M. Seidel, *Chem. Ber.* **1961**, *94*, 1555–1562.

[251] R. Cristiano, H. Gallardo, A. J. Bortoluzzi, I. H. Bechtold, C. E. M. Campos, R. L. Longo, *Chem. Commun.* **2008**, *0*, 5134–5136.

[252] R. Cristiano, J. Eccher, I. H. Bechtold, C. N. Tironi, A. A. Vieira, F. Molin, H. Gallardo, *Langmuir* **2012**, *28*, 11590–11598.

[253] T. Rieth, S. Glang, D. Borchmann, H. Detert, *Mol. Cryst. Liq. Cryst.* **2015**, *610*, 89–99.

[254] T. Rieth, N. Röder, M. Lehmann, H. Detert, *Chem. – Eur. J.* **2018**, *24*, 93–96.

[255] T. Rieth, T. Marszalek, W. Pisula, H. Detert, *Chem. – Eur. J.* **2014**, *20*, 5000–5006.

[256] S. Glang, T. Rieth, D. Borchmann, I. Fortunati, R. Signorini, H. Detert, *Eur. J. Org. Chem.* **2014**, *2014*, 3116–3126.

[257] A. G. Dal-Bó, G. G. L. Cisneros, R. Cercena, J. Mendes, L. M. da Silveira, E. Zapp, K. G. Domiciano, R. da Costa Duarte, F. S. Rodembusch, T. E. A. Frizon, *Dyes Pigments* **2016**, *135*, 49–56.

[258] V. A. Tartakovsky, A. E. Frumkin, A. M. Churakov, Yu. A. Strelenko, *Russ. Chem. Bull.* **2005**, *54*, 719–725.

[259] M. Wahren, *Z. Für Chem.* **1969**, *9*, 241–252.

[260] J. Fan, L. Lin, C. Wang, *Chem. Phys. Lett.* **2016**, *652*, 16–21.

[261] H. Min Kim, J. Min Choi, J. Yeob Lee, *RSC Adv.* **2016**, *6*, 64133–64139.

[262] J. H. Kim, M. Eum, T. H. Kim, J. Y. Lee, *Dyes Pigments* **2017**, *136*, 529–534.

[263] M. Kim, S. K. Jeon, S.-H. Hwang, S.-S. Lee, E. Yu, J. Y. Lee, *J. Phys. Chem. C* **2016**, *120*, 2485–2493.

[264] Y.-C. Duan, L.-L. Wen, Y. Gao, Y. Wu, L. Zhao, Y. Geng, G.-G. Shan, M. Zhang, Z.-M. Su, *J. Phys. Chem. C* **2018**, *122*, 23091–23101.

[265] H. Tanaka, K. Shizu, H. Miyazaki, C. Adachi, *Chem. Commun.* **2012**, *48*, 11392–11394.

[266] M. M. Rothmann, E. Fuchs, C. Schildknecht, N. Langer, C. Lennartz, I. Münster, P. Strohriegl, *Org. Electron.* **2011**, *12*, 1192–1197.

[267] P. Schrögel, N. Langer, C. Schildknecht, G. Wagenblast, C. Lennartz, P. Strohriegl, *Org. Electron.* **2011**, *12*, 2047–2055.

[268] M. Numata, T. Yasuda, C. Adachi, *Chem. Commun.* **2015**, *51*, 9443–9446.

[269] J. Lee, N. Aizawa, M. Numata, C. Adachi, T. Yasuda, *Adv. Mater.* **2017**, *29*, 1604856.

[270] M. J. Frisch, G. W. Trucks, H. B. Schlegel, G. E. Scuseria, M. A. Robb, J. R. Cheeseman, G. Scalmani, et al., *Gaussian Inc*, Wallingford, CT, **2013**.

[271] J. A. Pople, J. S. Binkley, R. Seeger, *Int. J. Quantum Chem.* **1976**, *10*, 1–19.

[272]   "mCP   -   1,3-Bis(N-carbazolyl)benzene,"   can   be   found   under
https://www.ossila.com/products/mcp, visited at 01.11.2019.

[273] S. Ye, Y. Liu, C. Di, H. Xi, W. Wu, Y. Wen, K. Lu, C. Du, Y. Liu, G. Yu, *Chem. Mater.* **2009**, *21*, 1333–1342.

[274] Q. Wang, J. Ding, D. Ma, Y. Cheng, L. Wang, F. Wang, *Adv. Mater.* **2009**, *21*, 2397–2401.

[275] "PPT," can be found under https://www.ossila.com/products/ppt, visited at 01.11.2019.

[276] J. Nishide, H. Nakanotani, Y. Hiraga, C. Adachi, *Appl. Phys. Lett.* **2014**, *104*, 233304.

[277] J. W. Sun, K.-H. Kim, C.-K. Moon, J.-H. Lee, J.-J. Kim, *ACS Appl. Mater. Interfaces* **2016**, *8*, 9806–9810.

[278] "DPEPO," can be found under https://www.ossila.com/products/dpepo, visited at 01.11.2019.

[279] S. Wu, M. Aonuma, Q. Zhang, S. Huang, T. Nakagawa, K. Kuwabara, C. Adachi, *J. Mater. Chem. C* **2014**, *2*, 421–424.

[280] D. R. Lee, M. Kim, S. K. Jeon, S.-H. Hwang, C. W. Lee, J. Y. Lee, *Adv. Mater.* **2015**, *27*, 5861–5867.

[281] M.-H. Tsai, H.-W. Lin, H.-C. Su, T.-H. Ke, C. -c. Wu, F.-C. Fang, Y.-L. Liao, K.-T. Wong, C.-I. Wu, *Adv. Mater.* **2006**, *18*, 1216–1220.

[282] "CzSi," can be found under https://www.ossila.com/products/czsi, visited at 01.11.2019.

[283] S. Kun, G. Z. Nagy, M. Tóth, L. Czecze, A. N. Van Nhien, T. Docsa, P. Gergely, M.-D. Charavgi, P. V. Skourti, E. D. Chrysina, et al., *Carbohydr. Res.* **2011**, *346*, 1427–1438.

8 Appendix

8.1 Curriculum Vitae

Personal Information

Name	Fabian Hundemer
Born	22nd of April 1992 in Landau in der Pfalz
Nationality	German

Practical Experience

Nov 2019 – Jul 2020   **EVONIK INDUSTRIES AG,** *HANAU (GERMANY)*
Junior Consultant in Procurement Strategy as part of M.B.A. degree
- Development and implementation of projects for the advance of operational and strategic procurement
- Elaboration of approvals and briefings on executive board level

Dec 2016 – Dec 2019   **KARLSRUHE INSTITUTE OF TECHNOLOGY (KIT),** *KARLSRUHE (GERMANY)*
Doctorate on organic chemistry in the Bräse Group
- Member of SFB 1176 (DFG) "Molecular Structuring of Soft Matter"
- Member of KSOP – Karlsruhe School of Optics & Photonics

May 2013 – Dec 2015   **MULTIPLE INSTITUTES WITHIN THE KIT,** *KARLSRUHE*, several research assistant positions
- Investigation on Li- and Mg-Ion energy storage systems
- Synthesis of electrodes and fabrication of coin cells in glovebox environments
- Kinetic studies in a shock wave tube
- Size-controlled synthesis of gold nanoparticles for catalytic studies
- Investigation on carbon monoxide oxidation using various catalytic systems

International Experience

Mar 2019 – May 2019   **UNIVERSITY OF ST ANDREWS,** *ST ANDREWS (UNITED KINGDOM)*
Visiting Researcher in the Zysman-Colman Group
- Optoelectronic characterization of materials for organic electronics and OLEDs
- Working in a cleanroom environment

Jan 2016 – Mar 2016   **STOCKHOLM UNIVERSITY,** *STOCKHOLM (SWEDEN)*
Visiting Researcher in the Bäckvall Group
- Iron-Catalyzed Cross-Coupling Reactions for the Synthesis of Allenols

Education

Jan 2017 – Jun 2020   **COLLÈGE DES INGÉNIEURS,** *PARIS, MUNICH, TORINO (FRANCE, GERMANY, ITALY)*
M.B.A. Candidate in the Science & Management program (Part-time next to doctorate)
- Tri-national, highly selective management program 100% funded by industry
- Specially designed to take engineering and science graduates from the world's top universities and prepare them for the highest and most responsible positions in the world of business

Oct 2014 – Nov 2016	**KARLSRUHE INSTITUTE OF TECHNOLOGY (KIT)**, *KARLSRUHE (GERMANY)* MSc in Chemistry focused on organic chemistry in the Bräse Group (grade: 1.0 with distinction)
	• Thesis: Modular Acceptor Design of Nitrile-based TADF Emitters
	• Modules included: Organic Chemistry, Radiochemistry, Toxicology, Law for Chemists
Oct 2011 – Sep 2014	**KARLSRUHE INSTITUTE OF TECHNOLOGY (KIT)**, *KARLSRUHE (GERMANY)* BSc in Chemistry focused on inorganic chemistry in the Breher Group (grade: 1.6)
	• Thesis: Non-frustrated Lewis pairs to activate small molecules
	• Modules included: Inorganic Chemistry, Organic Chemistry, Physical Chemistry, Experimental Physics, Applied Chemistry
Aug 2002 – Mar 2011	**EDUARD-SPRANGER-GYMNASIUM**, *LANDAU IN DER PFALZ (GERMANY)*
	• German Abitur: Chemistry, German, Geography, Mathematics (grade: 1.9)

Trainings and Seminars

Sep 2018	**INHOUSE CONSULTING EVENT	GOGLOBAL TRAINEEPROGRAMME**, *DARMSTADT*, hosted by Merck KGaA
Apr 2018	**LEADERSHIP TRAINING**, *LUDWIGSHAFEN/KARLSRUHE*, hosted in collaboration with the KIT and BASF SE,
Mar 2018	**SMC INSIGHTS – STRATEGY WORKSHOP**, *BUDAPEST*, hosted by Siemens Management Consulting
Sep 2017	**WACKER CHEMIE SUMMER SCHOOL**, *BURGHAUSEN*, hosted by Wacker Chemie AG
Sep 2016	**MODERN METHODS IN AGRICULTURAL CHEMISTRY**, *MONHEIM/KARLSRUHE*, hosted in collaboration with the KIT and Bayer AG

Extracurricular Activities

May 2018 – Dec 2019	**SOCIAL MEDIA RESPONSIBLE**, Bräse Group (KIT), *KARLSRUHE (GERMANY)*
Sep 2017 – Dec 2019	**SAFETY ADVISOR**, Institute of Organic Chemistry (KIT), *KARLSRUHE (GERMANY)*
Jul 2017 – Sep 2019	**BOARD MEMBER OF THE PH.D. STUDENTS' COUNCIL**, Faculty of Chemistry and Biosciences (KIT), *KARLSRUHE (GERMANY)*

Awards and Scholarships

Jun 2018	KIT innovation award "Neuland"
Dec 2017	Rétey award for the best MSc degree within the Institute of Organic Chemistry (KIT)
Jan 2017	M.B.A. fellowship of the Collège des Ingénieurs
Jan 2016	Erasmus+ scholarship for an internship abroad
Apr 2013	e-fellows.net scholarship (foundation of Deutsche Telekom, Georg v. Holtzbrinck, McKinsey)

Skills

| Language skills | German (native), English (fluent), Turkish (basic), French (basic) |
| IT skills | MS Office, Adobe Photoshop, ChemOffice, MestreNova, Origin, Turbomole |

8.2 List of Publications

Reviewed Articles

1) F. Hundemer, E. Crovini, Y. Wada, H. Kaji, E. Zysman-Colman, S. Bräse, *in preparation*

   Tristriazolotriazines as TADF acceptor core for sky-blue OLEDs

2) F. Hundemer, M. Zhang, I. Wehl, E. Spuling, L. Graf von Reventlow, S. W. Münch, D. M. Knoll, D. Lorenz, S. Seifermann, D. Volz, M. Nieger, U. Schepers, A. Colsmann, S. Bräse, *in preparation*

   Nitrile Derivatization of the 4CzIPN TADF System: Color Tuning and Bioconjugation of Arylated and Alkylated Tetrazoles

3) O. Franco, M. Jakoby, R. V. Schneider, F. Hundemer, D. Hahn, B. S. Richards, S. Bräse, M. A. Meier, U. Lemmer, I. A. Howard, *Front. Chem* **2020**, 8, 126.

   Sensitizing TADF Absorption Using Variable Length Oligo(phenylene ethynylene) Antennae

4) F. Hundemer, L. Graf von Reventlow, C. Leonhardt, M. Nieger, S. M. Seifermann, A. Colsmann, S. Bräse, *ChemistryOpen* **2019**, 8, 1413.

   Acceptor Derivatization of the 4CzIPN TADF System: Color Tuning and Introduction of Functional Groups

5) C. Bizzarri, F. Hundemer, J. Busch, S. Bräse, *Polyhedron* **2018**, 140, 51.

   Triplet emitters versus TADF emitters in OLEDs: A comparative study

6) S. N. Kessler, F. Hundemer, J.-E. Bäckvall, *ACS Catal.* **2016**, 6, 7448.

   A Synthesis of Substituted α-Allenols via Iron-Catalyzed Cross-Coupling of Propargyl Carboxylates with Grignard Reagents

Conference Posters

1) F. Hundemer, *"Modular Modification of TADF Emitters"*, 16th –18th of July **2018**, Darmstadt, Curious2018 – Future Insight conference by Merck KGaA.

2) F. Hundemer, I. Wehl, U. Schepers, S. Bräse, *"TADF in Cells via Bioconjugation"*, 7th of September 2017, Frankfurt am Main, 1st International TADF Symposium.

3) S. W. Münch, F. Hundemer, S. Bräse, *"Synthesis of multifunctional transporter moieties and "smart dyes" for cellular uptake"* from 11th of April **2017**, Speyer, 9th Seminar day, 2nd retreat of the GRK 2039.

8.3 Acknowledgements

First, I want to express my deepest gratitude to my supervisor Prof. Dr. Stefan Bräse for giving me this excellent and challenging research topic and for his guidance and patience along the way. I would like to thank him for the freedom in research he enabled and for giving me unique chances for my personal development like the Collège des Ingénieurs, various conferences, workshops and my stay abroad.

I am very grateful to Prof. Dr. Eli Zysman-Colman for giving me the opportunity to visit his group and laboratory in St Andrews. The help of Dr. Nidhi Sharma, Dr. Tomas Matulaitis, Ettore Crovini and David Hall was inevitable for the success of this research stay. I would like to thank the whole Zysman-Colman group for their warm welcome and support during my stay.

I would like to thank all the collaborators in the context of CRC1176. Prof. Dr. Uli Lemmer and his co-workers Dr. Andreas Arndt, Dr. Olga Franco and Marius Jakoby. Also Prof. Dr. Michael Meier, Dr. Rebekka Schneider and Daniel Hahn.

My gratitude also goes to Dr. Alexander Colsmann and his group members Dr. Min Zhang and Lorenz Graf von Reventlow for their effort in characterization and OLED fabrication.

I also want to thank Prof. Dr. Ute Schepers, Dr. Ilona Wehl and Dr. Stephan Münch for their collaborative work on the bioimaging project.

I would like to thank the cynora GmbH for their analytical support, especially Dr. Daniel Volz, Dr. Stefan Seifermann and Nico Thöbes for their guidance and fruitful discussions.

I want to acknowledge the work of Prof. Hironori Kaji and his co-worker Yoshimasa Wada at the Kyoto University and their support in OLED fabrication and characterization.

Prof. Dr. Joachim Podlech is kindly acknowledged for the acceptance of the co-reference of this thesis.

A special thank goes to the Collège des Ingénierus. Being part of this extraordinary Science & Management MBA course greatly influenced my personal development and expanded my horizon.

I want to acknowledge the work of Dr. Martin Nieger for crystal structure analyses at the University of Helsinki, Dr. Olaf Fuhr for crystal structure analyses at KIT, Dr. Andreas Rapp, Tanja Ohmer-Scherrer and Pia Lang for NMR spectroscopy, Lara Hirsch, Angelika Mösle, Rieke Schulte, Danny Wagner and Carolin Kohnle for mass spectrometry, IR and elemental analysis.

I am grateful of being part of the graduate schools of KSOP and CRC1176. I am not only thankful for financial support in the form of positions or travel funding, but also for the numerous workshops and networking opportunities.

A special thank goes to Dr. Eduard Spuling and Gloria Hong for spending their time to proofread this thesis. I am very grateful for your comments and suggestions.

I am very grateful to be part of the AK Bräse team. Thanks to all current and past members, keep up the pleasant work climate. A huge thanks to Dr. Christin Bednarek, Janine Bolz, Christiane Lampert and Selin Samur for their support in any organizational tasks and for their effort in keeping the group running.

I want to thank all the students who worked under my supervision either in the context of their thesis or during the advanced organic chemistry course.

A particular thank you to Dr. Eduard Spuling whose guidance along my career in the Bräse group cannot be neglected. He has always been a patient listener of many issues and a mentor with humorous but always deliberate suggestions.

I am very happy to always have Daniel Knoll accompanying me since the advanced chemistry course in school, throughout our studies and finally during our doctorate. Rarely one meets someone like-minded as him. He has always been a challenging companion, discussion partner and friend.

Finally, I am more than happy to always count on the support of my beloved parents Bärbel and Theo, my brother Steffen and of course Selin. Without your encouragement this thesis would not have been possible. Selin, you are my light.